A Disaggregate Freight Transport Model for Germany

Sascha Reiche

A Disaggregate Freight Transport Model for Germany

Springer Vieweg

Sascha Reiche
Magdeburg, Germany

Dissertation zur Erlangung des akademischen Grades Doktoringenieur (Dr.-Ing.) von Herrn Dipl.-Wirtsch.-Ing. Sascha Reiche, geb. am 28.10.1986 in Magdeburg genehmigt durch die Fakultät für Maschinenbau der Otto-von-Guericke-Universität Magdeburg

Gutachter:
Herr Prof. Dr.-Ing. Hartmut Zadek (OVGU, ILM)
Herr Prof. Dr.-Ing. habil Thomas Schulze (OVGU, FIN)
Promotionskolloquium am 12. Januar 2017

ISBN 978-3-658-19152-8 ISBN 978-3-658-19153-5 (eBook)
DOI 10.1007/978-3-658-19153-5

Library of Congress Control Number: 2017949513

Springer Vieweg

Printed on acid-free paper

This Springer Vieweg imprint is published by Springer Nature
The registered company is Springer Fachmedien Wiesbaden GmbH
The registered company address is: Abraham-Lincoln-Str. 46, 65189 Wiesbaden, Germany

Danksagung

Die vorliegende Arbeit entstand während meiner Tätigkeit als wissenschaftlicher Mitarbeiter am Lehrstuhl für Logistik des Instituts für Logistik und Materialflusstechnik der Otto-von-Guericke-Universität Magdeburg. Den Personen, die mich während dieser Zeit begleitet und unterstützt haben, möchte ich an dieser Stelle meinen besonderen Dank entgegen bringen.

Mein Dank gilt zunächst Herrn Prof. Dr.-Ing. Hartmut Zadek für die Betreuung dieser Arbeit sowie das entgegengebrachte Vertrauen und den konstruktiven Austausch während des Forschungsvorhabens. Herrn Prof. Dr.-Ing. habil. Thomas Schulze danke ich für die Übernahme des Zweitgutachtens und seine wertvollen Anregungen.

Weiterhin danke ich meinen Kollegen am Institut für Logistik und Materialflusstechnik für die freundliche Unterstützung und die stete Hilfsbereitschaft, die wesentlich zum Gelingen dieser Arbeit beigetragen haben.

Bei meiner Familie und meinen Freunden möchte ich mich ganz besonders herzlich bedanken für die uneingeschränkte und vielseitige Unterstützung während meines Studiums und auf dem Weg zum erfolgreichen Abschluss dieses Promotionsvorhabens.

Magdeburg, Januar 2017

Content

List of figures

List of tables

List of equations

List of abbreviations

Agg.	Aggregation
AGORA	AGORA-Group
BA Statistik	Statistik der Bundesagentur für Arbeit
BAFA	Bundesamt für Wirtschaft und Ausfuhrkontrolle – German Federal Office for Economic Affairs and Export Control
BAG	Bundesamt für Güterverkehr
BDA	Bundesverband der Deutschen Arbeitgeberverbände – Confederation of German Employers' Associations
Binnenreede-rei	Deutsche Binnenreederei AG
BinSchUO	Binnenschiffsuntersuchungsordnung
BMEL	Bundesministerium für Ernährung und Landwirtschaft – German Federal Ministry of Food and Agriculture
BMELV	Bundesministerium für Ernährung, Landwirtschaft und Verbraucherschutz – former German Federal Ministry of Food and Agriculture
BMVBS	Bundesministerium für Verkehr, Bau und Stadtentwicklung – former German Federal Ministry of Transport and Digital Infrastructure
BMVI	Bundesministerium für Verkehr und digitale Infrastruktur – German Federal Ministry of Transport and Digital Infrastructure
BMWI	Bundesministerium für Wirtschaft und Technologie – former German Federal Ministry for Economic Affairs and Energy
BNetzA	Bundesnetzagentur
BStatG	Bundesstatistikgesetz
BSV	Binnenschifffahrts-Verlag GmbH
Bundesbank	Deutsche Bundesbank
CPA	Statistical Classification of Products by Activity, 2008 version
DB Schenker	DB Schenker Rail Deutschland AG
Destatis	Statistisches Bundesamt – German Federal Statistical Office

Dev.	Deviation
EC	European Commission
EOQ model	Economic Order Quantity model
Eurostat	Statistical Office of the European Union
EWI	EcoTransIT World Initiative
FDZ	Forschungsdatenzentrum der Statistischen Ämter der Länder – German Research Data Centres of the Federal Statistical Office and the statistical offices of the Länder
Freq.	Frequency
GA	Genetic Algorithm
GDV	Gesamtverband der Deutschen Versicherungswirtschaft e.V. – German Insurance Association
HABEFA	Handbook Emission Factors for Road Transport
HPE	Bundesverband Holzpackmittel, Paletten, Exportverpa-ckung e.V.
hwh	hwh Gesellschaft für Transport- und Unternehmensberatung mbH
ifeu	ifeu Institut für Energie- und Umweltforschung Heidelberg GmbH – German Institute for Energy and Environmental Research
IfM	Institut für Mittelstandsforschung Bonn
IHK	Oldenburgische Industrie und Handelskammer
IPF	Iterative Proportional Fitting
IRPUD	Institute of Spatial Planning Dortmund
IWW	Inland Waterway Navigation as a mode of transport
Lohndirekt	Lohndirekt GmbH
LWKN	Landwirtschaftskammer Niedersachsen
MRCE	Mitsui Rail Capital Europe B.V.
MRIO data	Multiregional Economic Input-Output data
Nbr.	Number
NST	Standard goods classification for transport statistics, 2007 version
NUTS	Nomenclature of Units for Territorial Statistics
OD data	Origin/Destination data
OECD	Organisation for Economic Co-operation and Development
OSMF	OpenStreetMap Foundation

P.	Part
PC data	Production/Consumption data
PLANCO	PLANCO Consulting GmbH
Spec.	Specification
SZ	Süddeutsche Zeitung
tkm	tonne-kilometres
UNECE	United Nations Economic Commission for Europe
Uniconsult	Universal Transport Consulting GmbH
WSDO	Wasser- und Schifffahrtsdirektion Ost
WSDW	Wasser- und Schifffahrtsdirektion West
WSV	Wasser- und Schifffahrtsverwaltung des Bundes

Abstract

Although it is widely recognised that freight movements have significant impacts on economic, ecological and societal well-being, providing satisfactory freight traffic models to affected decision makers is still an open task. Previous developments dedicated to distinct aspects of freight transport analysis led to a large diversity of specific transport models with different operational value. The primary cause for an insufficient practicability can be seen in the scarcity of comprehensive specific data and the complexity of topical issues.

The model developed in this study addresses this obstacle by using public data to its best use. Therefore, a multi-modal commodity class specific freight model at the level of firms for the area of Germany has been developed, allowing an integration of macroscopic as well as disaggregate input data. Three modes of transport and 30 types of goods at the spatial level of 403 national as well as 29 international regions are considered, taking into account supply chain specifications of 88 different German business branches to provide fundamental insight into domestic freight transport organisation, which should prove useful to decision makers with reference to the subject.

Kurzfassung

Es ist weithin anerkannt, dass sich der Güterverkehr in komplexer Art und Weise auf das Wirtschaftsgeschehen, die natürliche Umwelt und damit die Allgemeinheit auswirkt. Dennoch mangelt es an geeigneten Güterverkehrsmodellen als Planungsgrundlage für Entscheidungsträger mit entsprechendem Themenbezug. Bisherige Konzepte münden in einer großen Vielfalt an Modellen zur Erschließung des Güterverkehrsgeschehens – jedoch mit eingeschränkter Praktikabilität. Als Hauptursache für einen mitunter stark eingeschränkten Anwendungsfokus ist nicht zuletzt der ausgeprägte Gegensatz zwischen Themenkomplexität und entsprechender Informationsverfügbarkeit anzusehen.
Die vorliegende Abhandlung zielt folglich darauf ab, öffentlich zugängliche Daten zum Güterverkehrsgeschehen möglichst weitreichend zu erschließen. Hierzu wird ein multimodales und zugleich gütergruppenspezifisches Modell für firmenindividuelle Güterverkehrsflüsse in Deutschland erarbeitet, welches sowohl eine Integration aggregierter als auch kleinteiliger Eingangsgrößen ermöglicht. Im Hinblick auf eine allgemeingültige und zugleich tiefgreifende Praktikabilität des Modells finden drei Transportmodi und 30 Gütergruppen sowie 403 territoriale Raumgliederungseinheiten mit Bezug zu 88 branchenspezifisch ausgeprägten Lieferketten Berücksichtigung.

1. Motivation and problem context

Freight transport is ubiquitous although accompanied by various societal challenges. It is closely interlinked with the industrial sector as well as with people's everyday life. Industrial products need to be transported from one production site to another and finally to a customer. On the one hand, this movement of goods is essential for an economy. On the other hand, it is the root of manifold negative impacts on the natural environment. This is why political decision makers have to cope with the challenge of providing an organisational framework for freight transport. A political decision related to the transport sector will be more efficient as it becomes founded on a thorough understanding of its impacts as well as its outreach.

However, when the focus is set on what is at hand for a structured freight transport organisation in Germany, the status quo is not satisfactory. The knowledge concerning freight transport activities within Germany and with its trading partners is very limited. Considering the tremendous volume of freight transports that are related to the German economy, each knowledge growth will have relevant effects on the governmental freight transport organisation capacities and in the end, the potential to meet the environmental and societal challenges.

For those in charge of providing the necessary information, e.g. transport planners and engineers, this is certainly a challenging task – at least due to the complexity of the topic. This is a fact that demands the interaction of several scientific fields, such as economics, engineering sciences or information technologies, a combination of what – in broad terms – is the subject of logistics engineering according to e.g. ARNOLD (2008, p. 4). This perspective is one way to evaluate the potential impacts and the scope of transport politics and, in return, to elaborate a freight transport analysis that will initialise profound political decisions.

From this point of view, the information at hand can be rearranged to a complete and more detailed depiction of domestic freight transport activities. Thus, the overall motivation of this study is to provide a fundamental insight into the German freight transport organisation. The subsequent evaluation of freight transport interactions from a market point of view as well as from the perspective of a logistics system aims at gathering a systematic understanding, respectively. Therefore, an identification of selected key steps affecting elementary activities as well as its encompassing system are indicated.

This insight may – at least in the long-term – likewise introduce the possibility to systematically assess the impact of general economic and societal trends on the freight transport organisation in Germany and vice-versa. In this context, for instance, the expansion of global or local sourcing strategies as well as particular aspects of social transformations, such as e.g. a demographic development with a different regional manifestations, can be discussed.

1.1. Research questions

Effectively accomplished transports in terms of *operational efficiency* and *ecological viability* are key success factors in a global economy, according to e.g. BMVI (2016b), BMVI (2015) and BMVBS (2010, p. 2)[1]. This is why an understanding of its *mechanism* is so important. An appropriate freight transport model promotes such understanding in preparation of deriving respective organisational measures. Thus, the representation of complex interactions for the *real-world* freight transport organisation is elementary for the presented study that aims at developing such model with pronounced explanatory modules which are conducive to understand and subsequently organise the process of the German freight traffic genesis, heading to the subsequent questions:

Q_1: Which data are appropriate to describe and analyse the process of freight transport commencing from its emergence up to its realisation?

Q_2: Which methods and techniques are appropriate to make use of these data for developing a comprehensive freight transport model for Germany?

[1] For the presented context, the *efficiency* addresses the avoidance of wasting resources, such as energy, materials, money, time or human resources for an intended transport organisation either on a national scale or downscaled to an individual operation. Since transport activities per se have consequences for the environment – as e.g. discussed in detail in OECD (2010) – an intended transport organisation should explicitly consider its corresponding *ecological viability*.

Q_3: How can complex multimodal freight transport path alternatives be sufficiently represented in a nation-wide model; at the same time reflecting real-world freight transport best possible?

Q_4: What practical recommendations can be derived from such a representation to improve the efficiency of the institutional German freight transport governance?

1.2. Scientific and practical relevance

Recent national freight transport models are mainly derived from aggregate input data, at least when dealing with the complexity of multiple commodity classes, regions and modes[2]. Within such a framework for Germany, for instance, accessible regional specific traffic volumes per mode are consolidated with commodity class specific data per mode – although both datasets are explicitly not published for this overlapping characteristics, resulting from an insufficient data validity[3]. However, within aggregate models this initial consolidation is essential to answer the questions of *where* and *how* freight traffic takes place within a focal *real-world* transport system. A subsequent disaggregation approach for the resulting data to e.g. firms with a corresponding regional and economic specification will then be used to refer to the question of *why* freight traffic takes place within the observed system – a question that is only insufficiently answered due to the inherent limits of a breakdown of aggregate traffic data to disaggregate data[4].
These shortcomings – among others – are the result of the methodological limits of aggregate data based freight transport models. Disaggregate freight transport models, in contrast, offer the opportunity to understand and disentangle drivers for the present state of a national, such as the German, freight transport system. This is due to the fact that disaggregate freight transport models derive traffic from an economic pattern – whereas basic aggregate

[2] Cf. overview on present modelling approaches in section 7.
[33] Cf. discussion on data discrepancies for German road freight statistics in section 4.4.3 as well as the share of mode road within the overall modal split in Fig. A-1.
[4] Cf. Fig. A-2 and the related discussion.

approaches work the other way around. Furthermore, they allow for the derivation of institutional opportunities of practical relevance for improving the efficiency and international competitiveness by:

- *decoupling* of the economic growth from a similar aligned freight traffic expansion through network effects – based on both public and private freight flow organisation measures[5]
- and ensuring a *better ecological compatibility* as well as the *cost efficiency* of domestic freight traffic, based on an improved transport infrastructure and/or enhanced freight transport activity organisation.

They thus reveal an insight into relevant steps necessary for the freight traffic genesis for which a respective action is needed. This is a goal that is not yet met in theory or practice.

1.3. Research Design and Scope

First of all, the German freight transport volume per annum is determined. Therefore, as for the remainder of this study, the reference year will be 2012. An identification of statistically reported total freight volume references will be required for subsequent calibrations and validations within the model as well as to outline adequate system boundaries.
Subsequently, the German freight transport system is analysed with a focus on relevant actors. On the one hand, there is a need to identify sources of a *transport demand* and related transport relevant decisions, respectively. On the other hand, it is necessary to develop an understanding of the way logistics decisions are made within the *transport supply* as well as how it is constituted. This step will not only be useful for a subsequent review of recent freight modelling advances but also for the following development of a framework for a disaggregate freight transport model for Germany and its implementation.

In general, there is no universal paradigm for freight transport models but rather individual concepts depending on the type of application and data available (CASCETTA, 2009, p. 239). That is why an overview on eligible accessible input data concerning freight transport in Germany is compiled. In

[5] The importance of *decoupling* as well as related obstacles are e.g. discussed in ALISES AND VASSALLO (2015), VERNY (2007) and TAVASSZY (2008, p. 48).

order to enable a complete transparency and reproducibility of the subsequent model, it is the goal of this study to use public data only. For one part this assures a high level of data quality and traceability of the proposed processing. For another part this guiding principle allows for an update of potential future rollouts of the model without financial burdens of commercial/private datasets.

From this starting point, the focus is set to manufacturing firms being a major source of the freight transport demand that, in turn, is at the heart of the nation-wide multimodal commodity shipment model based on a supply chain synthesis at the level of individual firms. In this line, a stepwise freight transport model will be developed. This approach will be in line with the 'classic' four-stage traffic concept, omnipresent for scientific traffic modelling. The aim of the resulting framework is to handle the complexity within the concept and its implementation as well as to identify linkages for the aspired governmental and private improvement measures in practice. The modelling sequence is given as follows:

- The first step identifies the set of examined firms. Hence, firms are listed based on firm size measures and a classification of in- and outgoing goods.
- A second step will help to estimate a likely supply chain configuration of these firms. It will be complemented by a shipment size determination for each respective pairing within the supply chain network.
- A third step determines the transport path design. This procedure involves a modal selection of one or multiple modes and subsequently a related transport network path.

The last part is one of the most critical in the proposed modelling approach. It aims at delivering an innovative and comprehensive solution in response to the central challenge of a large scale freight simulation – to represent the actual supply path that may, and in most practical cases will, differ from direct commodity trips starting from a point of production to a final consumption.

Within this framework, multiple logistics decisions are addressed. These are usually individual strategic decisions that can only be sufficiently captured in freight models based on the individual firm level. This, however, requires company specific information that, to a large extent, are not at hand. The goal of the proposed approach is nevertheless to derive a maximum benefit from the available data.

Therefore, concerning the general dilemma of scale and scope of models due to data restrictions, the following major limitations shall apply:

- spatial variations within the objective region are not considered,
- network changes are not performed within the modelled time period,
- interactions with passenger traffic are not considered in detail[6],
- only land-based transport modes are explicitly modelled[7] and
- the freight transport demand as well as the resulting traffic is modelled for commodities that are produced and/or consumed within the modelled time frame only[8].

The latter restriction refers to an exclusion of modelling transports of e.g. *secondary raw materials; municipal wastes and other wastes* (NST-14) that have a significant role within the German transport system, since about 8% of the overall transport volume relates to them[9]. Together with other excluded commodities, such as *goods moved in the course of household and office removals* (NST-17), *mail, parcels* (NST-15) and the large proportions of *grouped goods* (NST-18) as well as *unidentifiable goods* (NST-19) a total share of about 19% of the overall reported transport volume is not part of the model (cf. Table A-3)[10].

The reason for this limitation is that for waste and secondary raw materials or e.g. the shares of grouped and unidentifiable goods an allocation to the German economy is required other than that within the framework of immediate identifiable production- and consumption-related commodities.

---

[6] However, interactions in terms of traffic congestions and a derived impact on freight travel times and late arrivals are incorporated to a limited extent into the transport cost evaluation as presented in section 14. More detailed interactions, especially when it comes to traffic flow organisation, are out of scope of the presented study on a nation-wide freight transport modelling. For example, where there is a firm – whether it is a production site, a retail store or e.g. a library – there might be not only a freight transport demand, but also a distinct type of passenger transport demand.

[7] The considered modes of transport are: road, rail and transports on inland waterways. Cf. section 4.4 for details on selected limitation.

[8] Productions to or consumptions from stock are modelled.

[9] Cf. section 4.4.2 for details on utilised nomenclature.

[10] Note that the share of modelled commodities, when measured in terms of tonne-kilometres, exceeds 75% of the corresponding total transport volume.

In line with the overall objective, this study is structured in five major sections:

- an overview on *Freight transport modelling*
- the development of a *Framework for a disaggregate German freight transport model*
- the subsequent *Realisation of the disaggregate German freight transport* as well as
- a concluding *Résumé*.

A Contextual outline

A first overview on the German freight transport context will help to globally size the field of study. It is encompassed through different perspectives to measure the central element – the transport volume – as well as to identify suitable system boundaries and relevant actors.

2. Freight transport in Germany by volume

The total annual freight transport volume for Germany in 2012 and its modal split is evaluated by the *Federal Bureau of Statistics* in Germany, as e.g. depicted in Fig. A-1. The dominant road freight quantity, however, is only based on an estimation of the international road freight transport volume in Germany – a hindrance for more focused interpretations, as will be discussed in the following[11]. Nevertheless, it allows for the constitution of an overview on the overall land-based freight transport volume.

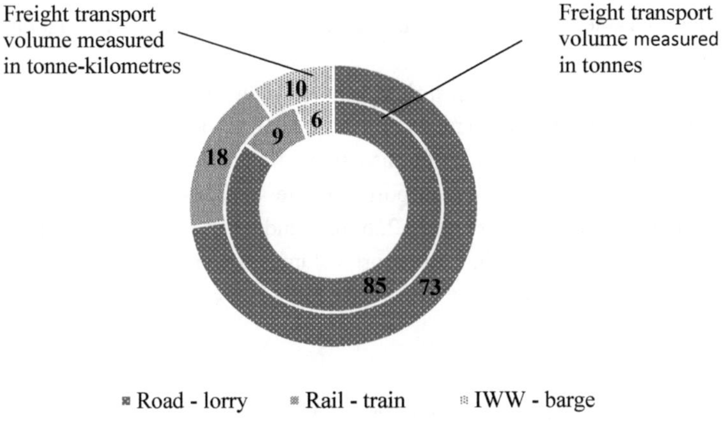

Fig. A-1 Shares of the modal split for a total 2012 land-based freight transport volume in Germany[12] [DESTATIS (2014m)]

[11] Cf. sections 4.4.3 for a discussion of the data base of corresponding statistical publications for Germany.

[12] See Table G-1 for absolute volumes.

In addition to the modal split, the overall freight transport volume in Germany can be specified by national and international transports as well as transports performed by domestic and transports of international freight forwarders. This first overview on the scope of freight transport activities related to Germany will be useful for calibrations of the presented model.
One of these specifications leads to the overall freight transport volume measured in tonnes while the other one leads to a total measured in tonne-kilometres. Most of the relevant information in this context is retrievable from the *Statistical Office of the European Union* (Eurostat).

Freight transport volume per mode

According to EUROSTAT (2014g), the total national *road freight transport*[13] volume in 2012 is estimated at 2,761,152 thousand tonnes. Another 29,185 thousand tonnes are transported within Germany by foreign forwarders[14], whereof about 97% are enrolled by forwarders from a EU-27 country (EUROSTAT 2014k; EUROSTAT 2014j). For international transports by road[15] the share of domestic and foreign freight forwarders is depicted in a consolidated format in Table G-2, together with a tonne-kilometre specific evaluation in Table G-3.
The equivalent national *rail freight transport* volume in 2012 measures about 247,117 thousand tonnes in total (EUROSTAT, 2014i). Cabotage by mode rail is not reported[16]. The volume of transnational freight flows by rail is given in EUROSTAT (2014e). This transport volume is consolidated to 45,286 thousand tonnes going out and 58,226 thousand tonnes directed to Germany (cf. Table G-4). Rail freight transport volumes for Germany measured by tonne-kilometres are depicted in Table G-5.

---

[13] See Table A-1 for references on volumes for road freight transports, rail freight transports and IWW freight transports.
[14] The share of road cabotage in terms of transported tonnes is about 1.1%. Measured in tonne-kilometre, the share of foreign freight forwarders for national transports by road is equivalent to 3.28% – the cabotage penetration rate for Germany in 2012.
[15] Freight transport statistics for Germany are set up on a different terminology – sending and receiving instead of import and export. The reason for this is that a receiving is not per se an import and vice versa, same as for outgoing loads and exports. For instance, a ship load arriving in the ARA area (ports of *Antwerp*, *Rotterdam* and *Amsterdam*) is usually not exclusively related to the Dutch, but also to imports by other European countries. Within the presented model, this effect will not be further differentiated apart from the context of country and port specific incoming and outgoing loads that are related to German imports and exports in section 13.4.
[16] Cf. section 14.3.

For inland waterways, the *IWW freight transports*, the total domestic transport volume in 2012 is about 54,569 thousand tonnes. For this volume a relevant cabotage rate is reported. 70% – that is 38,177 thousand tonnes – are transported by German flagged barges and 29% (16,392 thousand tonnes) are conveyed by freight vessels from other EU-27 countries (EUROSTAT, 2014n). International IWW transport volumes in terms of tonnes from or to Germany are presented in EUROSTAT (2014o) and EUROSTAT (2014m). Table G-6 and Table G-7 give a unified depiction this dataset.

Total freight transport volume

These mode specific transport volumes, measured in tonnes for Germany in 2012 – as reported in EUROSTAT (2014j, 2014k, 2014e, 2014n, 2014o, 2014m) – are displayed in the subsequent summary of Table A-1 by country of origin of related forwarders. A distribution of freight totals according to commodity classes is given in Table A-3[17]. As a result, a total freight transport volume of 3,815,014 thousand tonnes is identified to be related to Germany, whereof *3,706,181 thousand tonnes are directly related to either German origins and/or destinations*. This is an o1utline in terms of freight volumes for the subsequent freight model.

The goal of the subsequently presented model is to put in place a structure that allows one to understand the genesis as well as the related distribution of the identified overall German transport volume, accordingly.

[17] Note that statistical reports for Germany to EUROSTAT are basically submitted by DESTATIS. However, certain data discrepancies arise for variant data specifications when both reports are compared to each other for similar specifications – as e.g. given in DESTATIS (2014d, 2014c, 2014b). See also section 4.4.3 for more details on input data discrepancies as well as section 13.4 for a discussion of a commodity specific distribution of reported total freight volumes, respectively. For the presented determination of a *total freight transport volume* for Germany in 2012, results from EUROSTAT are decisive.

	National						International												Transit		
							sending						receiving								
	German forwarders			foreign forwarders			German forwarders			foreign forwarders			German forwarders			foreign forwarders					
	road	rail	IWW	road	rail	IWW	road	rail	IWW	road	rail	IWW	road	rail	IWW	road	rail	IWW	road	rail	IWW
BE				2%		9%	13%	4%	24%	8.3%		28%	15%	8%	14%	9%		12%	1%	16%	13%
BG				2%		0%		0%	0%			0%		0%	0%			0%		0%	0%
CZ				7%		2%	2%	8%	0%	7.3%		0%	3%	7%	0%	9%		0%	11%	3%	0%
DK				2%		0%	6%	1%		1.6%			7%	0%	0%	2%		0%	1%	4%	0%
EE				1%		0%													2%	0%	0%
IE				0%		0%									0%			0%	0%	0%	0%
EL						0%	0%	0%		0.5%			0%	0%		1%			0%	0%	0%
ES				1%		0%	1%	2%		3.2%			1%	1%		4%			0%	0%	0%
FR				1%		0%	19%	4%	6%	10.5%		3%	20%	3%	9%	9%		3%	2%	2%	27%
IT				2%		0%	6%	27%		5.1%			7%	15%		6%			0%	21%	0%
CY						0%			0%			0%			0%			0%	1%	0%	0%
LV				1%		0%									0%			0%	0%	0%	0%
LT				1%		0%		0%						0%					1%	0%	0%
LU				7%		2%	4%	2%	0%	2.6%		0%	3%	0%	0%	2%		0%	4%	2%	2%
HU				3%		0%	0%	2%	0%	2.0%		0%	0%	2%	2%	2%		0%	0%	0%	1%
MT						0%			0%			0%			0%			0%	4%	0%	0%
NL				23%		72%	19%	7%	61%	20.7%		63%	22%	31%	72%	21%		81%	12%	28%	31%
AT				7%		0%	11%	18%	3%	9.3%		0%	10%	12%	2%	8%		0%	8%	11%	5%
PL				33%		10%	2%	6%	0%	18.9%		0%	1%	9%	0%	18%		1%	39%	3%	0%
PT				0%		0%	0%	0%		0.4%				0%					0%	0%	0%
RO				2%		0%		0%	0%			0%		0%	0%			0%	2%	0%	0%
SI				2%		0%	0%	1%		0.9%				0%					3%	0%	0%
SK				2%		0%	0%	1%	0%	1.7%		0%	0%	4%	0%	2%		0%	4%	2%	0%
FI						0%		0%							0%			0%	0%	0%	0%
SE				0%		0%	1%	4%	0%	0.5%		0%	1%	4%		1%			1%	4%	0%
UK				0%		0%	2%	0%	1%	1.9%		0%	1%	0%	0%	2%		0%	0%	0%	0%
EU-27				98%		95%	87%	87%	96%	97.8%		95%	93%	97%	100%	99%		98%	98%	97%	78%
CH				2%		4%	13%	13%	4%	1.2%		2%	7%	3%	0%	1%		0%	2%	2%	21%
RU						0%		0%						0%	0%			0%		0%	0%
others				0%		1%	1%	0%	0%	0.5%		3%	0%	0%	0%	0%		0%			
Total	2,761,152	247,117	38,177	29,185	0	16,392	65,603	45,286	7,274	123,674	0	40,974	47,803	58,226	21,889	125,277	0	78,152	73,008	15,512	20,313
Totals	**3,706,181**																		**3,815,014**		

Table A-1 Identification of a total national, international and transit freight transport volume for Germany 2012 in thousands of tonnes[18]

[18] For a better readability entries with *no value* reported are indicated as blanks.

3. Freight transport in Germany as a general system

As the envisaged goal of freight transport planning policy measures is to be effective or, moreover, efficient, a profound understanding of the affected system is a prerequisite. This is especially true for complex systems, such as the outlined framework. For instance, at the socio-ecological level, impacts on global climate as well as on local air and noise emissions need to be considered for transport planning by public authorities. As a result, regional, national and international roadmaps and master plans are being released that call for a reorganisation of freight transport processes.

In response, models of the freight transport system are developed. In general terms, models serve as a tool to interpret a system's behaviour at a reasonable effort (BOSSEL, 2004, p. 15). In order to enable realistic representation of the transport system, it is crucial to identify its relevant components and respective interactions.

Therefore, a system analysis will be rolled out in the following. To analyse the freight transport system, general properties of systems and system states will be introduced first.

It is important to acknowledge that no specific modelling concept fits all facets of a complex system. However: '(...) *the key to effective advances is picking the appropriate categories within which to undertake analysis and subsequently build models*' (WIGAN AND SOUTHWORTH, 2006, 7 f.).

General properties of a system

In broad terms a system's essential is the integrity of a cause-and-effect structure that follows a certain purpose (BOSSEL, 2004, p. 35). Following Arnold et al. (2008, p. 76), systems may be:

- *defined as* a configuration of components or elements which are connected. These connections built upon specific attributes and related rules. In this sense, components represent a set of elements – the smallest and most basic unit of a system,
- *characterised by* a system state that contains the total of all state variables necessary to entirely describe the system at any time[19],

[19] Although static systems are purely hypothetic, they may be useful in a certain context. In such case they are described by constant state variables.

- *referred to* as a structure of components limited by distinct boundaries. Within these boundaries a system has interfaces to affect its environment and/or to be affected itself vice-versa. The system boundaries likewise define the range of values for the coupling of its components.

In the presented context, a freight transport system is regarded as a dynamic system, defined as the total of all interlinked components of the transport infrastructure, the transport demand as well as the transport supply. To understand the complex freight transport genesis – the inevitable starting point – a description of the interactions of a transport system's components in the context of an economic market contributes to this purpose.

4. Freight transport in Germany as an economic market

How does transport demand arise and how are transport services supplied? These questions may lead the way to a more specified analysis of the transport systems properties. In this sense, the general purpose of the transport system is stated as to bring together demanders and suppliers to trade freight transport services in a mutually beneficial way. This evolves from the perception that freight transport is formed by the effort that must be taken in order to bridge two spatially differentiated locations (BLAUWENS, BAERE AND VAN DE VOORDE, 2008, p. 21). This effort serves to convey products and goods between a number of suppliers and consumers – the transport demand[20]. The demand that meets a supply of transport services generates traffic on the corresponding transport infrastructure (NOTTEBOOM, 2013, p. 212).

Following this perception, the goal of the freight system analysis is to elaborate a system's cause-and-effect structure by identifying its components and their potential interactions. The focus is set on the general structure of the system as an economic market and more specifically on the interactions between different competences of logistics and relevant actors, respectively.

[20] See also BUTTON (2010) for further interpretations of a transport demand's evolution.

Components of the transport demand

Production facilities and retailers are considered to be the principles of a movement of goods, in accordance with e.g. BENDUL (2011, p. 48)[21]. As *shippers* they are located at the origin or source, as *receivers* at the destination or sink of a transport. Depending on the sourcing and distribution concept and the attributes of their manufactured goods, shippers demand a pick-up of certain products or goods, subject to logistics service requirements by transport service providers[22]. Similarly receivers attend a delivery specified by logistics quality attributes.

Shippers and receivers are distinct microeconomic market actors. They represent companies and their respective establishments – in the following referred to as firms – that can be grouped upon a varying resolution in accordance with:

- an economic activity and/or
- a spatial resolution of market elements.

Thus, the dominant economic activity of a firm can be helpful for a more aggregate classification of firms that, for instance, represent a common freight transport demand for *timber products* or the total transport demand of the *retail sector*. Thus, a distinction upon the economic activity either refers to an input or potential output of a firm.

Another typology of freight demand elements can result from a spatial differentiation into traffic cells. One example is the demand for freight transport in particular *urban areas*, another one is the *global* demand for freight transport.

Components of the transport supply

A transport market supply is organised by carriers and forwarders, such as road carriers, rail carriers and barge carriers. They make use of different

[21] See also section 6.2.2 for an interpretation of the role of households as a component of the freight transport market's demand.

[22] From an institutional point of view companies running production and/or retail facilities may be part of the market demand and supply at the same time in case they act as their own transport service provider. Their business activities and related business units may in most cases still be decomposed to fit the given context (PFOHL, 2010, p. 255 ff.). See also next section for continuation.

transport means, generally related to a mode of transport in a transport infrastructure[23]. Equivalent to shippers on the demand side, carriers and freight forwarders are regarded as microeconomic market actors/elementary institutions. According to NOTTEBOOM (2013, p. 214 f.), the market supply can be further divided into two categories:

- shippers/receivers operating their own fleet of transport means: the transport user deploys his own fleet of lorries, rail wagons and barges and
- third-party transports: specialised transport companies, such as trucking companies, railway or barge operators, offer a transport service to users[24].

However, as many production companies acknowledged that transport activities are not part of their core business, freight transport services are increasingly being outsourced (NOTTEBOOM, 2013, p. 214). In consequence, transport market demand and supply elements are evaluated independently in terms of potential interactions.

Market size

Neither transport service demand nor supply take place in a vacuum. Transport demand in a globalised business cannot be restricted by boundaries or walls of a particular nature. This also applies to transport services that, in order to meet the respective demand, perform operations on links as well as in nodal points within a complex transport system. Thus, the size of a transport system from a market perspective can hardly be limited. However, a segmentation of the transport system in time and space as well as for the nature of interactions allows for an exemplary qualitative market size limitation[25]:

---

[23] A transport infrastructure by region consists of roads, rail, sea, air, walkways etc. For further details see e.g. TAVASSZY AND BLIEMER (2013, p. 332). It can be represented by one or several networks (cf. next section).
[24] An overview on relevant companies for the transport market supply is e.g. given in KILLE AND SCHWEMMER (2013).
[25] Cf. section 6 for corresponding freight modelling analyses.

Scope of a freight transport market					
Allocation of ship-pers, receivers and ter-minals over time	Short-term	Long-term			
	Fixed location	Variable location			
Spatial extent and organisation of transports	Short-distance	Long-distance			
	Unimodal	Land-based		Non-land-based	
		Unimodal	Multimodal	Unimodal	Multimodal
Specification of sourcing and distribution interaction	Single-product	Multiple-product			
	Uniform vehicle	Uniform vehicle		Multiple-vehicles	

Table A-2 Exemplary morphologic transport market sizing

Market interactions

Starting from the perception that transport systems' interactions are performed as the result of logistics processes, the state of such a system is the result of logistics choices made by its subsystems and their decision making elements. In practice, these decisions are individually motivated. The system's state may – and in most cases will – differ from a theoretic system optimum then. From a market perspective this can be underlined.

For many markets the standard economic equilibrium framework for prices and quantities may provide only insufficient explanations because of vertical sub-markets and/or dynamic pricing phenomena (Ben-Akiva et al., 2012, p. 446). These features are typical for transport markets. As a result of logistics processes they are segmented markets, e.g. in terms of the physical process design of transport services. Individual transport units, shipment sizes, delivery times etc. are only a few of various criteria that limit the number of elements concerning demand and supply of a homogeneous transport market segment. Furthermore, the effect of economies of scale, achieved through logistics' process coordination in the market, results in dynamic prices. The effect of economies of scale describes the phenomena that, for instance, a single forwarder or shipper achieves cost advantages by increasing the respective transport volume on a certain network link. This equally holds true

for the observation of interacting suppliers of transport service, e.g. in the case of transport bundling. In contrast, by reaching a certain level, this effect is opposed due to capacity limits. These effects constantly change their relevance over time and result in a dynamic that sets limits to a transport system analysis based on a general market equilibrium approach[26].
As a result, the approach to analyse the equilibration of demand and supply within the standard economic equilibrium framework is not appropriate for understanding a complex transport market's behaviour.

4.1. Freight transport demand and supply interactions

A freight transport system is of such complexity, that no single scientific discipline would be able to encompass its functions and operations (RODRIGUE, NOTTEBOOM AND SHAW, 2013, p. 3). Nonetheless, it may be stated that for an analysis of informational and – even more relevant – physical flows, which might be identified as core processes of a freight transport demand and supply interplay, the consideration of logistics organisation capabilities is indispensable. Hence, the interdisciplinary nature of logistics science may be regarded as a suitable approach for an analysis of the broad range of interactions behind the physical movement of goods. This gives reason to an implementation of logistics concepts within the presented framework in order to adequately model relevant freight market interactions.

A line may be drawn between an evaluation of 'classic' logistics interactions and a more refined analysis of supply chain interactions. From a managerial perspective, supply chains, in essence, are built among independent organisations to extend the concept of elementary logistics interactions (cf. BAUMGARTEN, DARKOW AND ZADEK (2004, p. 2 ff.) to a level that explicitly seeks to improve cooperative competitiveness, thereby enhancing customer satisfaction (cf. BOWERSOX (2013, p. 30 ff.), WERNER (2013, p. 5 ff.), STADTLER AND KILGER (2008, 9)). Two broad dimensions for improving competitiveness are identified by LEE AND NG (1997, p. 191):

[26] An additional effect is the limit of information flows in the sector of transport bundling potentials, hence the imperfection of market interactions.

- the first one comprises a closer *integration of firms*. This strategy targets the organisational boundaries by working more closely with its suppliers and customers and
- the second dimension encompasses the *coordination of flows* in a supply chain. These flows represent the threefold elementary structure of logistics process analysis: materials, information, and finance.

The assumption may be stated that by today basically no firm will neglect the supply chain management potential[27]. That is why an understanding of the way freight transport market demand and supply meet each other is crucial for an elaborated freight transport planning.

4.2. Structural framework

Besides market demand and supply, the transport network is a third significant component of the freight transport system. Physical interactions follow a network structure, which may be defined as a set of nodes and a set of connecting edges.

- Transport nodes serve as access points or as intermediary locations for the overall transport system or a particular subsystem. The latter function is mainly serviced by terminals. Terminals represent starting, ending and transhipment points for transport flows (RODRIGUE, NOTTEBOOM AND SHAW, 2013, p. 4).
- Transport edges connect transport nodes. Edges primarily represent transport infrastructure elements such as roads, rails and waterways.

A network link is defined in this context as a pair of nodes connected by an edge. In the centre of such a transport system is the shipment of goods and commodities. These shipments are realised by certain transport modes, either uni- or multimodal that, in turn, are accessible via transport terminals.

Transport network nodes

Transport is realised between origins and destinations, namely two transport nodes. A terminal – another major representative of an analytic transport

---

[27] See for instance STADTLER AND KILGER (2008, 1 f.) for an exemplary description of the supply chain management potential in practice.

node – may be defined as any facility where freight is being handled in a transport process (RODRIGUE, COMTOIS AND SLACK, 2009, p. 127). This may be a sea or an inland port as well as a freight yard. In a broader sense, each access or release point for trailer loads may be considered a terminal and likewise each origin or destination for transports. These nodes within a transport system are interlinked by edges to build up a transport network.

Transport network edges

Arcs within a transport network connect the network nodes. They are also referenced as arcs or links. Along them, a transport of goods and commodities takes place.
From a market perspective there is a demand for transports from an origin to a destination. These two representatives of network nodes are interlinked by transport paths that consist of one or the aggregate of multiple edges or links.

Shipments and commodity flows

In this context, the realisation of a transport demand is referenced as *a shipment.* A shipment, in turn, is defined as the amount of commodities shipped between two locations or transport nodes in a transport system – specified by shipment size, frequency etc. Shipments are mainly formed by market demand elements[28] based on individual lot size calculations. A further bundling with other shipments, induced e.g. by transport logistics service providers, and/or their realisation along similar freight transport network links, leads to *commodity flows*. In this context, a commodity flow refers to a realisation of a shipment within a transport network.

Shipment size determination as well as commodity flow configuration are individually motivated actions within a transport system. They are the outcome of calculations for at least the underlying logistics service costs. From this perspective, the overall result of transport market demand and supply interactions is a distinct commodity flow configuration.
A commodity flow configuration may – and often will – differ from production/consumption flows. Such a production-consumption trade flow (*PC*

[28] Less frequently also the market supply side within a transport system may be responsible for a lot sizing. Significantly more often a commodity flow bundling will take place on the supply side.

flow) describes a direct relation between two network nodes that are not necessarily equivalent to the origin/destination flow (*OD flow*) between certain network terminals. This phenomena is also considered as a micro-macro gap in freight transport systems[29].

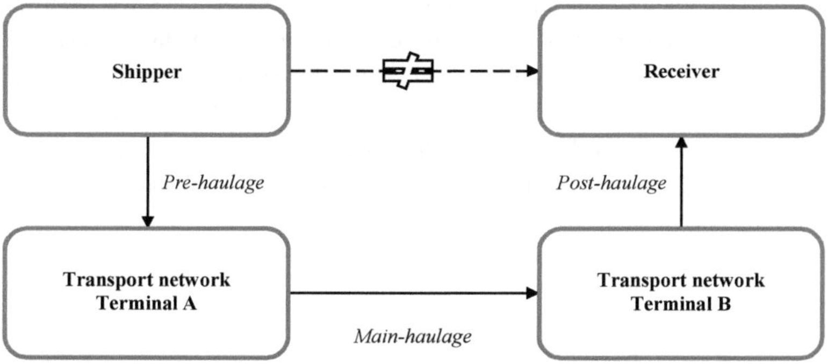

Fig. A-2 Micro-macro gap for intermodal transports [adapted from BLAUWENS, BAERE AND VAN DE VOORDE (2008, p. 32)]

Transport modes

Within a transport system usually multiple modes are relevant. Typical *land-based transport modes* are:

- Road
- Rail
- IWW (Inland Waterway Transports)

A transport via pipelines is only viable for selected commodities, e.g. crude oil and natural gas and furthermore merely part of transport systems that seek to analyse long-term market demand and supply interactions. Another alternative mode, the transport by sea, is almost exclusively bound to international transport relations and therefore restricted to certain locations and port access of international origins or destinations.

[29] For further discussions of this problem context see also section 15.1.

Multimodal or intermodal freight transport

A multimodal transport is performed with at least two different means of transport from the point where a good is originated, to a designated point of delivery. Multimodal transport may be further specified by the use of containers, swap bodies, road lorries and also their separated trailers as transport units to combine a transport between the road, rail, barge and sea infrastructure. For those transports where moved goods and commodities remain in one and the same loading unit or road vehicle for the entire transport operation, the wording *multimodal* is preferable to *intermodal* transports. The selected terminology is also in line with the definition of intermodal transports in UNECE (2012, p. 2). Intermodal and combined transports are applicable for the same context.

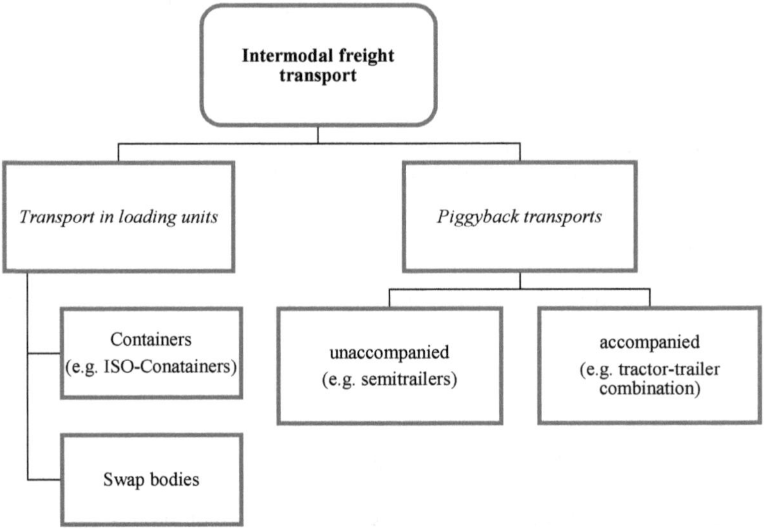

Fig. A-3 Types of intermodal transports in freight transport systems

Although transported by multiple modes, the non-land-based transport modes air and sea are not per se denoted as multimodal transport freight transport systems. Instead, the *land-based* transport sequence may be separately evaluated. Since for air freight hardly any of the transport units, as specified before, are deployed in practice along the entire transport chain, the role of intermodal transports including air-freight main hauls is often negligible. This differs from sea shipments to arrive in – or be sent from – national ports. Here, multimodal transports are realised with e.g. ISO-containers as a considerable alternative to unimodal transports.

Goods and commodities

Within a transport system's arrangement the general focus is usually set on the transported goods – a term mostly used in economic considerations that is also applied as a more general term for commodities. The notion commodity is used to mutually examine raw materials as well as semi-finished and finished products that have common characteristics.
Commodities can be classified by producers and/or consumers together with distinct characteristics, as exemplarily indicated for three basic economic sectors in Fig. A-4. Hence, they are related to a sector of an economic activity and further specified by individual characteristics such as weight, volume and value.

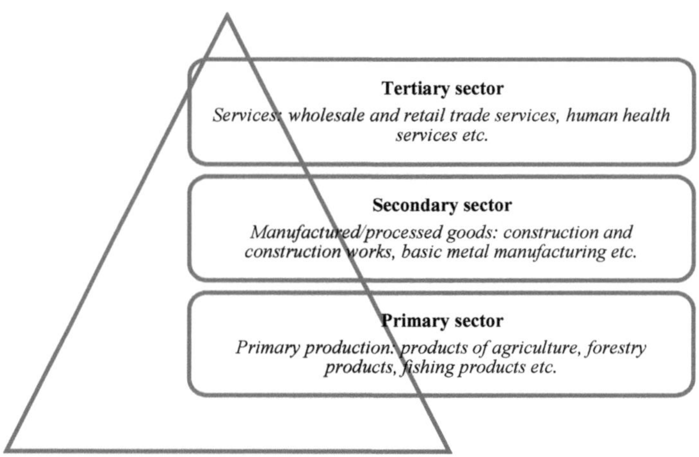

Fig. A-4 Exemplary classification of commodities according to their producers

The industry standard classification systems – such as the International Standard Industrial Classification (ISIC), sponsored by the United Nations

Statistics Division, the North American Industry Classification System (NA-ICS) of the Statistical bureaus of the U.S., Canada and Mexico as well as the Statistical Classification of Economic Activities (NACE) in the European Community – are also designated to classify economic activities and their respective outcomes within statistical reports for freight transport systems.

4.3. Synopsis

The result of the given interpretation of a transport system as a market with its components and their logistics interactions is a *commodity flow configuration.* One way to describe this specific setup is to depict a generic structure and its components as follows:

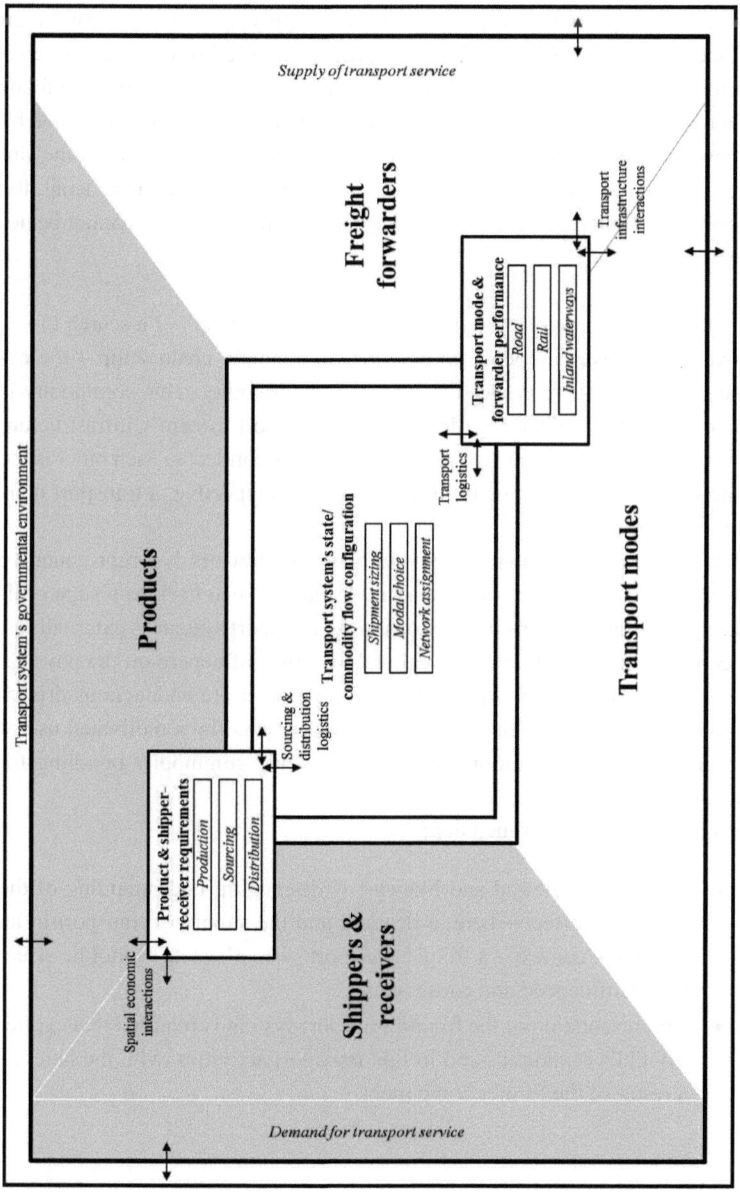

Fig. A-5 A generic setting of a freight transport market's setting and its driving forces

From this perspective, transport demand and supply can be observed between shippers or receivers and freight forwarders. In broad terms, they are connected to each other due to a demand or the supply of a product. The focus of a freight transport model is mostly set on the transport demand formed by economic interactions of spatially separate shippers and receivers. If the aim is to interpret a transport system's demand configuration in more detail, the supply of transport services on a given transport infrastructure cannot be neglected[30].

When this connection is interpreted in detail, as the result of a search to improve cooperative competitiveness – hence, a supply chain setup – the exchanged product is referenced as a specific commodity. This commodity is transported by freight forwarders along a transport system's infrastructure via multiple modes. The way these driving forces connect to each other is the transport systems' state, or from an economic perspective, a transport markets' setting.

Within this understanding, the governmental framework is a representative for exogenous influences to the market. It may be seen itself as a superordinate system that performs interactions to the transport system as externalities. In contrast, the economic, technical and logistics influences on the physical as well as informational organisation of the market are endogenous drivers for a change of the system. This is mainly described by a motivated use of certain routing configurations, transhipments and commodity bundling for each mode of transport.

As a result, it may be argued that:

- without a profound and likewise wide-ranging understanding of the cause and effect – here, a demand and the supply of transport in an economic context – a freight transport system's state cannot be effectively influenced and consequently,
- a representation of the freight transport system is required that explicitly links economics and freight transport activities, with the latter as a result of the former component.

[30] Cf. discussion on the role of alternative transport costs per path in the context of a modal selection in section 9.

4.4. Specific statistical data

Since an understanding of the way freight transport market demand and supply meet each other is crucial to elaborated freight transport planning, it has to be a feature of a model for the German freight transport system. However, the more detailed the model of a complex system is set up, the more informational content is required. A trade-off that can be dealt with when light is shed on individual firms' *supply chain interactions* and the related information at hand.

These interactions have be evaluated within a structural framework of *multiple modes*, namely road, rail and IWW. This selection is the result of a breakdown of the total annual transport volume in terms of tonnes for Germany in the base year of 2012, as depicted in Fig. A-1 and Table G-1. The total annual transport volume, including transits, is estimated to a total of about 4.3 billion tonnes[31]. Thereof, more than three quarters were transported by mode road, 9% by train followed by 7% for costal sea shipping and 5% for transports on inland waterways. A share of 2% of the aggregate transport volume in tonnes is reserved for crude oil transports in pipelines. The proportion of only a 0.1%, equivalent to 4 million tonnes, is transported by cargo flights.

Referring to the latter, the share of domestic air freight relations in 2012 is only 0.1 million tonnes, equivalent to an overall proportion of 0.002% (DESTATIS, 2013e, p. 12). Although of minor impact on the national modal split, cargo flights play a non-negligible role in international trade relations, especially with a view to oversea regions. In consequence, a freight model for Germany should consider air freight transports in addition to the land-based alternatives – at least by implementing synthetic network nodes of production and consumption.

Since statistical reports on national airports are limited in terms of their informational content, the share of domestic air freight transports is not appropriately separable from international transport volumes. With regard to a share of only 0.002% for the overall national transport volume, this impact is assumed to be negligible.

[31] Thereof 3.8 billion tonnes are related to mode road, rail and IWW as depicted in Fig. A-1.

For sea shipping the reported annual total transport volume of about 300 million tonnes is likewise spread to a large extent among international transport relations. According to DESTATIS (2013g), only 7.3 million tonnes are transported between domestic ports, that is about 0.2% of the overall national sea shipping transport volume. In contrast to air freight related statistical data, the national as well as the international sea freight transport volumes can be unambiguously distinguished[32].

Furthermore, for crude oil may be assumed that it is dominantly destined to refineries in Germany, each of which is connected to a pipeline system or an oil port (BMVI, 2014a, 16 f.). Thus, it is further assumed that no significant alternatives to pipeline transports are considered for German crude oil refineries[33]. Additionally, for natural gas similar assumptions might apply. As a result, neither the pipeline network will be part of further evaluations for a model of the German freight transport system, nor will be the product *extractions of crude petroleum and natural gas* of (CPA-06).

For the remainder of this study, only transports by mode road, rail and IWW will be evaluated for national transports as well as international transports. Additionally transports by sea and air are considered as options for international transports.

4.4.1. Regional specific information

These transport modes are accessible via *different types of transport nodes*. These are either locations of production and consumption or transhipment nodes. All together they are classified by a geographic referencing unit, the Nomenclature of Units for Territorial Statistics (NUTS) for European countries. The first level of this hierarchical system is referred to the current 28 EU member states (EFTA and EU-candidate countries are separated) with a two-letter country code. Further subdivisions into regional levels are represented by additional digits. For Germany according to the NUTS 2010 nomenclature, as given in EUROSTAT (2011, p. 29 ff.), these are:

[32] Cf. section 13.3.

[33] Apart from that for coking plants, as a second part of the commodity group CPA 19, the remaining modes are genuine options for a modal choice within a freight transport system.

- NUTS-1: major socio-economic regions (19 German *Regierungsbezirke*)
- NUTS-2: basic regions (39 German *Regionalbezirke*)
- NUTS-3: small regions (402 German *Landkreise/ kreisfreie Städte*[34])

To identify a centroid of each of the modelled international NUTS-0 and national NUTS-3 administrative units, valuable indications are given in EUROSTAT (2015a), STÄDTESTATISTIK (2012) and IRPUD (2005).
Each transport considered node of the German transport system is assigned to either of these regions. Inland nodes are distributed within the centre of one of the NUTS-3 regions and international nodes are assigned to a NUTS-0 centroid. This setup likewise provides a distinct restriction criterion for system analysis – the spatial outreach of transport activities under consideration.

4.4.2. Commodity specific information

These nodes are connected via multiple transport links, enabling multimodal transport options of various commodities. For a German transport system a useful classification scheme for commodity classes is the European NACE industry standard classification system using a four-digit code for a classification of economic activities (EUROSTAT, 2014u). The economic origin of a commodity is referenced to this classification via the Statistical classification of products by activity (CPA), a universal classification of products at the level of the European Union (EUROSTAT, 2012). CPA product categories are related to activities as defined by NACE standards. The CPA classification has a hierarchical structure with six levels of detail. Up to the fourth level of detail the NACE and CPA coding are identical with very few exceptions and both may be used in the same manner in practice to express an economic activity by characteristic commodities (DESTATIS, 2008b, p. 49)[35].
The current NACE Rev. 2 standard is, according to EUROSTAT (2008, p. 61 ff.), classified as follows:

[34] The latest European statistics publication on regions in the EU does however not consider recent local administrative reorganisations, e.g. for small regions of Mecklenburg-Vorpommern by late 2011 in northern Germany, leading in consequence to declaration discrepencies of 429 NUTS 3 regions and the German equivalent of 402 Landkreise/ Kreisfreie Städte.
[35] Since only two-digit NACE and respective CPA codes are considered in the following, this proposed unification will be applied for a more convenient readability.

- 21 Sections – coded by an alphabetical letter
- 88 Divisions – identified by a two-digit numerical code
- 272 Groups – identified by a three-digit numerical code
- 615 Classes – identified by a four-digit numerical code

The CPA systematic is organised as (EUROSTAT, 2013):

- 21 sections (alphabetical letter)
- 88 divisions (two-digit numerical code)
- 261 groups (three-digit numerical code)
- 575 classes (four-digit numerical code)
- 1.342 categories (five-digit numerical code)
- 3.142 subcategories (six-digit numerical code).

Besides economic activities, the European NACE industry standard classification system also covers their outcomes. One classification scheme is Prodcom statistics by Product (EUROSTAT, 2010). Its name is derived from the French *PRODuction COMmunautaire* (Community Production). Prodcom statistics contain about 3900 different types of manufactured products for mining, quarrying and manufacturing. Therein, products are identified by an eight-digit code that, for the first four digits, corresponds to the NACE and, for the first six, to the CPA system (ibid.).

In addition, the *Standard goods classification for transport statistics* (NST) in its current version from year 2007 is a specific statistical nomenclature for transported commodities – by road, rail, inland waterways and sea. The latter classification scheme considers the economic activity from which the commodities originate, thereby connecting related commodities to the CPA standard and the NACE framework[36].

In principle, a transformation of statistical information from one standard to another is feasible according to the correspondence table given in EUROSTAT (2009). However, depending on the level of detail and the direction of a transformation, the correspondence is not unambiguous. On a two-digit level for the CPA classification and a three-digit level of detail for the NST systematic, a conversion from NST data to CPA leads to a direct comparison as depicted

[36] Cf. Table G-9 to Table G-11 for details on the European commodity classification scheme.

in Table G-12. Therein, double entries that require further adaptions are marked. They originate from subgroups of the NST scheme that, in turn, are related to multiple subgroups of the CPA scheme, for instance group NST-048 that relates to CPA-10 and CPA-12.

In more detail, the NST standard groups together *other food products n.e.c. and tobacco products* (NST-048) that are related to either *food products* (CPA-10) or *tobacco products* (CPA-12). It is assumed that the volume of food products from NST-048, which is not classified elsewhere within the group of CPA 10, is rather small, compared to the volume of *tobacco products* (CPA-12). The remaining double entries are treated accordingly to obtain a basis on which to work within further modelling stages[37]. For this system and its state or market configuration, an initial assumption is stated, whereupon:

- the setup of commodity flows is assumed to be constant and equally distributed over a certain period of time – here, one year.

Although this assumption limits the ability of the model to represent the reality, it is inevitable since the goal is to analyse the German transport market interactions at the most detailed level possible. Seasonal effects and other time-related changes are neglected in that context.

Furthermore, commodity specific information is related to firms that in fact may also represent multiple sites or establishments of a single firm in terms of its legal structure. That leads to the following simplification that:

- within the presented context, each organisational unit that has a distinct spatial location as well as a distinct object of business organisation is modelled separately, addressed as an individual firm.

In total, more than 2,000,000 firms as both an origin and destination of freight transport activities are modelled within the borders of Germany. Another 29

[37] Note that a transformation of CPA formatted data into the NST structure on a similar level of detail is less trivial. A significant number of divergent relations between subgroups of *Chemicals and chemical products* (CPA-20) and *Basic pharmaceutical products and pharmaceutical preparations* (CPA-21) hinder a direct transformation. Fortunately, this drawback is of minor importance for the presented model since mostly a correspondence from NST-3-digit data to CPA-two-digit information is required.

synthetic firms are modelled as aggregates for neighbouring countries. This set is completed by 44 extra origins and destinations with significant relevance for land-based transports in Germany. Along with this understanding another assumption is stated for the model's representation of a final consumption. Thereafter:

- the final consumption of a product is related to retailing firms within a region.

In other words, private transports for e.g. shopping of *electrical equipment* (CPA27) in a hardware store are not modelled, whereas the commercial transport from its origin of production to the hardware store is part of the freight transport model.

4.4.3. Mode specific statistical information on the German freight transport market

Apart from overall mode specific transport volumes as given in Table A-1, certain data sources are accessible to set up a distribution of transport volumes for Germany by commodity class[38].

For mode rail, national freight volumes are listed in EUROSTAT (2014h). For international rail freight transports commodity specific transport volumes are determined according to an unpublished data source of the German *Federal Bureau of Statistics* (DESTATIS, 2013b). A detailed distribution of the overall volume of goods transported by IWW can be determined accordingly, including data from DESTATIS (2013a).
Deviant to road freight volumes for Germany, each shipment by mode rail and IWW is statistically recorded. These annual commodity flow counts are published on demand by the *Federal Bureau of Statistics*. Reports on rail and IWW transports are accessible and contain for instance:

[38] Originally the NST two-digit classification scheme applies in this context. As presented before, this setting can be transformed to the CPA standard.

- a regional specification for origins (NUTS-2)[39] and
- a regional specification for destinations (NUTS-2) of a transport path,
- a related commodity classification according to the current *standard goods classification scheme for transport statistics* (three-digit NST) and therefore:
    - a total annual transport quantity in tonnes (including weight of transport unit) and
    - the share of quantities transported in vehicles, containers, swap bodies (without weight of transport unit)[40].

In contrast, statistically reported freight volumes for mode road are the result of a survey without a comparable spatial and commodity class specification. Hence, the results are only useful to a limited level of detail.

Data discrepancies for German road freight transports

The German *Federal Motor Transport Authority* collects a dataset on road freight activities as a monthly survey. However, this survey only represents about 5 out of 1000 vehicles of a preselected tranche of vehicle classes (KBA, 2012a, p. 1). Furthermore, for confidential reasons detailed statistics on regional origins and destinations for differing commodities transported by lorry are not open to the public.

Considering the small sample within the statistical population of road freight transports, the given data need to be considered as an approximation to the *real-world* data, only. In consequence, the indicated road freight total is questionable. Moreover, with a higher level of detail the validity of survey data is further limited, for instance a specification of transport volumes by CPA. That is why additional data specifications, such as a spatial allocation of transport volumes, are not published for road freight transport volumes.

Nevertheless, at least an allocation of mode specific annual transport totals to groups of goods – hence, commodity classes – is feasible along with the presented rail and IWW data as well as with the survey results for mode road.

---

[39] Although an even more detailed request is possible, a trade-off between a more detailed dataset in contrast to additional anonymised sections applies (cf. section 12.2).

[40] This element will be valuable for a modelling of combined transport alternatives in section 14.

In EUROSTAT (2014g), the volume for national road freight transport estimates (cf. next section) is presented. EUROSTAT (2014p) and EUROSTAT (2014q) offer the volume of goods that are sent and received in Germany by domestic freight forwarders.

The counterpart of sent and received commodity classes transported by international freight forwarders can be determined from a consolidation of EUROSTAT (2014b), EUROSTAT (2014p) and EUROSTAT (2014q) [41]. Results of this data collection are depicted as follows[42].

[41] Here, from EU-27 countries. Volumes from other countries are not presented.

[42] Note that not only road freight transports are performed by national as well as foreign forwarders, but also IWW transports. These transports are sufficiently reported to the Federal Bureau of Statistics for this differentiator. The respective total transport volume in tonne kilometres is depicted in Table G-8.

	National					International											
						sending					receiving					
	German forwarders		foreign forwarders		IWW total	German forwarders		foreign forwarders		IWW total	German forwarders		foreign forwarders		IWW total	
CPA	road	rail	road	rail		road	rail	road	rail		road	rail	road	rail		
1	6%	2%	(6%)		12%	(5%)	2%	(6%)		12%	(7%)	1%	(12%)		8%	
2	0%	6%	(0%)		8%	(1%)	0%	(1%)		0%	(0%)	15%	(0%)		27%	
3	10%	14%	(10%)		22%	(7%)	10%	(10%)		19%	(6%)	10%	(3%)		16%	
4	17%	1%	(17%)		5%	(12%)	1%	(10%)		6%	(14%)	1%	(10%)		4%	
5	1%	0%	(1%)		0%	(1%)	0%	(1%)		0%	(1%)	0%	(1%)		0%	
6	7%	4%	(7%)		1%	(8%)	6%	(9%)		1%	(6%)	5%	(10%)		2%	
7	3%	14%	(3%)		20%	(2%)	3%	(3%)		10%	(2%)	5%	(1%)		18%	
8	7%	12%	(7%)		10%	(13%)	7%	(10%)		18%	(14%)	5%	(11%)		12%	
9	9%	4%	(9%)		3%	(8%)	1%	(8%)		3%	(6%)	1%	(5%)		1%	
10	7%	11%	(7%)		5%	(8%)	9%	(10%)		9%	(9%)	12%	(13%)		4%	
11	3%	0%	(3%)		0%	(3%)	0%	(4%)		1%	(3%)	0%	(4%)		0%	
12	4%	6%	(4%)		0%	(4%)	4%	(5%)		2%	(5%)	4%	(6%)		0%	
13	1%	0%	(1%)		1%	(1%)	0%	(2%)		0%	(1%)	0%	(3%)		0%	
1 – 13	**77% 188,003**	**72% 40,423**	**(77%) (6,368)**		**87% 9,540**	**(73%) 19,118**	**42% 8,558**	**(79%) 60,334**		**81% 10,309**	**(75%) 14,991**	**60% 13,799**	**(80%) 65,926**		**93% 20,738**	**76% 458,107**
14	6%	3%	(6%)		11%	(4%)	3%	(2%)		6%	(5%)	2%	(3%)		2%	
15	3%	0%	(3%)		0%	(1%)	0%	(1%)		0%	(1%)	0%	(0%)		0%	
16	3%	1%	(3%)		0%	(2%)	1%	(3%)		0%	(4%)	1%	(3%)		1%	
17	2%	0%	(2%)		0%	(1%)	0%	(0%)		0%	(1%)	0%	(0%)		0%	
18	9%	1%	(9%)		0%	(7%)	1%	(8%)		0%	(7%)	0%	(8%)		0%	
19	0%	22%	(0%)		1%	(0%)	40%	(2%)		9%	(0%)	32%	(2%)		3%	
20	0%	1%	(0%)		0%	(0%)	0%	(3%)		0%	(0%)	0%	(3%)		0%	
Non-EU-27						(13%)	13%	(2%)		3%	(7%)	5%	(1%)		1%	
Total	245,267	56,326	8,308	0	10,912	26,352	20,310	76,424	0	12,688	20,030	23,159	82,369	0	22,228	**604,373**

Table A-3 Total transport volume per NST commodity division for Germany 2012 in thousands of tonnes

To further specify these transport totals within a spatial setting – a prerequisite for a qualified freight transport policy – the available statistical input data is not sufficient.

B Freight transport modelling

A starting point of modelling is the presence or expected occurrence of a system's condition which a planning institution wants to influence (ARNOLD, 2008, p. 36). In this sense, the aim of a transport model is to make transparent the interacting or interdependent components of a real transport system. More specified, the purpose of a transport model is to describe and explain the behaviour of a transport system (BEN-AKIVA, MEERSMAN AND VAN DE VOORDE, 2013, p. 4). This general purpose can be further specified.

Specific purpose of a freight transport model

The classification of their application for different planning purposes is a useful general classification of freight transport models. Since a model's scope per se is limited, only an evaluation of the suited class of applications – formed by expectations of a planning institution – reveals the degree of quality to represent a real system. According to WIGAN AND SOUTHWORTH (2006, p. 4) model applications may be:

- traffic management studies,
- infrastructure investment planning,
- supply chain operations analysis,
- shipper mode and/or carrier choice studies,
- analysis of vehicle load factors and vehicle transport distances (notably for environmental impacts analysis),
- light duty commercial vehicle/service industry activity analysis (notably for land use planning),
- location or facility specific freight traffic generation/attraction estimation
- warehouse terminal locations modelling,
- modal and intermodal market competition modelling and
- modelling the effects of alternative user charges, subsidies, standards and/or regulations (on carriers, shippers, storage companies, third party logistics agents, customers).

Although this is only a general overview, it exhibits the potentially wide range of different transport models. That is why different general modelling types and related modelling techniques are discussed in the following – without claiming to be exhaustive. Subsequently, an overview on existing freight transport models will be presented.

5. General types of freight transport models

Transport modelling types for the aforementioned application purposes may be differentiated as:

- commodity vs. vehicle flow models,
- static vs. dynamic models,
- international, national, regional and firm-level models,
- optimisation vs. descriptive models,
- aggregate vs. disaggregate transport demand models and
- microscopic, mesoscopic and macroscopic supply models.

Commodity vs. vehicle flow models are clusters that refer to a general perception of freight transport systems according to WERMUTH AND WIRTH (2005, p. 296). If freight transport activities are evaluated as movement of goods, commodity based models are favoured. Otherwise, in case of vehicle movements in focus, vehicle flow models apply. The first concept requires transformation procedures to obtain data on transport infrastructure by vehicles but allows for a larger scope of freight transport analysis. Vehicle flow models generally are mode specific and used for detailed studies of limited spatial scope.

According to TAVASSZY AND BLIEMER (2013, p. 333), the separation of *static vs. dynamic models* follows the concept of time dependent transport modelling. For certain applications the dynamic nature of transport is of special interest and dynamic modelling is uncompromised. However, not all transport studies rely on different variables such as departure times for certain modes of transport or hourly peaks in infrastructure use. The latter holds true, for instance, for more strategic long-term studies.

A separation into *international, national and regional models* is one of the major classification methods in the context of freight system analysis differencing the spatial coverage of the model.

A distinction of *behavioural vs. descriptive models* which can be found in CASCETTA (2009, p. 239) is based on different perceptions of the market elements' behaviour. The former suggests explicit assumptions for transport patterns in the market. The latter refers to a reproduction of reality based on empirical data.

Clustering general modelling concepts into *aggregate vs. disaggregate transport demand models* as TAVASSZY AND BLIEMER (2013, p. 333) propose, is based on assumptions about the homogeneity of transport market demand elements. Either individual actors are analysed in detail in disaggregate studies[43] or their macroscopic counterpart, one aggregate or a few homogeneous components of transport demand. Although disaggregate models are able to represent more accurately the real transport system, i.e. preferences of transport demand elements such as an individual decision made in a manufacturing facility, it increases the model's structural complexity and data needs.

Macroscopic, mesoscopic and microscopic supply models are categories of general modelling concepts by TAVASSZY AND BLIEMER (2013, p. 333) that refer to the type of modes and means of transport considered in a model. Whereas macroscopic models only represent homogeneous and continuous streams of traffic, microscopic models in most cases consider specific transport vehicles in dynamic models. Again, the model purpose determines the level of detail. In this sense, the advantage of macroscopic models is their applicability to large scale transport networks, while the high level of detail in microscopic approaches is often limited to smaller networks. Mesoscopic models are applied for intermediate network scales with a higher level of detail than macroscopic.

---

[43] Microscopic transport models explicitly set focus on the logistics interactions framework and the decision units in a freight transport system that are individual shippers and receivers as well as related freight forwarders and/or logistics service providers.

Classification criteria for transport model concepts					
General perception of freight transport activities	*Scope of model in time*	*Scope of mode in space*	*Concept of modelling technique*	*Level of demand aggregation*	*Scale of supply representation*
Commodity flow models	Static models	International models	Behavioural models	Aggregate models	Macroscopic models
Vehicle flow models	Dynamic models	National models	Descriptive models	Disaggregate models	Mesoscopic models
		Regional models		Firm-level models	Microscopic models

Table B-1 Overview on general transport model concepts [adapted from MEST (2011, p. 23)]

The purpose of a model influences the general concept of the model and likewise the related techniques. That gives reason to the subsequent overview.

6. An overview on freight modelling techniques

In broad terms, different modelling concepts refer to variations of one or several steps of the 'classic' four stage transport planning process. Although it is a very basic structure, originally applied for passenger traffic models, the four stage concept and its variations[44] are still a guideline for freight transport models. The four stage concept is therefore deployed to structure the presented modelling techniques as described in the following.

6.1. The four stage transport modelling concept

This concept comprises a *trip generation* as the first stage to determine a total of transport trips originated at a source or origin and designated for a sink or destination. The second step of *trip distribution* is performed to combine relations of these nodes as trips. The application of the *mode choice* description

[44] See e.g. CLAUSEN ET AL. (2015, p. 44).

in the third step is followed by the fourth and last step, the *route choice* planning. At this stage, traffic flows are assigned to the modelled network. From this point of view, the four stage model is a specific form of a transport system analysis (see Fig. B-1). This is one reason why according to e.g. TAVASSZY AND BLIEMER (2013, p. 339) and WIGAN AND SOUTHWORTH (2006, p. 6) this general structure is still in use for many transport planning purposes. Related to a generic transport market's setting as given in Fig. A-5, the four stage modelling sequence can be illustrated as follows:

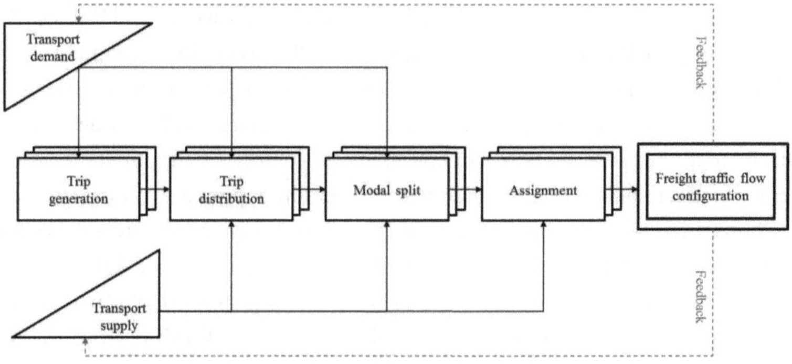

Fig. B-1 The four stage model as an application of a transport system analysis [adapted from MCNELLY (2000, p. 39)]

Although the application of this structure may be disputed, it provides not only a guideline to address specific transport planning issues, but also a typology of global categories of modelling concepts (cf. section 6).

6.2. Transport trip generation models

The first out of four stages of the basic transport planning architecture aims to predict the number of freight transport trips entering and leaving so called traffic cells. Traffic cells refer to zonal aggregates in the area of investigation. The objective of this stage can be approached in a number of ways. One process is based on direct observations or surveys. An alternative technique emphasizes the underlying role of economic trade among countries, regions or even firms to derive trip data. This leads to a separation of two basic types of models that are examined in the following to represent different trip generation modelling concepts:

- direct surveys or trip counts and
- input/output based models.

6.2.1. Direct surveys or trip counts

An elementary method to obtain estimates for freight generation and attraction volumes by zone is to make use of basic transport data collection techniques. Two of the most practical representatives are surveys and link traffic counts. Especially the latter technique requires the taking into account of specific inconsistency considerations[45]. Empirical data from freight surveys – gathered e.g. amongst shippers and receivers or carriers and forwarders in the designated area – as well as trip counts may be further extrapolated or used in regression analysis being adequate for certain applications. Factors such as household size, car ownership, employment, income etc. are of use for extrapolations and growth factor methods (DIOS ORTÚZAR AND WILLUMSEN, 2011, p. 139 ff.). An example for this type of trip generation method is implemented within the urban commercial traffic model WIVER from MEIMBRESSE AND SONNTAG (2001).

Explicit disaggregate trip based models analyse e.g. the use of specific classes of truck types, rail wagons or IWW vessels, mostly in a limited region such as local freight transport models or intra company vehicle fleet models. They then may suffice traffic management application purposes for urban areas studies (cf. introduction section 0). Aggregate demand models are put in place to describe accumulated in- or outgoing trips, e.g. in tonnes per day and mode between zonal aggregates, such as those being used in national statistics.

---

[45] For further information on data collection methods, types of surveys and technical limits see e.g. DIOS ORTÚZAR AND WILLUMSEN (2011), AMBROSINI AND ROUTHIER (2004) or RICHARDSON, AMPT AND MEYBURG (1995).

However, the *derived* nature of a freight transport demand within the economic context hinders a direct transfer of these techniques for more comprehensive modelling applications[46]. Another drawback for a direct implementation of forwarder survey data or trip counts within comprehensive freight transport models results from the micro-macro gap for intermodal transports (cf. Fig. A-2). Nevertheless, within specific concepts, direct survey or trip count data might be useful in the context of comprehensive freight transport models (c.f. section 15).

6.2.2. Input-output data based models

Input-output based models rely to a large extent on an economic input-output table as formally introduced by LEONTIEF (1936). Within this context, an input-output analysis describes relationships between different national or regional economic sectors. These sectors are linked – for both a customer of outputs and a supplier of inputs – upon the exchange of goods and services, measured in monetary units. In this framework, economic activities by sector being complemented by imports and exports if necessary, are linked to a final consumption by households. Input-output based models intend to reproduce a transport demand by the level of economic activities. A basic notion for the calculation of a total production x per economic sector, related to a final consumption c, would be:

$$x = (I - A)^{-1} \cdot c \tag{1}$$

Therein A represents a matrix of technical coefficients (indicating how much input is used for the given consumption) and I denotes the identity matrix of the same dimension[47].

In the following, two applications of an input-output analysis in the context of freight movements will be discussed in more detail, the *multiregional input-output models, spatial computable general equilibrium models* and *firm-to-firm transport trip generation models.*

---

[46] Cf. discussion of the present German *Federal Transport Infrastructure Plan* (BVWP 2030) in section 6 as well as the complementary context of Table A-2.

[47] For a comprehensive overview on the input-output research field and its applications see e.g. MILLER AND BLAIR (2009).

Multiregional input-output models

A related zonal trip generation for input-output tables can be derived from a total sectoral production in relation to spatial distribution parameters. This procedure allows for a translation of economic forecasts into transport forecasts[48]. Examples for operational models and explicit modules of comprehensive transport models that perform systematic procedures to make input-output data usable at the level of regions are multiregional input-output models (MRIO) such as:

- the multimodal elastic trade coefficients multi-regional input-output model for freight demand for Europe at the level of countries (CASCETTA ET AL., 2013),
- the PECAS framework of a generalised system for allocating economic production, exchange and consumption quantities for North American regions and certain US states (HUNT AND ABRAHAM, 2005) and the
- RUBMRIO analysis of production and trade patterns for Texas state counties by a random-utility-based multiregional input-output model (KOCKELMAN ET AL., 2005).

Due to the aggregate nature of trend and time series for economic developments as impetus, input-output tables are usually only available at the national or a limited regional level.

Spatial computable general equilibrium models

Another descendent from the elementary economic input-output concept is the class of computable general equilibrium (CGE) models. In the context of transport trip generation multi-regional or spatial computable general equilibrium (SCGE) models evolved. More recent representatives treat trade, transport and economic activities in integrative manner following the ideas of the new economic geography (NEG) that amongst other ideas takes into account an imperfect economic competition along with transport costs for

[48] Vice versa, input-output based models may be used to anticipate an economic development as a result of transport projects. An historical overview on applications is given for instance in WEISBROD (2008).

trade activities as well as the related effects of economies of scale[49]. Examples of these concepts applied in practice, are:

- the CGEurope concept of a multiregional spatial computable equilibrium model for Europe (BRÖCKER, 1998) and the
- RAEM framework in its latest version, a spatial computable equilibrium model for the Netherlands with a an explicit dynamic structure (IVANOVA, 2007).

These SCGE models represent a valuable concept to generate transport trip data on a large scale. They are – e.g. in comparison to traffic count or survey based trip rate models – to be seen as an advantageous promising approach to elaborate comprehensive freight transport models.
The broad spatial coverage may be a property to serve the purpose of an integrative analysis of transport and economic development, such as in cost-benefit studies (cf. introduction section 0). However, due to usually aggregate nature of national input-output data, SCGE models might be limited to a coarse zonal level of spatial detail[50]. This is where an alternative concept of input-output based models comes into play: the class of firm-to-firm transport trip generation models.

Firm-to-firm transport trip generation models

These transport trip generation models or modules start from a disaggregate transport demand structure. This is a prerequisite to capture logistics interactions at the level of microeconomic market actors (BEN-AKIVA AND JONG, 2013, p. 75). As a consequence, detailed interactions between firms may be modelled instead of trade between zones. This approach combines aggregate input-output information with disaggregate data on firms, e.g. a size measurement, a location as well as a sectoral activity. The latter information is at

[49] See e.g. KRUGMAN (1996) and LAFOURCADE AND THISSE (2011) for further elaborations of the concept of new economic geography as well as WEGENER (2011) and TAVASSZY (2009) for its deployment in SCGE models.
[50] One exception is a hybrid freight transport model impetus of aggregate input-output tables and mode specific traffic counts as well as commodity flow survey data, as proposed HANSEN (2011) for the PINGO SCGE model of Norway from VOLD AND JEAN-HANSEN (2007). This concept includes a transformation procedure to be valuable for disaggregate transport studies. Note that Norway is one of the few countries that fulfils the essential condition of being provided with national commodity flow survey data.

the core when firm level data is connected to input-output data. Examples of applications of this concept can be found in[51]:

- the disaggregation procedure of the aggregate-disaggregate-aggregate freight transport framework for Norway and Sweden (JONG, 2005), or the
- InterLog model's generation modules for an actor-based approach to commodity transport modelling in Germany (LIEDTKE, 2006) and the
- framework of the freight activity micro simulator for the U. S., in the firm generation module (SAMIMI, MOHAMMADIAN AND KAWAMURA, 2010).

6.2.3. Comparison

An overview on the examined procedures to generate transport trip data, including advantages and disadvantages for an inclusion in contemporary freight transport models, is given in the following figure:

		Advantage	*Disadvantage.*
Direct surveys or trip counts		• limited data requirements for local studies	• little insight into causality of policy effects • limited spatial coverage
Input-output based models			
	MRIO models	• link to economy • policy effects may be included	• require specific input-output table • limited spatial resolution • no explicit transport cost consideration
	SCGE models	• link to economy • policy effects may be included • time specific effects may be considered if dynamic	• limited spatial resolution
	firm-level models	• link to economy • policy effects may be included • high level of spatial resolution	• require firm specific data and/or separate allocation algorithms

Table B-2 Summary of freight transport production and attraction concepts [adapted from JONG, GUNN AND WALKER (2004, p. 111)]

[51] The given examples are further discussed in section 6.

The common output of trip generation models or modules is an undirected volume (for a certain time interval) of freight production and consumption as P_i and C_i per traffic cell i in the area of interest.

6.3. Trip distribution models

The trip distribution section of the four stage transport model concept determines trade flows[52] between origin and destination zones based on measures of production and consumption – usually obtained from a preceding transport *trip generation* model. In a model these trade flow relations are representable as x_{ij} by consecutive indices for transport network nodes of production $i = 1,2, \dots, N$ and consumption $j = 1,2, \dots, M$. The result of this transformation can be displayed in a trip distribution matrix.

Consumption / *Production*	1	2	3	…	J	*Total freight produced*
1	x_{11}	x_{12}	x_{13}	…	x_{1j}	$M1$
2	x_{21}	x_{22}	x_{23}	…	x_{2j}	$M2$
3	x_{31}	x_{32}	x_{33}	…	x_{3j}	$M3$
…	…	…	…	…	…	…
I	x_{i1}	x_{i2}	x_{i3}	…	x_{ij}	M_i
Total freight consumed	$N1$	$N2$	$N3$	…	N_j	X

Table B-3 Generic structure of trip distribution model results

In Table B-3, x_{ij} displays the total of trips (in tonnes of goods or vehicle movements) from i to j. The total amount of sent goods or equivalent departing trips in i – that corresponds to the transport *trip generation* output of the previous step of the 4 stage concept – is displayed as:

$$M_i = \sum_{J=1}^{i} x_{ij} \qquad (2)$$

The total of goods or trips in a consumption zone is formulated as:

[52] Note the subsequent discussion of trade vs. commodity flows in section 6.3.3.

$$N_j = \sum_{i=1}^{j} x_{ij} \tag{3}$$

The total of overall trips or transported goods in the investigated area is represented as:

$$X = \sum_{i} M_i = \sum_{j} N_j \tag{4}$$

To obtain such formal description of trade flow relations, the following two general concepts[53] are introduced, the *growth factor method* as well as the *synthetic trip distribution concept.*

6.3.1. Growth factor models

The growth factor method applies systematic multipliers to existing trade patterns, e.g. a sample trip matrix. A further distinction of this particular method is made between the *average growth factor method*, the *Detroit growth factor method* and the *Furness* or *Fratar approach*[54] (BLAUWENS, BAERE AND VAN DE VOORDE, 2008, p. 264). Their common prerequisite is to rely on an already available trip matrix. This, for instance, can be an outdated matrix, while having growth indicators at hand or a mode specific distribution matrix accompanied by information on the overall total freight consumption and distribution rates.

6.3.2. Synthetic trip distribution models

In case none of this data is available or the intention linked to a trip distribution model's scope is different, synthetic trip distribution models apply. Within this context, trade flows are considered as the result of decisions made by specific shippers and/or receivers. For the related sourcing and/or distribution of their exchanged goods the question of the underlying rationale for

[53] For a discussion of more specified problem types, e.g. regional specific empty vehicle *trip generations*, see e.g. (HOLGUÍN-VERAS, THORSON AND ZORRILLA, 2010).

[54] The Detroit method is also referenced as a single constrained growth factor method, whereas the Furness and Fratar method are similar representatives of the doubly constrained growth factor concept (DIOS ORTÚZAR AND WILLUMSEN, 2011, p. 180).

such trip distribution evolves. A formal framework to represent such a process is often based on descriptions for the perceived (in terms of the models representatives for shippers and/or receivers) transport resistance to bridge a spatial separation of demand and supply. One of the most common synthetic models is the application of the *gravity approach* for freight transport systems, see e.g. DIOS ORTÚZAR AND WILLUMSEN (2011, p. 182), BLAUWENS, BAERE AND VAN DE VOORDE (2008, p. 265) and HENSHER AND BUTTON (2000, p. 47).

The gravity model for freight traffic applications

The gravity model is analogous to Newton's law on a gravitational force between two planets that is proportional to their masses and inversely related to the squared distance in between. Transferred to freight trip distribution models, the number of trips or amount of transported goods between production and consumption nodes or zones is assumed to be proportional to their attraction indicators and reversely related to the functional form of transport resistance to bridge their spatial partition. This may be expressed in mathematical terms as:

$$x_{ij} = k \cdot M_i \cdot N_j \cdot f(c_{ij}) \tag{5}$$

where x_{ij} denotes the traffic flow from i to j, with k as a shape parameter to fit the functional form to an observation, an attraction expression of zone M_i as a point of production as well as N_j as a point of consumption and a resistance function $f(\cdot)$. In such context the transport resistance function may cover transport costs c between i and j, following e.g. the concept of total logistics cost (TLC) evaluations[55], including components for travel times, transport expenses in monetary terms, general appropriateness of trade relationship etc. Examples of functional terms in the given context are for instance given in TAVASSZY AND BLIEMER (2013, p. 335) by an exponential cost function $f(c_{ij}) = \alpha \cdot \exp(-\beta c_{ij})$ and top-lognormal cost function

[55] For an exemplary application of the TLC concept cf. chapter 14. Alternative or additional factors apart from costs may be e.g. variables for regulatory limits such as so called border resistances in international trade. For further discussions on an application of the gravity model for freight traffic applications see also BEN-AKIVA AND JONG (2013), BEUTHE, VANDAELE AND WITLOX (2004) and BAUMOL AND VINOD (1970).

$f(c_{ij}) = \alpha \cdot \exp(-\beta \ln(c_{ij}/\gamma))$ with α, β, γ as positive shape parameters. Freight traffic gravity approaches for freight trip distribution within comprehensive applications can be found e.g. in:

- the Dutch national transport Model for Integrated Logistic Evaluations SMILE (TAVASSZY, SMEENK AND RUIJGROK, 1998),
- the NEMO network model for freight transport within Norway and between Norway and other countries (VOLD ET AL., 2002) and the
- the Transport forecasting model TRANS-TOOLS (version 2) for Europe (MONZÓN ET AL., 2010).

A more general approach for distribution modelling is given through the principles of entropy-maximisation. Formally, the gravity transport model is a representative of the entropy-maximisation model class, albeit with a number of specific assumptions. In broad terms, an entropy-maximisation procedure seeks for the most likely configuration of the elements within a system. It is assumed that the trip distribution x_{ij} is proportional to this particular configuration, which, according to WILSON (1969, p. 111), is the same as the one obtained from a gravity model.
Such generalisation will be necessary for applications where the gravity model can no longer be written in the format of a regression equation – as displayed in the former section – due to additional constraints. This, however, is out of scope of the presented context[56]. An overview on advantages and disadvantages of growth-factor models and gravity models as a representative for basic synthetic models is given as follows:

[56] For further details see e.g. WILSON (1970) .

	Advantage	*Disadvantage.*
Growth-factor models	• limited data requirements (e.g. no specific transport resistance data needed)	• only applicable when transport relations are already reported • little insight into causality of policy effects
Gravity models	• limited data requirements (e.g. sound trip pattern observation not necessary) • insight into causality of policy effects through resistance (cost) functions	• limited scope for analysing policy effects on specific resistance (cost) factors • requires specific calibrated resistance (cost) factors

Table B-4 Summary of freight distribution modelling concepts [adapted from JONG, GUNN AND WALKER (2004, p. 112)]

6.3.3. The OD-PC-gap in freight distribution modelling

Growth factor models as well as basic synthetic trip distribution models require detailed input data. For national freight transport models this involves e.g. complete regional freight *trip generation* results – a prerequisite that might cause challenges to their application. For instance, within the German context a complete *trip generation* is not directly derivable from the total of all mode specific transport *trip generation* volumes and the related trip attractions per zone, since trips are not analogously observed for all modes[57].

Given these conditions, for a modelled transport system valuable supplementary data might be at hand, for instance nearly-complete records for corresponding transport relations might be at hand – or equivalent firm-to-firm transport trip estimates – as an initiation of the growth factor model. Alternatively, a selection of the respective information is available to calibrate an implementation of the gravity model. In case one of these inputs is provided, the starting situation is much improved but the challenge persists. The application of a basic synthetic trip distribution model is limited to mode specific distributions. Intermodal freight trip distributions, instead, require a concept that bridges the gap between origin-destination relations and production-consumption pairings.

[57] Cf. discussion of statistical road freight data reports within the context of the present German *Federal Transport Infrastructure Plan* (BVWP 2030) in section 6.

The OD to PC gap describes a divergence of general trade flow and detailed transport flow interactions. Both relations fall apart e.g. when transhipment nodes in a transport network are used. The capability of elementary gravity models that rely on trade-related transport trip data to capture this phenomena, is limited. In case a regional and mode specific freight *trip generation* is incomplete – such as for the German context – a bridging of the OD-PC gap is not only required to determine intermodal transport trips, but also to a mode specific trip distribution for the unknown mode. That is why models that aim to capture multiple modes in detail – here, a more advanced synthetic trip distribution model – require, at least, an additional data processing to identify the respective freight distribution.

6.4. Modal split models

The aim of modal split models is to evaluate the share of transport modes for zone to zone flows. Hence, mode choice models transform trip distribution tables into mode specific tables[58].

A huge variety of research projects has been carried out in this area, as modal split models are crucial for many transport studies[59]. Often political interventions in freight transport activities intend to influence modal choice processes that – in contrast to measures related to the *freight generation* and *distribution* – are assumed to have less burdening effects on the economy (BLAUWENS, BAERE AND VAN DE VOORDE, 2008, 269 f.). In this context, both aggregate as well as disaggregate models were established. A common modelling technique in this context – for aggregate as well as disaggregate studies – is the application of general principles of choice modelling. These principles are discussed within the following excursus.

6.4.1. Excursus on choice modelling techniques

Following the concepts of utility maximisation, choice modelling techniques attempt to describe the underlying behavioural process for a particular decision of an individual or a population of sufficient homogeneity (TRAIN,

[58] The general format of mode specific *trip distribution* tables may be comparable to Table B-3.

[59] DIOS ORTÚZAR AND WILLUMSEN (2011, p. 207) argue for instance that the choice of transport mode – at least for public transport – is probably the most important stage in transport planning and policy making.

2009). A specific case is the discrete choice model framework for analysing a decision that has to be made among a finite and countable set of alternatives – such as in mode choice decisions[60]. Discrete Choice analysis may be linked to utility theory, as it was done in the seminal work of MCFADDEN (1974). Further elaborations of this interpretation can be found e.g. in BEN-AKIVA AND LERMAN (1985).

In this context, factors that collectively determine a choice are regarded. These factors are classified as factors observable to the modeller and those that remain unobserved. A functional expression for a behavioural process would be $y=h(x, \varepsilon)$. Therein, ε represents the unknown factors, for which specific functions may be selected as candidates to model an expected outcome. A basic assumption posited therein is that:

- with a choice of individuals or populations, there is an underlying ratio, which is to strive towards an economic utility maximisation.

If the unknown choice component is assumed to be randomly distributed, and as such following e.g. a Gumbel distribution, the common multinomial logit model (MML) is an adequate candidate to study the decision problem. The parameters of these models may accordingly be estimated from available data using various regression techniques. For relaxed assumptions on homogeneity of grouped decision makers, defects in the interpretation of the *rationale for decision-making*[61] or for choices among specific alternatives, e.g. dependent choice sets, other functional forms with different levels of tractability may be used or combined. In total, discrete choice models are favoured when a sound knowledge (measured e.g. as statistical relevant number of observations) for choices concerned with sets of mutually exclusive alternatives is at hand. For transport system analysis, this is for instance supported by shipper or commodity surveys.

---

[60] Theoretically the *trip distribution* process, using the gravity approach, can be interpreted as a derivative of the discrete choice model for interpreting a firm's choice of trading relations and likewise of the vehicle choice process in the last step of the four stage concept (TAVASSZY AND BLIEMER, 2013, p. 340 ff.).

[61] Cf. MANSKI (1977).

6.4.2. Disaggregate modal split models

Disaggregate modal split models use data from shippers, carriers and/or external logistics companies[62] surveys or, alternatively, from commodity flow surveys. Both surveys of individual transport market components as well as surveys following shipments of specific transport orders are established in the format of *revealed* and *stated choice analyses*[63] in order to gain information on the mode choice procedure. Exemplary applications of discrete choice models in a superordinate freight transport modelling context are:

- a joint SP/RP model of freight shipments from a region in northern France (JONG, VELLAY AND HOUPÉE, 2001),
- a stated preference survey among shippers and 3Pls in Canada (PATTERSON, EWING AND HAIDER, 2010),
- a survey among shippers and freight forwarders to determine modal split functions for a Swiss national freight transport model (FRIES, 2008) and
- a freight forwarder survey for mode choice analysis for Germany (BÜHLER, 2006).

In contrast to disaggregate concepts that deal for instance with data at the level of firms, aggregate mode choice models are based on modal split data for accumulated vehicle or commodity flows. Although strong restrictive assumptions have to be made to apply the aforementioned choice modelling framework to aggregate datasets and consequently significant drawbacks results – such as the calculation of model outputs only valid for a restricted context – these models are still in practice. One reason may be the broad spatial coverage of aggregate statistical data, e.g. for national freight models[64]. Examples of such models are:

---

[62] Such companies which organise market interactions are usually referred to as third party logistics companies or 3PLs.

[63] Some models are exclusively applied to study mode choice c in detail, some are also applied in operational models. See discussion and classification on *revealed preference* (RP) and *stated preference* (SP) choice studies and their application to obtain time and cost information, as e.g. given in JONG (2008).

[64] See e.g. RICH, HOLMBLAD AND HANSEN (2009, p. 1007) for a discussion on aggregate models and suggestions for appropriate applications of analytic choice modelling techniques

- the European transport model Transtools of version 1 and 2 (MONZÓN ET AL., 2010),
- the weighted logit freight-choice model for the Oresund region (RICH, HOLMBLAD AND HANSEN, 2009),
- the Worldnet model for European and intercontinental long-distance transport (NEWTON, 2008) and
- the dynamic macroeconomic Assessment of Transport Strategies model ASTRA (SCHADE, 2005).

A summarizing comparison of the presented mode choice concepts is given in the following table:

	Advantage	*Disadvantage.*
Disaggregate choice evaluation	• Sound theoretical foundation	• Large datasets (surveys) required for representative analysis of larger scale (e.g. national level)
Aggregate choice evaluation	• Limited data requirements (e.g. national or regional freight statistics may be used)	• Limited insight into causality of policy effects due to restrictive assumptions • Validation challenges in case of aggregate data used for calibration

Table B-5 Summary of models for modal split evaluations within a superordinate freight transport modelling context

6.5. Assignment models

The final step of the four stage process seeks to allocate selected modes to routes, respectively links, within the modelled transport network. This includes a unimodal as well as multimodal assignment for either freight traffic only or together with passenger freight flows. When joint freight and passenger models are built, usually assumptions on truck and car drivers' behaviour have to be made, e.g. both having uniform routing preferences (TAVASSZY AND BLIEMER, 2013, 341 f.). Furthermore, conversion rates for space occupancy are determined, e.g. trucks expressed in passenger car equivalents (JONG, GUNN AND WALKER, 2004, p. 16). Examples for assignment modules in comprehensive freight transport models are:

- the traffic assignment module of the nation-wide macroscopic freight traffic model for the area of Germany from MÜLLER, WOLFERMANN AND HUBER (2012) and
- the tour building module of the INTERLOG simulation model for German road freight transports from LIEDTKE (2006).

6.6. Résumé on the four stage transport modelling procedure

The specific model design – including the theoretical specification and applied modelling technique – is the consecutive result of available input data in the context of a modelling purpose (see introduction of section 0). That is why many of the concepts presented for a certain stage of the four stage transport modelling procedure are transferred to other steps as well.
For instance, choice modelling techniques also may apply for *trip distribution* processing. The gravity model is likewise applicable for multiregional input-output based *trip generation* models. Certain techniques are also applicable for more than one stage, e.g. within a combined *trip generation* and *trip distribution* model or a *combined modal split and network assignment* model[65].
Although notation and terminology are specific, for instance if the generalised logistics cost concept is applied for mode choice studies where each alternative requires a unique indexing, the 'classic' four stage processing is an adequate procedure to globally categorize the different approaches in contemporary freight transport models in relation to their output. A brief summary is given as follows:

---

[65] Cf. section 6 for more examples.

	Input	*Output*
1. Transport trip generation	• Extrapolated traffic counts or survey data • Location of economic activities or spatial distribution of agglomerations • …	• Outgoing traffic per cell (M_i) • Incoming traffic per cell (N_J)
2. Trip distribution	• Output of transport *trip generation* process • Transport costs • …	• Trade flows (Xij) • Commodity flows (Xij)
3. Modal split	• Output of *trip distribution* • Cost and rates per mode • Travel times • Choice elasticities • …	• Modal share
4. Assignment	• Output modal split • Time savings and distance savings	• Freight flow assignment to infrastructure

Table B-6 Input and output data examples for modules of the four stage freight modelling concept [adapted from BLAUWENS, BAERE AND VAN DE VOORDE (2008, p. 259)]

New impetus to the 'classic' four stage traffic modelling concept is given through various developments in transport modelling research in terms of new combinations of the presented techniques, further data collections or advanced data integration. An overview will be given in the following section.

7. Present modelling approaches

Recent freight transport models on a national scale represent the trade-off between limited input data and their overall purpose. Some are aggregate models while others also include disaggregate elements. Instead of groupings of the reference units – such as trade flows, shipments and firms – the more complex disaggregate freight traffic models aim to represent *real-world* observations that are referenced as logistics modelling elements[66].

[66] See e.g. discussion in JONG ET AL. (2013), LIEDKE, SCHRÖDER AND ZHANG (2013), TAVASSZY, RUIJGROK AND DAVYDENKO (2012) and COMBES AND LEURENT (2007).

Logistics modelling elements

Logistics considerations are a paradigm to identify an underlying individual decision process for mainly mode choice models, with outreach to *trip distribution* and network assignment procedures (see subsequent examples). The inclusion of logistics coordination mechanisms as endogenous choice determinant results, for instance, in combined shipment size and mode choice models based on survey data. '*This, however, requires an exceptional amount of input data and represents a model complexity level completely different from what has been seen so far in freight transport modelling*' (RICH, HOLMBLAD AND HANSEN, 2009, p. 1007). This may give reason to the persisting use of aggregate freight transport models, especially on a national scale in a multimodal, multi-commodity transport system analysis.

Nevertheless, certain models or at least concepts are developed in response to a political appeal for more specific freight transport assessment tools. This development makes an assessment of the quality of existing freight models considerably less straightforward (WIGAN AND SOUTHWORTH, 2006, p. 7). As a result, logistics considerations are discussed along with the 'classic' four stage concept within the subsequent review.

Characteristics of reference freight transport models

The intention of this section is to give an overview on important freight transport modelling concepts from the recent past. The application of any of the four stages of the corresponding concept will be a major guideline. Apart from that, the overall *motivation*, the selected *scale* as well as the applied *methodology* will be evaluated for each of the selected reference models.

7.1. Characteristics of disaggregate national freight transport models

ADA concept for Norway and Sweden

Motivation: The Aggregate-Disaggregate-Aggregate freight model system (ADA) is a concept that was developed for Norway, Sweden, Denmark and the region of Flanders by BEN-AKIVA AND JONG (2013, p. 69 ff.). It is designed to help public authorities evaluate various freight traffic related policy measures with regard to individual aspects of logistics processes.

Scale: In order to meet its overall goal, the ADA freight model system is set to represent freight traffic on a national and partly international level. To likewise model logistics interactions, the reference values are commodity class specific shipments between individual firms. This approach also comprises multimodal transport options, albeit their related transport costs are not represented individually. The most advanced implementation, the one for Norway, covers 400 regional zones, 32 different commodity classes and about 100,000 sending and 400,000 receiving firms being modelled according to JONG (2005, p. 135).

Methodology: The general framework of the ADA approach is divided into three sections. By analogy with the four stage model structure, the first step includes the *freight generation* as aggregate flows of goods between zones of production and consumption. This section uses production and consumption statistics, input-output tables and trade statistics as primary inputs.
Subsequently this aggregate flow configuration is disaggregated to firms and shipments in order to obtain a distribution of freight flows between production origins and consumption destinations. At this point a specific logistics model is introduced covering the respective disaggregation as well as individual decision making processes. The latter refers to a shipment size selection, the use of consolidation centres, the use of containers and finally, a mode specific network allocation. Within the third stage the network traffic assignment is processed on an aggregate level to allow for an overall freight related policy evaluation.
Although it is part of the theoretic concept, the detailed elementary arrangement of aggregate commodity flows between sites of production and consumption within the practical implementation is only partly published. Instead, as indicated in JONG AND BEN-AKIVA (2004, p. 18), third party inputs from public Norwegian and Swedish institutions, private firms and their consultants are incorporated without a scientific discussion, adequate to the well-elaborated concept.
For an implementation in Norway commodity flows between sites of production and consumption are given as a multi-regional input-output distribution that defines a commodity specific annual trade flow volume between specified spatial regions. As given in BEN-AKIVA AND JONG (2013, p. 89), this setup together with an estimated number of receivers per sender – based on

expert interviews – results in a random assignment of firm pairs within a gravity model.
Within the well-defined logistics model potential transport paths are allocated to these firm-to-firm flows, including transhipment locations in order to identify alternative paths to reduce the time- and distance-related transport costs. Subsequently, a shipment size is determined accordingly. Based on this shipment size module the paths with the lowest logistics costs – here, the trade-off between costs for inventory holding and consolidation on the one hand and transport on the other hand – is assigned to each firm pair. Finally, firm-to-firm flows are aggregated to zone to zone flows, with an option to include empty return shares.
These mode and commodity specific zone to zone flow rates are then calibrated with related observed data. A validation is subsequently processed based on link flows from traffic counts.

Résumé: The ADA freight model system is one of the most sophisticated concepts to model nation-wide freight flows as a basis for profound freight policy measure evaluations. If a multi-regional input-output distribution and a commodity flow survey as well as an expert survey on potential trade flow configurations are at hand, many aspects of the methodology are transferable to other regions with reasonable efforts. However, a significantly higher number of nodes and links and, moreover, an increased total annual freight volume within the model leads to a different context.
A higher number of firms within a region leads to an increase of the impact of the random component within the disaggregation of regional trade flows to firm pairings. An effect that could be compensated with a higher zonal resolution of a multi-regional input-output distribution as well as the related commodity flow survey and expert interview data. This, in turn, requires an increasing number of professional resources for large numbers of nodes and links and the corresponding trade flow volumes. It may also result in the total absence of a multi-regional input-output distribution, a commodity flow survey or a related expert survey. This is a circumstance that is applicable for the German case.
Nevertheless, various aspects of the ADA approach serve as a benchmark for the German context. Although not explicitly labelled the same way, the procedure of a disaggregation of data from economic statistics to individual

firm-pairs is inevitable. Also the mode and path choice concepts are considerable as state of the art. At the same time it is in the nature of a comprehensive freight modelling framework to aggregate individual commodity flows between sites of production and consumption to flows between network origins and destination. For a German freight model this procedure should likewise include as many aspects of individual logistics decisions as possible within the context of less detailed input data.
Unfortunately, the ADA approach is missing a résumé on an implementation of the very promising concept of integrating consolidation centres[67]. Beyond that, no explicit application of the model is presented so far, giving an indication on the performed calibration and the final goodness of fit of the overall result in a comparison with e.g. link flow counts.

FAME for the United States

Motivation: The *Freight Activity Micro simulation Estimator* (FAME) for the United States, designed by SAMIMI, MOHAMMADIAN AND KAWAMURA (2010) introduced a behavioural freight movement microsimulation framework for the United States. This concept explicitly focusses on the role of firms and decision makers within the context of supply chain management elements.

Scale: The reference unit for the U.S. microsimulation framework is the shipment between firm-types. Firm-types aggregate firms with a common location, sector of industry and establishment size. This complexity reduction enables the bimodal FAME simulation to span about 330 industry classes for 50,000 firm-types in 130 geographic zones within the U.S., as indicated in SAMIMI (2010, p. 95).

Methodology: In order to model supply chain management decisions the modelling framework has a five-stage structure. Within the first section information on business establishments is gathered. This includes a dataset on

[67] Although consolidation centres in Norway and Sweden might differ from their German counterpart, e.g. those major centres located at either and of a ferry line allowing for consolidations of various different commodities, the concept might be relevant for certain coastal and inland network nodes in Germany as well. This could result e.g. in the capability of a German freight model system to represent wagon-load-specific train configurations instead of block trains only (cf. discussion in section 14.3).

location, employee size and industry type of establishments on a disaggregate level. Since the publication of this data includes a considerable number of details subject to confidential issues, an iterative proportional fitting (IPF) method[68] is applied.
The second modelling step addresses the estimation of trade flows between firm-types, the supply chain replication. According to SAMIMI (2010, p. 53 ff.) inputs for this trade flow assignment are on the one hand *input-output statistics* for an initial commodity class specific general allocation. On the other hand are the results of a web-based survey of a hired marketing company among 316 establishments within the U.S. Intention of the survey is to bridge the gap between data from aggregate statistics and individual specific data. One part of the survey was set up to test the likelihood of supply chain partnerships in relation to two criteria, size and distance. Finally, a general trade flow allocation is specified as flows between firm-types according to an application of the fuzzy set theory in line with ZADEH (1965). Analogously, as the third step, a shipment size between these firm-type pairings is processed based on fuzzy rules.
Stage four of the Freight Activity Micro simulation Estimator covers a modal selection module, using a binary probit mode choice model for mode road and rail as another outcome of the survey data processing. The aggregated result, the overall modal split, is than validated with the total road freight share from national statistics. The fifth step of the model design, the network analysis is still work in progress. Similarly, a validation and potential calibration of the commodity and mode specific sections as well as the supply chain synthesis – as part of the overall target – is not yet documented.

Résumé: A particular focus on supply chain configuration aspects is inevitable for state-of-the-art freight models. However, as the concept of a Freight Activity Micro simulation Estimator for the United States demonstrates, assembling the required individual specific input data is a challenge. If the individual specific input database is not sufficient, a calibration and validation of the data processing may require another external dataset. Nevertheless, an application of methods such as an IPF will be helpful to get a maximum amount of information from available datasets to achieve a microscopic freight modelling system at the best possible rate.

[68] The IPF algorithm helps to estimate unknown cell values of distribution tables (as e.g. depicted in Table B-3) in a way that either row or column totals remain fixed.

Unfortunately, the FAME model does not present any mode specific transport costs between firm-type pairings or a cost-related shipment size estimation. This could be helpful to validate parts of the commodity and mode specific sections as well as supply chain synthesis.
Such a setup of mode and shipment size choice modelling techniques following the concepts of utility maximisation on the one hand and the application of deterministic cost-minimising logistics modules on the other hand could be of interest for the German context.

INTERLOG for Germany

Motivation: LIEDTKE (2006) presented the simulation model INTERLOG for the German road freight transport market with a focus on microscopic modelling elements – here, synthetic firms – to represent the *real-world* decision-making of shippers and forwarders. Its aim is to analyse reactions of these actors as a response to policy measures and infrastructure upgrades.

Scale: With its distinct microscopic focus the reference unit within INTERLOG is the shipment between individual firms, specified by economic sector, size and location. For shippers this specification is broken down to a three-digit CPA level – equivalent to 261 commodity groups. They are allocated to about 440 national regions, based on a NUTS-3 spatial segmentation. Forwarders are then related to shippers to perform transports on the national road freight infrastructure. Although transport costs are not modelled, aggregate transport resistance functions are incorporated into a tour building procedure, including partial and empty loads.
The simulation time lapse is not fixed as it may include several reruns of the market interaction modelling procedure. This procedure is implemented to cope with different scenarios of a selection of 1000 senders and receivers as well as 200 forwarders at maximum.

Methodology: INTERLOG is divided into three sections:

1) The generation and location of shippers and transport companies that are heterogeneous in terms of economic sector, size and location
2) Simulation of a commodity flow configuration between suppliers and receivers

3) Integrated simulation of a shipment specification and routing as transport market interactions between shippers and forwarders

Statistical employment data serve as an input for the shipper generation module. This dataset is available at a form where the number of employees per region is included only at a NUTS-3 spatial resolution level together with a two-digit CPA level of detail on economic activities. Additionally statistical reports for Germany are available indicating the aggregate number of employees for certain company size classes on a three-digit CPA level. On this basis a Monte Carlo simulation[69] is performed, a data distribution algorithm that predominantly relies on repeated random sampling, complemented by certain assumptions on marginal totals – here, the three-digit CPA level employment distribution.
Within the second section of the model production rates are assigned to these synthesised firms. Therefore sectoral production data are obtained from German statistics. Missing data are filled by product specific weight density rates that are complemented by estimates or a deduction from production rates of related aggregate product classifications. Subsequently, a sectoral destination of a producing unit within the set of synthesised firms is estimated by using input-output tables. Since the level of detail within German *input-output statistics* is limited to two-digit CPA equivalents, a transformation is deduced from expert interviews and sectoral production data.
The supplier choice for an assignment of microscopic trade flows is modelled with yet another application of a Monte Carlo simulation. The role of randomness is reduced by weighted probabilities as the result of a distance sensitivity estimation as well as of sectoral trade flow rates.
The third section of INTERLOG simulates a road freight tour building. Therefore elements of the fuzzy set theory – as e.g. formalised by ZADEH (1965) – are implemented based on expert interview information on relationships between number of shipments, length of trip, number of stops and regions, empty shares as well as loading weights. The subsequent transport market simulation represents the contracting of shippers and forwarders to

[69] The Monte Carlo algorithm in the context of a cell value estimation for distribution tables is in broad terms comparable to the IPF algorithm. Both seek to refill missing values of a contingency table such as the example given in Table B-3. However, instead of a fitting to given marginal totals that e.g. result from additional observations, the Monte Carlo relies on a stochastic distribution of cell entries.

meet the predefined transport demand specifications in a way that road freight trips are estimated for the synthetic landscape.

Résumé: Due to the focus on individual behaviour within the shipper and forwarder contracting as well as to computational limits, only segments of the entire German road freight volume are modelled. Therefore, elements of a calibration and partial validation are performed based on empty trip shares within the modelling results. A validation of the generated artificial economic landscape is not performed. This may be the result of an absence of adequate reference data. However, INTERLOG presents a comprehensive structure on how to establish a database on producers and consumers within a German freight model. With an adaption of the number of shipments, the model also includes explicit logistic decision modelling elements. The INTERLOG model indicates potentials and limits of the applied techniques for further developments. The two most important lessons learned might be:

a. Instead of attempting to break down the available data to a maximum amount, a deterministic instead of stochastic refilling of the already existing gaps within the two-digit CPA level data might be preferred – especially within the context of a limited level of detail in the German *input-output statistics* and the absence of validation data.
b. This will be in line with the goal to displace the use of any random input factor to the latest possible step within further developments of the presented modelling structure.

It is indisputable that for policy measure evaluations as the central goal, a multimodal freight modelling framework is inevitable. In this respect, the role of international trade flows within a German freight model is not irrelevant. Similarly a limitation to selected freight flows is not target-oriented from this point of view, nor is the absence of a corresponding validation.

SYNTRADE for Germany

Motivation: The initial goal of the SYNTRADE model from FRIEDRICH (2010) is to apply formal concepts from the scientific field of *operations research* that originally focus on individual firms to an overall economic sector, estimating its transport demand. The German food retail sector is selected as an application example.

Scale: The model is split into a periphery of an artificial spatial company distribution including flows of goods between producers and retail stores – the reference unit – and a focus on warehouse structures for food retailers. Therefore, the model represents 439 German regions based on a NUTS-3 spatial segmentation as well as 41 regions representing European countries. 78 economic sectors are modelled within the periphery that are complemented within the core model by specific retail companies, types of logistic service providers and groups of wholesalers. With its focus on food retailers, the simulation reflects a unimodal road freight framework only.

Methodology: The twofold structure of SYNTRADE is reflected within its three modelling stages. The stage comprises a data collection and transformation for the core model on the food retailers (type and location of stores, article specific turnover) and the peripheral model for a distribution of different consumer goods from synthetic shippers to related receivers that is in large parts analogous to section one of the INTERLOG model from LIEDTKE (2006).
The second stage in SYNTRADE estimates supply paths between producers of consumer goods on the one hand and food retailing firms as well as consuming regions on the other hand. Subsequently, a transformation from *trip generation* to *trip distribution* is processed, comparable to the basic four stage transport model structure. At this stage, trade flows between regions and food retailers and their suppliers are set up. For the food retailing segment a number of suppliers per retailer size class is assumed and a random assignment that is guided by an evaluation of the size class is processed. For the remaining regional specific consumption a gravity model is implemented, based on assumptions for distance and size evaluation parameters.
The third stage within the model simulates a national warehouse structure for food retailers and all other transport paths, including the number, spatial location as well as the location within the food supply chain (central and regional warehouses). Therein a supply path decision is modelled for each type of consumption that incorporates a flow bundling to lot sizes along alternative transport links. A subsequent total logistics costs minimisation procedure finally leads to a warehouse distribution for food retailers and the remaining transport flows. The core model for food retailers is calibrated with company specific data, opening the way for a validation of the peripheral model.

Résumé: The SYNTRADE model presents an attempt to apply optimisation procedures that originate from *operations research* on a sectoral level. With its focus on one of the most important German economic sectors that integrate consolidation centres within its transport chains, it presents a guideline in this field. In this regard, the modelling concept also reveals the corresponding challenges. Above all there is the huge data demand to represent the heterogeneity of the German economic landscape in a microscopic model. As a result, random based data transformations are applied for various instances within SYNTRADE. A second major challenge is the complexity of modelling detailed logistics decision elements within a national model. The methodology that is chosen for the core model on food products is not directly transferrable to other consumer products. Without comparable product specific input data – e.g. type and location of stores, article specific turnover, exemplary distribution channel data – the remaining segment of consumer products is difficult to model on a large scale. Another burden arises for an adequate solvability of basic optimisation methods from *operations research* that gets increasingly affected by a larger scope of the model.

7.2. Characteristics of aggregate national freight transport models

ASTRA for Europe

Motivation: The macroeconomic model ASTRA, as presented in Schade and Krail (2006), is developed for an *Assessment of Transport Strategies* (ASTRA). It intends to provide a tool for long-term assessments of the European transport policy.

Scale: The ASTRA model is organised in eight modules that represent the population (POP), macro-economy (MAC), regional economy (REM), international trade (FOT), transport (TRA), vehicle fleet (VFT), environment (ENV) and welfare measurement (WEM) within 29 European countries. These modules are arranged to interpret wide-ranging causal relationships of passenger and freight transport related policy measures. Therefore, the REM module comprises 53 European regions, 15 goods producing sectors relevant for an aggregated *freight distribution* of three goods categories to three freight modes along five distance categories within the TRA module.

Methodology: ASTRA is a *system dynamics* model representing an approach to explain the behaviour of complex systems over time as a result of the dynamics of its interlinked components. According to SCHADE AND KRAIL (2006, p. 2) it is implemented in a software environment as one of the most complex *system dynamics* models worldwide.
To result in functional cause and effect relations, several statistical aggregates are transformed through a combination of different modelling techniques. Within the REM module *input-output statistics* are transferred from monetary values to volumes of tonnes to interpret sectoral trade flows for each modelled economic sector. As a result, national and international trade flows are generated and distributed within the modelling environment. Logit choice equations are implemented to break down sectoral flows to regional freight flows. A similar technique applies within the TRA module to estimate a mode choice along the distance bands.

Résumé: The goal of this model is to represent a wide scope of affected areas of a transport related policy. That gives reason to an absence of explicitly microscopic elements within ASTRA, representing decisions on the level of firms and shipments. For instance, supply chain related choices or modal selections are modelled on an aggregate level only. For an explicitly freight flow oriented national model the resolution of the modelling details should be significantly higher in order to estimate the magnitude and scope of policy measures in more detail.
The macroscopic ASTRA model helps to answer questions such as: How much would a fuel price change presumably affect the European freight transport in terms of vehicle miles travelled? However, a more disaggregate approach is required if the goal is to answer this kind of question with a specific regional or sectoral focus. The same applies when there are more specific questions to be answered, for instance: How much would the deployment of a certain infrastructure section affect the freight flow on this particular segment as well as the overall freight transport system?

SMILE for the Netherlands and Europe

Motivation: TAVASSZY, SMEENK AND RUIJGROK (1998) developed the Strategic Model for Integrated Logistics Evaluations (SMILE) as a national freight transport model for the Netherlands. Its aim is to estimate the effects of different scenarios on future freight flows operating as a decision support system. The model thus includes a set of specific logistics choices.

Scale: To forecast freight flows related to the Netherlands the SMILE model is structured into three segments. The first step describes the national production and consumption and derived trade flows on a regional level. Therefore 40 national and 37 international regions are encompassed. The subsequent calculations result in a spatial distribution of freight flows for commodities, grouped into 50 *logistic families*, along different distribution channels including three configurations of distribution channels, the direct distribution or the use of either one or two distribution centres. The last step involves an assignment of freight flows to one out of six modes comprising: road, rail, inland waterway, air, pipeline and sea.

Methodology: With its three level structure the SMILE model relates to the basic four stage concept of a complete freight model system. To generate freight flows within the first section, make-use tables are included. To transform the resulting trade flows to freight flows, the *trip distribution*, alternative distribution channels together with aggregate characteristics (e.g. transport and warehousing costs, lead times, available modes) of the selected homogenous commodity groups are processed within a multinomial logit choice model. A separate subsection forecasts the location of regional distribution centres in relation to the identified trade flow relations. Distribution centres are therefore assumed to be accessible within each modelled region. In line with the *path choice sequence*, a mode specific transport path is determined in the final section according to logistic costs and time factors specified for each logistics family class.

Résumé: According to JONG ET AL. (2013, p. 352) the SMILE freight transport model is the first national model where logistic decisions along with the transformation of trade flows into freight flows are explicitly included. It comes along with a number of innovative general concepts that may give reason to its implementation within parts of the European transport model Transtools (MONZÓN ET AL., 2010). However, compared to more recent publications (e.g. the ADA model, the FAME concept or the Interlog model) certain limitations for further adaptions can be identified.

For the German context the use of make-use tables instead of input-output tables is questionable. German make-use tables both rely on different prices (purchaser price and sales price) and require a transformation to apply these in a common context. Since this transformation on the level of commodity classes leads to the structure of input-output tables, their immediate implementation into a freight modelling context is reasonable. Using production factors instead of a distinct link between economic sectors of production and consumption creates an option to include them as variables within the modelling framework. This may be an advantage for forecasting capabilities of a freight model. However, since their transformation according to the principles of an input-output table is required in any case, a separate adaption of inter-sectoral trade flows within the input-output configuration is favoured.
The SMILE model is an aggregate freight transport model. Logistic decisions for supply paths and modal selection are solely modelled for aggregate flows instead of being firm specific. Likewise, shipment sizes and the use of distribution centres are modelled for large commodity aggregates instead of specific commodity classes as e.g. given in make-use tables and input-output tables, respectively.
An estimation for the potential usage of consolidation and/or distribution nodes could also find its place within the German context. Its execution on a microscopic scale is best placed at the end of the modal choice and transport path selection for each OD-pairing.

BVWP 2030 for Germany

Motivation: The overall aim of the German *Federal Transport Infrastructure Plan* (BVWP 2030) is to resolve problems within the national transport network. In this context the transport plan is designed as a governance tool control instrument to achieve greater efficiency within the expansion of existing infrastructures and new buildings. Its actual version with a planning horizon in the year 2030 is presented in BMVI (2016a).

Scale: Within the German infrastructure plan projects for an expansion and new builds within the public road, rail and IWW infrastructure are designated. This selection, to a large extent, results from a prognosis on interconnected passenger and freight traffic flows, as given in SCHUBERT (2014). The *land-based* freight traffic related prognosis considers 20 commodity types, based on the NST classification scheme. It derives from regional and commodity specific OD transport matrix per mode for the year 2010 and partly 2008 as the reference units of this study.

Methodology: The inland freight vessels transport matrix is based on a project specific statistical evaluation. As a result, zones for a NUTS-3 level of detail are included in this particular transport matrix for containerised and non-containerised transports.

For mode rail a similar level of detail is established within the related OD transport matrix. Therein the share of commodity specific containerised transports is estimated. Additionally, transports of hard coal and lignite are specified based on separate observations. International transport relations are evaluated as aggregates to adopt the structural data format within the prognosis. Furthermore, together with an informal dataset from German Railways Company, a segmentation for block and single wagon load trains is implemented as well.

For mode road an equivalent NUTS-3 level OD freight traffic matrix for 20 specified commodity types is estimated. Because of the contrast to modes rail and IWW, the largest freight traffic segment by mode road is statistically specified by a national freight forwarder survey that, in turn, is only accessible for certain aggregate attributes. These attributes only distinguish the total road freight traffic for German NUTS-1 regions into e.g. regional or ingoing and outgoing[70] – a challenge for a breakdown to NUTS-3 level specific flows, especially with an additional commodity class specification. Another relevant drawback for a processing of German road freight survey data within a road freight OD matrix estimation is the limited focus on national lorry drivers. That is why international transports are difficult to estimate, especially in terms of tonne-kilometres.

As a consequence, a national road freight traffic estimationfor the year 2007 is processed together with a breakdown of aggregate reference values from the year 2010 German road freight survey. A comparable data projection is executed to obtain partly disaggregate data for international road freight transport relations for the year 2008. In order to obtain 2030 data from the

[70] The further use of the German freight traffic survey report is not only limited due to its aggregate nature, but also the significant number of missing entries per attribute as a result of inadequate related sample sizes.

2010 cell to cell flows a *trip distribution* and *modal split* estimation in line with the 'classic' four stage traffic modelling sequence are processed.
The *trip distribution* is a regression based approach that involves aspects of the German economic landscape – here, selected locations of coal and lignite mines, crude oil refineries, metal fabrications and the representatives of the automotive industry. A gravity model is applied to estimate a future *trip distribution.*
For a modal specification of these zonal OD transport flows a choice estimation model that follows the concepts of utility maximisation[71] is applied. Therein, an unpublished[72] *stated* and *revealed preference interview* is processed. The *revealed preferences* analysis is related to transport cost, time and time punctuality variables. The resulting functional terms are then related to global transport time punctuality estimates as well as OD specific transport cost and time parameters. These specific parameters are the result of the study of BMVI (2014a).

Résumé: Although freight traffic flows for some parts are represented for a high level of detail, the BVWP 2030 freight matrix estimation procedure is a representative of an aggregate freight modelling approach. The model does not focus on specific shipments per firm pairing but annual traffic flows between traffic cells. This elementary top-down approach targets a breakdown of aggregate traffic flow data to industry-related trade flows in order to estimate the magnitude of impacts from economic variances on freight traffic flows in turn. Although this breakdown of aggregate reference values is challenging for a high level of spatial detail within a multi-commodity framework, its outcome might be helpful for a validation of corresponding microscopic models.
However, the usability of an aggregate top-down freight model within the context of freight flow forecasts is limited. Instead of individual firms and their related logistics cost minimisation principles, only freight flows between traffic cells are estimated. As a result, the *trip distribution* and modal choice are only represented for aggregate zonal flows. In other words, each producer or consumer within a zone can no longer individually choose adequate trade partners or the respective mode of transport and transport path.

---

[71] See also section 6.4.1.

[72] According to SCHUBERT (2014, p. 141) the survey is not commodity specific. Neither the sampling selection nor the number of participants is documented and so is the related response rate.

The underlying logistic decisions – such as a minimisation or the overall storing and transport costs for different modal alternatives – are not examined, although they are crucial for an assessment of potential modal shifts on a national level.
Even on this aggregate level the validity of the crucial *trip distribution* and modal choice modelling stages is challenged within the context of an only estimated regional distribution of commodity class specific road freight traffic data. In addition, the scope for a validation of aggregate input data based freight transport models is very limited.
The lack of an explicit link between freight traffic on the one hand and an economic pattern for a distinct system on the other hand, such as for Germany, is yet another obstacle for an interpretation of various potential economic changes within the context of a freight forecast. Without a sector-specific economic framework as a basis, a forward projection of specific local and global economic trends cannot be unambiguously associated to freight traffic – neither in detail, e.g. seaports and their inland counterparts, not in general to the overall national freight flow volume.

7.3. Conclusion

Different nation-wide freight transport models exist. They vary in terms of their overall objectives and therefore within their scale and the implemented modelling techniques. As a result, freight transport models are difficult to compare. Some freight models mainly intend to introduce different perspectives or formal methods to the context, others claim to serve as a governmental decision support tool. Microscopic concepts can be identified focusing on firm-to-firm transport flows and individual logistic decisions whereas also aggregate traffic flow models are still state of the art. A simplified comparison of the selected models and their capabilities is depicted within the following table[73]:

[73] The format for a depiction of the literature review synthesis is inspired by the work of FRIEDRICH (2010, p. 100). It is an attempt to compare complex concepts and models that are structurally very diverse.

		Aggregate models			Disaggregate models				
		ASTRA	*SMILE*	*BVWP*	*ADA concept*	*FAME*	*INTERLOG*	*SYNTRADE*	*present study*
scale	multi-commodity	Partly modelled	Partly modelled	Explicitly modelled	Explicitly modelled	Explicitly modelled	Explicitly modelled	Explicitly modelled	Explicitly modelled
	multimodal	Explicitly modelled	Explicitly modelled	Explicitly modelled	Explicitly modelled	Explicitly modelled			Explicitly modelled
	international transport	Explicitly modelled	Partly modelled	Partly modelled	Partly modelled		Partly modelled	Partly modelled	Partly modelled
depth	economic network representation	Partly modelled	Explicitly modelled		Explicitly modelled	Explicitly modelled	Explicitly modelled	Explicitly modelled	Explicitly modelled
	firm-to-firm shipments				Explicitly modelled	Partly modelled	Explicitly modelled	Explicitly modelled	Explicitly modelled
	transport path cost evaluations			Partly modelled	Partly modelled				Explicitly modelled
logistic decisions	trade flow configuration	Partly modelled	Explicitly modelled		Explicitly modelled	Explicitly modelled	Explicitly modelled	Explicitly modelled	Explicitly modelled
	shipment size selection				Explicitly modelled	Explicitly modelled	Explicitly modelled	Explicitly modelled	Explicitly modelled
	mode choice	Explicitly modelled	Explicitly modelled	Explicitly modelled	Explicitly modelled	Explicitly modelled			Explicitly modelled
	use of consolidation centres		Explicitly modelled	Partly modelled	Explicitly modelled			Explicitly modelled	Partly modelled
	location of consolidation centres		Explicitly modelled					Explicitly modelled	
	lorry tour building						Explicitly modelled		

Explicitly modelled
Partly modelled

Table B-7 Comparison of scale, depth and the role of logistics decisions within reviewed freight transport models

C Framework for a disaggregate German freight transport model

The conceptual framework, as proposed hereinafter, follows the identified role of logistic interactions in a national freight transport market. The result is a disaggregate microscopic modelling concept that focusses on specific decision makers – an encompassing set of *individual firms*[74]. Therefore, a framework is designed to cope with the major hindrance of detailed national freight transport models: the absence of specific information on individual firms' logistic interactions.

This modelling framework aims at representing the German *real-world* freight transport system as an economic market for a:

- *short-term* period of one year with fixed locations for all considered transport network nodes,
- *long-distance* nation-wide scale, including international linkages, where the physical result of spatial economic interactions is realised by *multimodal* alternatives of land-based transport,
- a sourcing and distribution of *multiple-products*, whose transport is arranged with *multiple vehicles*[75].

As a consequence, the model is stepwise implemented. This concept will be in line with the four-stage traffic modelling framework, as e.g. presented in section 6.6. The goal of the resulting framework is to handle the defined complexity as well as to identify linkages for structured freight transport market interventions.

Thus, the proposed German freight transport modelling framework contains three basic modules – with a specific combined modal split and assignment module – as depicted in Fig. C-1. Each sequence comprises a set of public data sources as an input and/or a separate validation. For necessary details within these inputs that are subject to confidential issues – the data gaps – the corresponding processing shall, as far as possible, follow the principle of a deterministic reconstruction. In order to further raise the level of modelling

[74] See Table B-1 for a related classification scheme of transport model concepts.

[75] Cf. Table A-2.

realism, the proposed structure also includes recursive relationships (see dashed lines in Fig. C-1).

8. The Freight generation module

The first module seeks to generate firms as trip sources and sinks. Focal subjects are individual firms, specified by a type of economic activity, firm size and location. Additional information such as regional interaction patterns, production design, supply chain configuration, turnover, firm size in terms of square footage etc. could be helpful for the present as well as further modelling stages.
However, sources of data for a German firm-based modelling to the extent of an entire country are limited. Most company specific data are of a confidential nature and hence not available to the public.

By contrast, the critical part of firm specific data may be accessed through national statistics. Such data are given e.g. in labour market statistics. Therein, firms are classified by number of employees, geographic location and the predominant economic activity up to a certain level of detail. Again, much information within this particular context, necessary for a detailed nation-wide virtual economic landscape, is subject to confidential issues. Nevertheless, German national labour statistics provide probably the most comprehensive input data for this module.

> *As a consequence, a firm generation procedure, based on selected statistical data that enable a mostly deterministic treatment of potential data gaps, is considered to be particularly suitable for this module.*

9. The freight distribution module

The consecutive *trip distribution* module consists of two sub-sections, namely a supply chain synthesis and the subsequent allocation of supplier-consumer pairings for a final trade flow generation. Both stages, in an allied context, are regarded as adequate instruments to enable a representation of firms that are part of one or multiple supply chains in *real-world* transport systems. The focus at this stage is therefore set on two global dimensions of a supply chain organisation:

- the *integration of firms* within a supply chain synthesis and
- the *flow coordination* within a shipment size determination[76].

Since both complementary firm management streams include more company specific decisions than present public datasets allow one to represent, assumptions for the modelling framework have to be made. This may give reason to the observation that this stage of transport system analysis is one of the most underdeveloped in transport literature (cf. section 0).

> *As a result, this module aims at representing a generic structure for individual real-world supply chain compositions within a freight transport modelling context. This procedure provides interfaces for additional data inputs in future that are not at hand at present.*

9.1. Supply chain synthesis

Within this stage, a primary twofold procedure for a supplier supply chain configuration synthesis is included to represent the trade flow network *integration of firms* for Germany in practice. This procedure includes a supplier pre-selection as well as a final allocation of supplier-consumer pairings.

Supplier pre-selection

The outcome of this part of the supply chain synthesis sub-section is a distinct set of potential national and international sources of production for each consumer of a specified commodity class.

> *In order to reach this target, trade statistics as well as national input-output calculations are consulted as an initial classification of potential trade partnerships between commodity class specific firms.*

Based on this initial distribution of source of production – here, suppliers – and sinks of consumption – here, intermediate and final consumers – a second procedure is applied to obtain a final supply chain allocation of firms, relevant for the German freight transport market.

[76] See also section 4.1.

9.2. Allocation of supplier-consumer pairings

The second part of the supply chain synthesis module aims at answering the modelling question: Which one of the potential suppliers per firm is finally selected in practice? Since sinks and sources – which are also addressed as suppliers and customers or producers and consumers within the context of a supply chain analysis – interact explicitly inhomogeneous when setting up partnerships along one or multiple supply chains, the context of this question is complex.

A starting point to handle this complex task would be a commodity flow survey. In the format of a shipper-based survey, transport paths of certain commodity classes per mode could be interpreted to reconstruct an underlying supply chain organisation within the reference area. But unlike e.g. Sweden, Norway (BEN-AKIVA AND JONG, 2013) and France (COMBES, 2012b) this source of information is not available for Germany. This type of information on exemplary commodity flows could be helpful to estimate likely supply chain configurations by product. For instance, when a distinct pattern within the setting of suppliers and consumers is identifiable, a more refined modelling is feasible compared to cases without a survey.

Nevertheless, in order to conduct a model of the commodity flow system for Germany, each considered firm needs to be represented within its unique supply chain configuration as best as possible. Therefore, a deterministic approach is favoured – a structured allocation of firm pairings that is not part of the reviewed freight transport models[77]. This gives reason for a brief review on supplier choice selection modelling techniques that are applied outside the context of freight transport models.

According to a literature review on formal supplier choice selection problems performed by HO, XU AND DEY (2010, p. 16), the following contemporary alternative solution paradigms can by identified – either in individual or in combined approaches:

---

[77] Compare with discussion in section 7. For instance, the ADA freight model system as one of the most sophisticated concepts to model nation-wide freight transport has a regional specified input-output dataset as an input.

- *data envelopment analysis* (DEA),
- *mathematical programming,*
- *the analytic hierarchy process* (AHP),
- *scoring models,*
- *fuzzy sets theory,*
- *genetic algorithms* (GA) and
- *neural network approaches.*

The *data envelopment analysis*, as introduced by CHARNES, COOPER AND RHODES (1978), can be used for supplier selection procedures performance using empirical measures to evaluate the efficiency of potential suppliers[78]. The efficiency evaluation is defined as a ratio of a weighted sum of outputs – e.g. quality, customer satisfaction – to a weighted sum of inputs – e.g. total cost of shipments, supplier reputation. The efficiency criterion leads to an allocation that represents a formal frontier along which the relative performance of all assessed suppliers can be compared. Examples can be found in SAEN (2007) and TALLURI AND SARKIS (2002).

According to HO, XU AND DEY (2010, 17 f.), various types of *mathematical programming* methods also apply for supplier choice modelling. These include linear programming, integer linear programming, integer non-linear programming, goal programming and multi-objective programming. They are commonly applied in search for an optimal problem solution, for instance to solve a supplier selection problem involving multiple objective functions for total costs, the quality rejection rate, the late delivery rate, such as in WADHWA AND RAVINDRAN (2007) or additionally the flexibility rate as in LIAO AND RITTSCHER (2007).

The *analytic hierarchy process* as proposed by SAATY (1987) is a procedure that decomposes a supplier selection decision problem into a hierarchy of sub-problems, each of which is evaluated independently by paired comparison, with respect to a given criterion that is stated as an upper hierarchy level element. These evaluations are converted into numerical values that can be

[78] A prerequisite for most final selection procedures – such as the AHP – is a pre-selection of potential suppliers with common attributes. A procedure that is comparable to a large scale clustering of potential firm-to-firm relationships according to input-output tables from national statistics. For a detailed discussion of formal pre-selection procedures see e.g. AISSAOUI, HAOUARI AND HASSINI (2007, p. 3520 ff.).

consistently processed allowing for a straightforward supplier selection by priority. AHP supplier choice literature examples are CHAN ET AL. (2007) and KAHRAMAN, CEBECI AND ULUKAN (2003).

A *scoring model*, such as formally introduced by ZANGEMEISTER (1971) – in broad terms – is related to an AHP. The method uses basic arithmetic calculations to approximate a supplier choice ranking. In contrast to the standard AHP, the ranking criterion is not determined by pairwise comparison, thus each alternative supplier is only weighted with a representative of a given set of criteria. Multiplied by a criterion specific weight factor, the procedure directly bears estimates for a ranking among the alternatives. Due to the fact that the scoring model is advantageous in terms of processing time, but dominated by the analytic hierarchy process regarding consistency, the model is rather used in hands-on applications than in supplier selection literature.

Fuzzy sets theory is applied in supplier selection procedures following e.g. the principles of ZADEH (1965). Fuzzy sets generalize basic Boolean logic, in which an observed element is either included or not included in a set. Expressed by membership functions, fuzzy sets theory permits a gradual membership instead. This property may be useful for a supplier evaluation and selection problem when e.g. a vague judgement of preferences is more confident to a decision maker than crisp values. Examples for such applications are GUNERI, YUCEL AND AYYILDIZ (2009) and SARKAR AND MOHAPATRA (2006).

Genetic algorithms techniques, as raised predominantly by HOLLAND (1992), are useful to solve supplier selection problems. They are applied as search heuristics inspired by selection mechanisms of biological evolution. Starting with a randomly generated candidate solution, superior solutions sets are usually searched for through iterative recombination and mutation such as in genetics and afterwards assessed through fitness functions to further work out a suitable solution. Examples for such processing of a supplier choice selection are given in CHE AND WANG (2008) and DING, BENYOUCEF AND XIE (2006).

Résumé

For the secondary stage of the supply chain synthesis various techniques are generally considerable. However, due to limited data availability for the nation-wide context, most discussed supplier choice selection formalisations are not applicable for a German freight transport model system.

> *In this context and with regard to limited references for practical supply chain evaluations, a scoring model is implemented within the disaggregate German freight transport model as the second component of a supply chain synthesis.*

9.3. Shipment size determination

In addition to a supplier choice modelling procedure, a shipment size configuration is required for a partial consideration of informational as well as material and financial aspects of a *real-world* freight *flow coordination*[79]. The shipment size and frequency determination for every single firm pair is designed according to pull-principles. Hence, the sourcing unit – be it a firm or a final consumer – takes control of the shipment size definition. Useful methods for such detailed microscopic shipment size determinations are:

- *the basic Economic Order Quantity model* (EOQ),
- *extended lot size determination models.*

The EOQ model, as introduced by HARRIS (1913) and further formalised by WILSON (1934), is a basic production logistics model applied in the field as well as in theoretic analysis. The model generally determines an order quantity in terms of lot sizes that minimize total inventory holding costs – usually increasing – and ordering costs that usually decrease with lot size.

Extended inventory theoretic models for shipment size determination consider additional variables such as demand fluctuations as well as for considering multiple products in a joint optimisation procedure. Usually, dynamic versions of the EOQ model for the one- or multiple-product case are in line

[79] In absence of nation-wide information on individual firms and their supply chain network coordination procedures, these dimensions – especially the financial element – can so far only be modelled very sparsely, e.g. in terms unrestricted financial flows in the shipment size planning.

with the proposition of WAGNER AND WHITIN (1958) and corresponding further developments[80].

An assessment of the empirical validity for a large heterogeneous population – as such given by shippers and receivers in this analysis – for most of these microeconomic formulations is, however, mostly pending. An exception is e.g. presented in COMBES (2012b), who used the national French shipment database ECHO[81] to econometrically assess the basic EOQ model. The study concludes that this particular lot size model is an effective approximation to consider a real-world firm's shipment size decision in a comprehensive freight transport model (COMBES, 2009, p. 290).

> *This gives reason to the application of the basic microeconomic EOQ model in the given context. That is in line with the twofold character of the proposed model – to set focus on the behaviour of firms in an encompassing transport market modelling not without designating an adequate data input*

10. Combined modal split and network assignment model

The third module that is proposed to model individual firms within a German freight transport model is a composition of stage three and four of the 'classic' four stage transport modelling concept. The goal of a *combined modal split and network assignment* model is a representation of one of the most important logistic considerations in a transport system, the individual mode choice decisions, without relying on complex survey data (cf. Table B-5) that are not at hand for Germany[82].

One possible way of setting up such a *modal split* module is an evaluation of transport costs and time for each potential mode specific transport network path alternative with respect to each particular transport relation of firm pairings that are identified within the *freight distribution* module. Subsequently,

---

[80] For further discussion of advanced lot size determination procedures and potential solution algorithms see e.g. (SIMCHI-LEVI, BRAMEL AND CHEN, 2005).

[81] The ECHO database is based on a commodity flow survey for approximately 10,000 observed shipments in France for the year 2004 (COMBES, 2009, 62 f.).

[82] Note that not only the absence of such data gives reason to a consideration of alternative approaches but also the limited appropriateness of survey data itself for many model applications (cf. argumentation of RICH, HOLMBLAD AND HANSEN (2009), as given in introduction of section 7).

the modal selection is the result of two decision affecting components – a cost and time factor.

The latter component is part of the shipment size determination model in terms of an estimation of the value of time per commodity class. This means that e.g. *computer, electronic and optical products* (CPA-26) are potentially less bundled and more frequently ordered in smaller shipping units between a firm pairing than – ceteris paribus – *products of forestry, logging and related services* (CPA-02). For this comparison shipments by road freight lorries may than seem to be a more cost efficient option for CPA-26 than for CPA-02. Vice versa, higher transport costs affect the shipment size determination.

> *This gives reason to the implementation of a feedback loop between the combined modal split and related network path selection on the one hand and the shipment size determination of the freight distribution module on the other hand.*

Although interlinked, the *value of time* estimation is mostly part of the *EOQ modelling* section. The *evaluation of costs* and a derived mode choice preference, in contrast, is merely part of the *combined modal split and network assignment* model.

This concept requires a complete evaluation of all potential transport path alternatives for the German freight transport system as well as the respective transport costs. Each of these specific evaluations as well as the final modal selection corresponds to a distinct transport network path choice and vice versa.

> *The transport cost related modal selection module directly refers to a path selection and the other way around – that gives reason to a combined modal split and network assignment processing.*

The proposed setup assumes a modal selection in favour of minimum transport costs.

> *This concept gives reason to a detailed transport cost evaluation for each mode specific transport path – the result of a combination of multiple links – for each modelled commodity class specification.*

Nevertheless, this concept is not directly applicable to the *real-world*, where a cost effective bundling, together with the choice of a non-road based transports, is not per se observable. At this point, an additional corrective element is required.

> *To enhance the validity of a minimum transport cost based combined modal split and network assignment model, sets of mode specific observations are included within the respective processing.*

11. Model structure overview

An overview on the proposed framework for a firm-based German freight transport model is given in the following. Therein, the processing stages as well as the related input data are depicted.

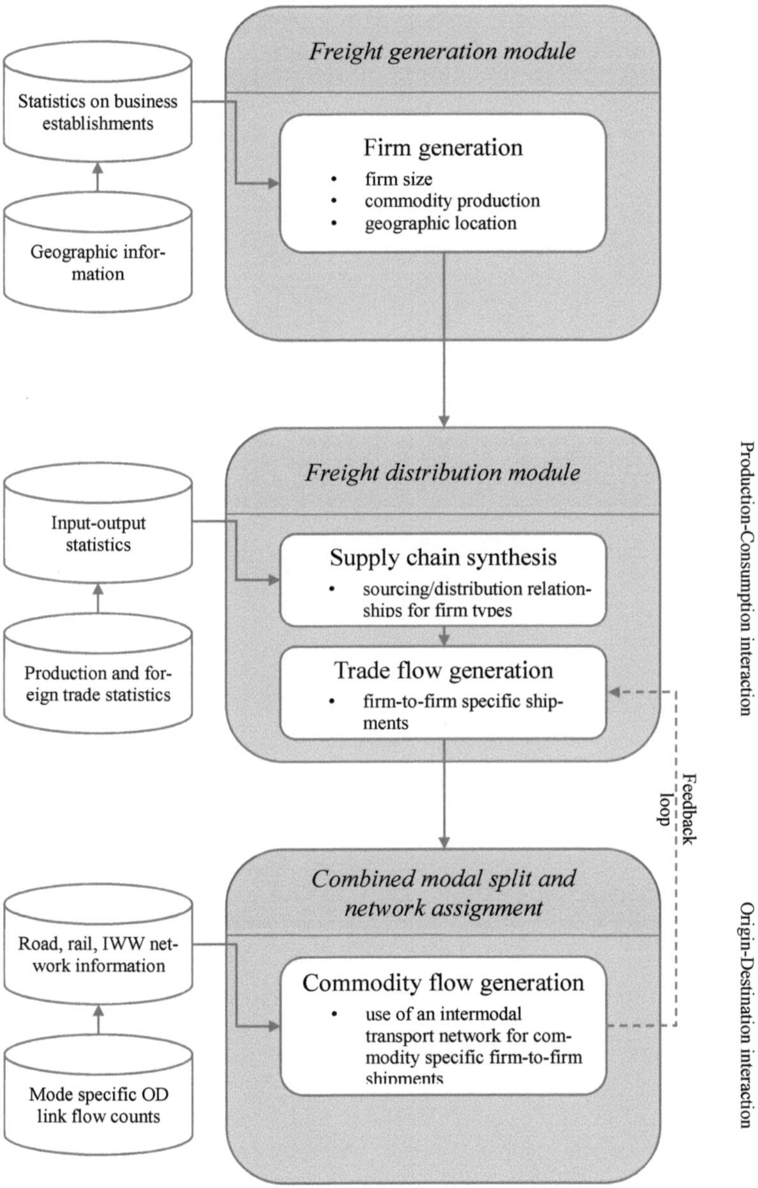

Fig. C-1 Structural framework of a disaggregate German freight transport model

D Realisation of the disaggregate German freight transport model

The realisation of the German freight transport model follows the proposed structure, including a section for *freight generation*, *freight distribution* and *combined modal split and network assignment.*

12. Modelling freight generation

The *freight generation* module generates the transport system's components that represent the transport market demand[83]. Therefore, national labour statistics are proposed to serve as input data to generate a spatial distribution of business establishments.

For Germany there is a valuable NACE-based dataset on firms. It is published by the national employment agency as a report on the number of employees per firm, including specifications on the firm's location and the corresponding statistical economic sector[84]. Both specifications relate to the classification schemes as discussed in section 4.4.1 and 4.4.2.

12.1. Input data selection

The German labour statistics on business establishments are collected on a detailed regional level higher than a NUTS-3 resolution and a corresponding statistical economic sector classification up to a five digit code. These datasets, however, are limited in terms of their information content. With a higher resolution a considerable number of details is not published due to legal restrictions[85].

This may give reason to the implementation of additional commercial data on business establishments in Germany, such as e.g. implemented within the comprehensive study on modelling an economic structure for freight models

[83] See also Fig. A-5 and Fig. B-1.

[84] Note that the German labour statistics only takes into account employees liable to social security contributions (BA Statistik, 2013b). This excludes about 25% of all gainfully employed persons in Germany, such as civil servants, self-employed, unpaid family workers, professional soldiers and those serving their community or military service. This, however, is regarded as of negligible impact on a manufacturing business establishment synthesis for freight traffic generation purposes.

[85] The reason is basically to disclose indications that directly point to one particular firm within a region. For more details see the subsequent discussion.

of BOCHYNEK ET AL. (2009) that is consulted e.g. for the nation-wide macroscopic freight traffic model for the area of Germany by MÜLLER, WOLFERMANN AND HUBER (2012).
However, without additional private data that significantly limit the model's transparency and replication opportunities for future research, an adequate firm specific *freight generation* module for Germany is feasible. This is shown, for instance, within the INTERLOG freight transport modelling system of LIEDTKE (2006) as well as the SYNTRADE model worked out by FRIEDRICH (2010). Here, a synthetic firm dataset is presented as the result of the processing of only public labour statistics[86]. In both studies, however, public labour statistics are not exploited to their full extent as they are only consulted for data on firms and employees per economic sector as a base for derived functional expression of firm size classes. At this point, significant statistical information on the spatial location of German firms get lost. Consequently, a random based spatial distribution of firms is applied in both models.

The presented study, however, aims at modelling an economic structure for the freight demand that is deterministic as far as possible. This principle needs to be considered when the initial level of detail is selected. The level of detail, in turn, is influenced by the trade-off between a high resolution for spatial and economic specifications on the one hand and the number of confidential entries within the related labour statistics on the other hand. As a result, the following conditions apply for the implementation of statistical firm data from BA STATISTIK (2012) and BA STATISTIK (2013a) within the presented context:

- economic activity is represented according to two-digit CPA codes (88 sections),
- a spatial resolution of NUTS-3 districts (402 regions)[87] is chosen and
- eight classes to obtain an estimate on the number of employees per firm for the year 2012 are selected[88].

---

[86] Cf. section 7.
[87] Cf. Table G-46 to Table G-57 for details and the corresponding model specific encoding.
[88] Note that selecting the company size-classes requires again consideration of the trade-off between the number of confidential cells and the higher level of research provided. The narrower the range of requested classes, the more data is anonymised. Therefore the chosen format applied for the given

This selection aims at obtaining the most appropriate level of informational detail on business establishments in Germany. The resulting dataset requires handling approximately 20% of anonymised entries – compared to 65% in case of a three-digit CPA classification. In terms of higher regional resolution and a narrowed company size range, comparable conditions apply.

The processed input dataset consists of two parts – both being interlinked. Part one contains a total *number of firms* in region a, a corresponding economic sector b and a firm size within range c. Part two comprises a total *number of employees* in region a, related to the economic sector b, and firm size-class c[89]. Therein, the chosen first two company size-classes represent small firms, the subsequent three classes represent medium-sized and the remaining classes span large firms. This typology relates to two different definitions of small and medium-sized firms. First, a recommendation released by the European Commission handles companies according to the number of employees as well as the annual turnover and/or an annual balance sheet total (EC, 2003a). According to this – and neglecting the financial status – companies of fewer than 50 employees are considered to be small. A medium-sized company, however, is a company of less than 250 employees. This upper limit is substituted in the following by the definition of the Institut für Mittelstandsforschung (IfM), a German institution dedicated to research on small and medium-sized enterprises (IFM, 2014). According to the IfM, Germany hosts a significant number of companies in the range of 250-500 employees that are still different from what is defined as a large company.

12.2. Firm generation

The firm generation module is implemented to derive an economic structure of shippers and receivers from the selected input data. Initially, all confidential entries need to be eliminated. Even though the anonymisation within German labour statistics is intended to avoid data reconstruction, certain data recovery procedures may apply.

context contains eight classes, representing employees by number of (1 to 10), (10 to 50), (50 to 100), (100 to 200), (200 to 500), (500 to 1,000), (1,000 to 5,000) and (5,000 and more).

[89] The basic format of both input data sets is sketched in Table G-13.

In general, hidden entries in employee-related statistics are the result of the German jurisdiction (Bundesstatistikgesetz 1/22/1987) to keep confidential all entries of values lower than 3 employees and/or 3 establishments per region and economic sector. Additionally, dominant concentrations are anonymised in case that the largest firm per region contributes to a majority[90] of all employees allocated to the respective regional and sectoral totals. An example for the latter case is an automobile firm that manufactures *motor vehicles, trailers and semi-trailers* (CPA29) in the city *Wolfsburg*.

Although confidential cells comprise only 6% of the missing information in terms of the total number of employees, they cover more than 20% of all data entries. Furthermore, most important information on large business establishments in Germany – hypothetically eligible for large amounts of transport bundling, hence with significant impact on the German commodity flow configuration – are covert[91]. To reveal as much information as possible within all eight firm size class groups for both employees and firms an additional overall distribution pattern is consulted. As a result, 18 datasets are incorporated in the following processing.

Within the overall datasets M_i^{Tf} and M_i^{Te} display row totals – hence, the total number of firms and respective employees for each region. Equivalently, the overall totals of firms and employees per economic activity are represented by N_j^{Tf} and N_j^{Te}. The corresponding un-classified distribution of firms and employees per region and economic activity is denoted as x_{ij}^{Tf} and x_{ij}^{Te}. All eight classified datasets for firms are formalised by:

$$M_i^{cf} \text{ and } N_j^{cf} \text{ as well as } x_{ij}^{cf} \text{ with } c = 1,2,\dots,8$$

The eight classified datasets for employees are formatted by:

$$M_i^{ce} \text{ and } N_j^{ce} \text{ as well as } x_{ij}^{ce} \text{ with } c = 1,2,\dots,8$$

[90] The criterion is defined as for less than 10 firms per region, no firm belongs to more than 50% of the total of employees – and vice versa – no firm per cell representing more than 10 firms per region binds more than 85% of employees.
[91] See distribution of anonymised cells for employees and firms for firm size classes 1000 to 5000 as well as 5000 and more employees in Table D-1.

where:

$$M_i^T = \sum_{c=1}^{8} M_i^c \tag{6}$$

and:

$$N_j^T = \sum_{c=1}^{8} N_j^c \tag{7}$$

and:

$$x_{ij}^T = \sum_{c=1}^{8} x_{ij}^c \tag{8}$$

According to the rules of confidentiality for German labour statistics, all restrained entries are symmetrically distributed in the firm and employee data. Furthermore, it should be taken into account that all M_i^T and N_j^T are entirely published for firms and employees, whereas several M_i^c and N_j^c are subject to confidentiality.
The confidentiality rules furthermore lead to a share of anonymised x_{ij}^T that is significantly lower than most corresponding x_{ij}^c for the eight groups, since limiting the number of firms per region and sector increases the amount of entries that fall within the scope of anonymisation.

Using this as a starting point allows for a cascade comparison of all row and column totals as well as for all matrix entries with their complement overall totals for both establishment and employee data and in consequence a considerable reconstruction of missing entries according to the following operations:

- First, for all eight employment size groups the missing row and column totals representing the number of employees and respective firms per region and economic sector are refilled. This may be done since the specific row and column totals for the total of all groups is entirely published. The first data processing reconstructs all hidden totals per employment size group with a unique formal degree of freedom. As a result only 109 compared to originally 334 out of 3397 entries for row and column totals of all eight datasets remain anonymous.
- In the second step the amount of matrix entries featuring one formal degree of freedom are reconstructed according to their new row and column totals.
- A third computation is performed to reveal all matrix values that are at least represented in the overall total and at the same time only hidden in one of the eight grouped datasets. This may be done both row and column wise. Thereby, substantial information is revealed by only using dataset inherent information.

Along with the previously introduced formalisations, the following structure gives a more detailed description of the processed data recovery:

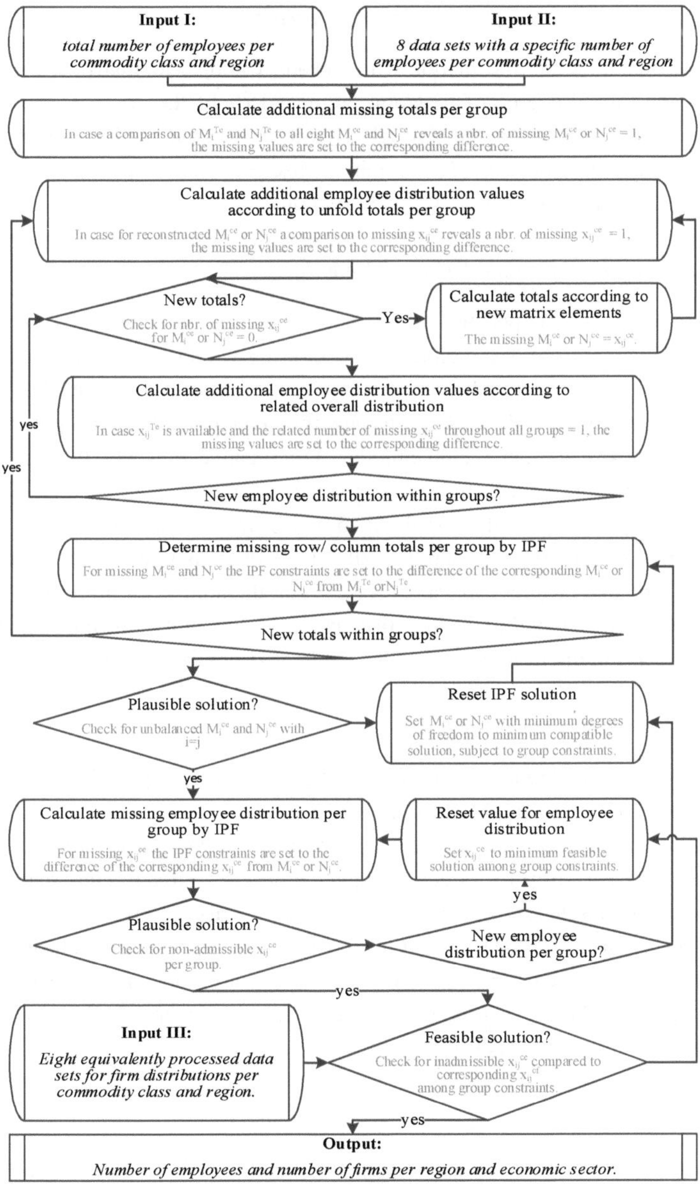

Fig. D-1 Simplified structure for reconstruction of confidential cells in firm generation dataset

Firm generation through basic data reconstruction

Along with inherent information – that means dealing with matrix entries of only one formal degree of freedom – almost 15% of the originally anonymised data is reconstructed through a basic data reconciliation within the size-class specific datasets. The remaining disclosed matrix entries are reconstructed by an IPF procedure (cf. next section).

Firm generation through data complements

The latter processing performs a distribution of entries to empty cells according to size-class specific regional and sectoral row and column totals. This concept becomes unsuitable for extreme outliers within a dataset. For the previous example of the city *Wolfsburg*, this would result in a proportional distribution of less than 25,000 employees. In practice, however, the resident largest firm that manufactures motor vehicles employs nearly 60,000 employees. According to this, for about 25 dominant firms within a region and economic sector, an additional manual data recovery applies, based on a review of related business statistics. As a result, the number of outliers is significantly reduced within the total firm dataset. A box plot of the missing data percentages in the format of quartiles in Fig. G-1 depicts this outcome[92].

Firm generation through IPF

Finally all remaining unknown matrix entries are refilled by applying an IPF algorithm for all eight size-class specific datasets.
An IPF, such as the one proposed, is employed in various scientific disciplines. A basic form of its current applications was already elaborated by DEMING AND STEPHAN (1940). This procedure is also referred to as Fratar (1954) or Furness (1965) method for traffic and transport matrix adjustment problems[93] or as RAS technique in economic input-output analysis (BACHARACH, 1970). An IPF is used to adjust data matrix elements to fit in total to given row and column constraints.

---

[92] The lower quartile is a 25th percentile, splitting off the lowest 25% of all undistributed employees by regions and sectors. The upper one is equivalently a 75th percentile. Since the given firm data and employment data are characterized by similar missing entries, both yield into comparable results in terms of perceptual data reconstruction.
[93] Cf. section 6.

This is done iteratively so that the input – e.g. given by an outdated distribution or an initial uniform cell value distribution – and each further elaborated distribution of the matrix cell values is rescaled to converge to a solution that finally satisfies the constraints. In practice, convergence is taken to be reached when no cell value will change by more than a predefined precision criterion.

For a two-dimensional table x_{ijn} – with matrix elements in row *i*, and column *j* and an iteration *n* as well as M_i and N_j as row and column constraints – new cell values for rows are estimated according to the following equation:

$$x_{ij(k+1)} = \frac{x_{ijk}}{\sum_j x_{ijk}} M_i \quad (9)$$

and equivalent for columns, by:

$$x_{ij(k+2)} = \frac{x_{ij(k+1)}}{\sum_i x_{ij(k+1)}} N_j \quad (10)$$

The former processing is done for both employee data and company data. As a final result an entire distribution of companies as well as employees per NUTS-3 region and two-digit CPA for eight firm size groups is obtained.

Since the IPF-procedure is based on fractions, it generates results that need to be rounded. Consequently, all IPF treated data need to be validated with their own size class restrictions as well as to their counterpart of firms and, vice versa, to employees. The latter necessity becomes more transparent for the last procedures of data processing, when both firm and employee datasets are put together (step 2 in Fig. D-1).
At this point, the nearest integer for the number of firms is processed. In consequence, the final amount of distributed firms is 0.05% lower than the actual total value, whereas the number of employees deviates by 0.001% compared to its respective reported total.
This option is selected with regard to the next processing of distributing sectoral production rates to firms according to their number of employees. Therein, the accuracy of the number of employees per cell is highly important.

The result of the basic data reconstruction by a data complementation and the IPF application is summarised within the following table:

		non-adjusted		adjusted to inherent information		adjusted to IPF for column & row totals	
*firm size class**	*number of firms*	*number of anonymised cells*	*share*	*number of anonymised cells*	*share*	*number of anonymised cells*	*share*
1) 1 to 10	1,666,997	13,573	1%	13,214	1%	13,214	1%
2) 10 to 50	340,396	23,021	7%	14,937	4%	14,937	4%
3) 50 to 100	50,349	11,175	22%	8,521	17%	8,517	17%
4) 100 to 200	25,593	12,670	50%	10,110	40%	10,097	39%
5) 200 to 500	13,525	8,527	63%	7,067	52%	7,059	52%
6) 500 to 1000	3,516	2,592	74%	2,259	64%	2,236	64%
7) 1000 to 5000	1,656	1,358	82%	1,132	68%	1,099	66%
8) 5000 and more	112	108	96%	53	47%	35	31%

*by number of employees

Table D-1 Distribution of anonymised cells within firm generation input data and within deterministic reconstruction stages [incl. data from BA STATISTIK (2013a)]

Firm generation through random assignment

A subsequent procedure is applied to distribute the remaining employees per size class to their respective employers. This is done according to the following random-based structure:

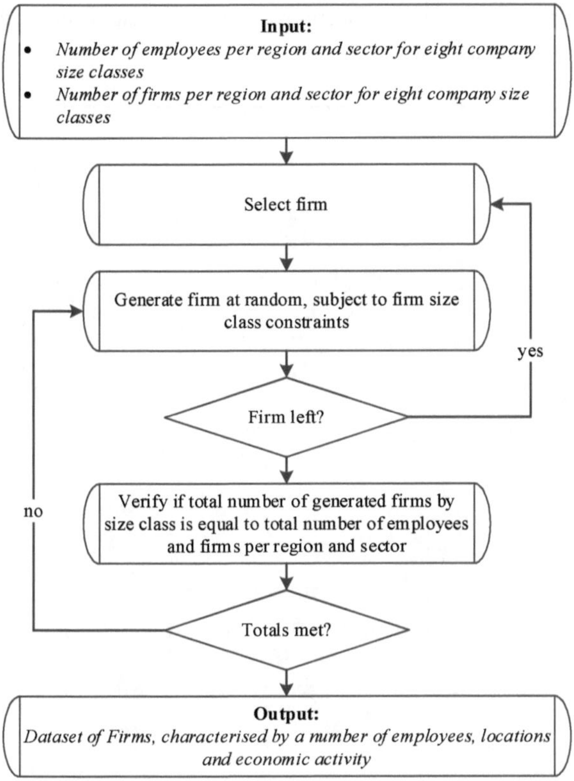

Table D-2 Simplified procedure for random-based reconstruction of confidential cells for firm generation dataset

Within the size class boundaries, a firm size is randomly generated since no further information is available. At this stage, the trade-off between increasing data anonymisation and selecting size classes as narrowly as possible to limit the random distribution per size class becomes more transparent. In case per region and economic sector more than one firm is present – according to the corresponding dataset on the number of firms – a random distribution applies within the particular size class range.

Finally, around 30,000,000 employees are distributed among about 2,000,000 firms, each of which is categorised by a NUTS-3 equivalent location (out of 402 regions) and assigned to an economic activity represented by a two-digit CPA code (of 88 sections). The aggregate result of this section can be visualised as follows[94]:

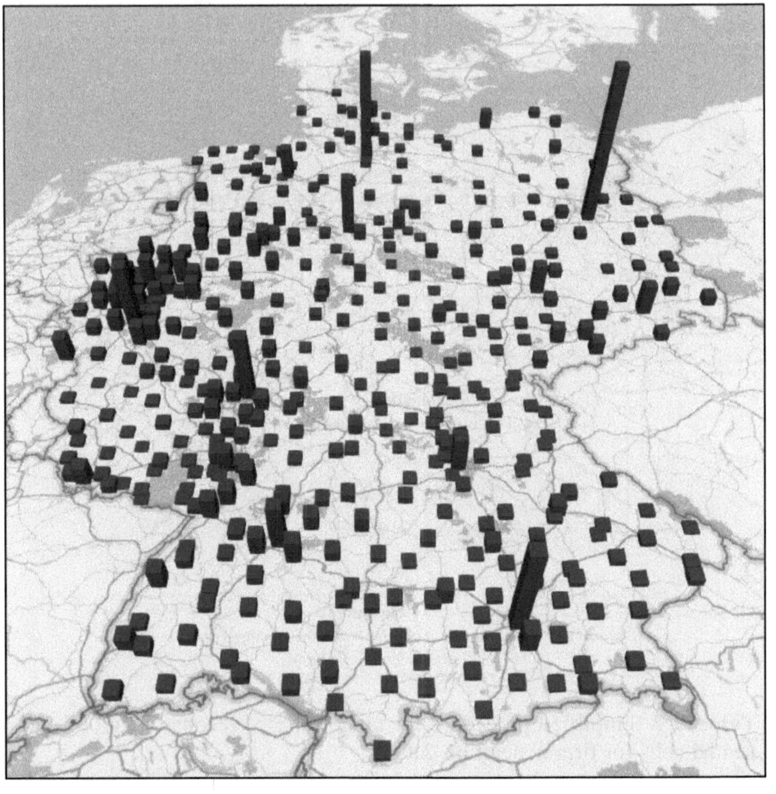

Fig. D-2 Spatial Distribution of employees in German NUTS-3 regions in total for Germany in 2012 [own results, depicted with BING MAPS (2016)]

[94] Note that this simplified depiction of the spatial distribution of employees in Germany that are subject to social insurance contributions does not take into account the allocation to firms and their related economic activity, although it is included in the resulting dataset to be further processed. The aim of this depiction is to introduce the segmentation of the domestic part of the spatial coverage of the model as well as to illustrate the data results.

12.3. Sectoral output generation

If a complete dataset for firms in Germany is at hand, the level of production for each element can be estimated. Afterwards, the transported tonnes for production-consumption interactions within the German transport system can be determined.

To estimate the level of production per firm, the total output per sector of economic activity – here, coded by two-digit CPA – is identified first. Therefore, German *production statistics* (DESTATIS, 2013f) and German *foreign trade statistics* (DESTATIS, 2012) as well as certain complementary statistical reports are incorporated. Subsequently, the total production will be distributed among corresponding firms according to their firm size in terms of the number of employees[95].

The incorporated *production statistics* are classified by a specific code, the GP 2009 – named: *Güterverzeichnis für Produktionsstatistiken, Ausgabe 2009* (DESTATIS, 2008a)[96]. This nomenclature is based upon the European PRODCOM-standard. Thereafter, the first four digits per heading refer to an equivalent class within the NACE-system, allowing for a final conversion to the two-digit-CPA standard of the presented context (cf. section 4.4.2). The GP 2009 comprises nine digits at a maximum resolution. Information for the year 2012 on quantities per GP 2009 are generally published for this resolution only. German *production statistics* from DESTATIS (2013f) are specified by:

- a value of production for sale per production class in EUR,
- the number of firms per production class and
- a corresponding amount of production in tonnes and/or kilograms and/or other specific units and/or pieces[97].

[95] A detailed breakdown of the level of productivity for each modelled firm according to an annual monetary turnover would be favourable, but is not implemented due to a lack of adequate references.
[96] From EUROSTAT (2014t) an equivalent eight-digit level of detail data is obtainable for Germany as well.
[97] Specific units are reported for instance for liquids or gases. Pieces apply for example for listings of animals or building materials.

Due to the high level of resolution per heading, a significant number of required information is subject to confidential issues[98]. In consequence, a significant number of dataset entries is partially or even completely anonymised. To allow an incorporation of this valuable input, certain transformations and the consultation of additional data inputs apply.
The first addition to this primary set of input data is taken from German *foreign trade statistics* from DESTATIS (2012), being consulted as proposed in LIEDTKE (2006) and BOCHYNEK ET AL. (2009). Therein, for a significant number of entries, a volume in tonnes is indicated as well as in specific units and/or the related number of pieces. Furthermore, the value of exports and/or imports as well as the weight are reported.
Along with these relations, a transformation of indications on specific units, pieces or a measure of the value of the level of production within the *production statistics* can be transformed into a volume of produced tonnes[99]. The underlying conversion factors between the nine-digit-GP2009 classification and the eight-digit standard of goods for foreign trade is given by the German *Federal Bureau of Statistics* (DESTATIS, 2013d).

Subsequently, a first basic calculation scheme applies to the remaining anonymised weight-values for the disaggregate nine-digit level of detail d. These missing headings are compared with data entries of same set that are at hand, defined by an aggregate eight-digit-code a. This procedure also iteratively applies for further levels of aggregation up the final two-digit-classification. It is formulated as:

$$W_d^m = V_d^m * \frac{\sum_a W_d(V_d)}{(\sum_a V_d - V_d^m)} \tag{11}$$

with:

W_d^m missing disaggregate weight of GP 2009-heading d of an aggregate set a,

W_d weight of disaggregate heading d,

[98] See discussion on statistical confidentiality regulations in section 12.3.
[99] Each conversion is derived from export relations first and, only if necessary, from import relations, since both are not congruent.

V_d^m value of production for disaggregate heading d with weight missing,

V_d value of production for disaggregate heading d with weight at hand.

A second calculation scheme is included for the four-digit level as well as the two-digit level of detail. These intermediate steps incorporate additional volumes of production for sale into four- and two-digit-classifications[100], both given in DESTATIS (2014i). The aim of this inclusion is to further enhance the validity of the missing-weight estimation procedure. The corresponding calculation to determine additional values of production for anonymised aggregate four- and two-digit-sets by disaggregate five- and three-digit-headings is defined as:

$$V_d^m = \frac{V - \sum V_d^o}{n} \tag{12}$$

with:

V_d^m missing value of production for disaggregate GP 2009-heading,

V at hand value of production for aggregate set of corresponding headings,

V_d^o at hand value of production for disaggregate heading,

n number of missing disaggregate values per aggregate set of corresponding headings.

The inclusion of the German *foreign trade statistics* and the application of the aggregation procedure as well as the incorporation of aggregate production volumes are depicted in the following in Fig. D-9. The sequential information gathering for each nine-digit-heading is subsequently listed in Table D-3.

[100] Therein, no indications on tonnage or specific units and/or pieces are given. However, the value of production per aggregate two- or four-digit heading is a valuable input for the final production estimation per two-digit CPA.

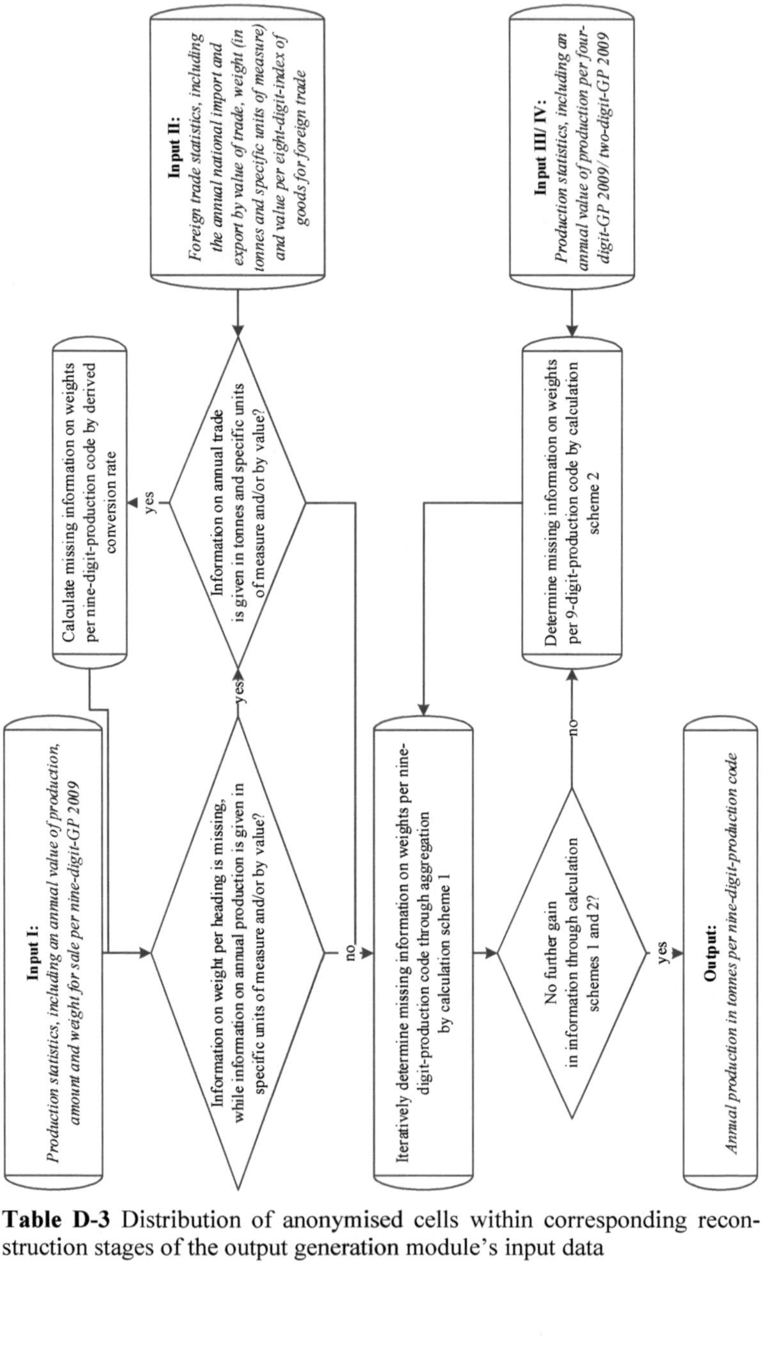

Table D-3 Distribution of anonymised cells within corresponding reconstruction stages of the output generation module's input data

CPA	*non-adjusted*	*value-weight transformation*	*1st agg.*	*2nd agg.*	*3rd agg.*	*4th agg.*	*5th agg.*	*1st agg. with 4-digit values*	*2nd agg. with 4-digit values*	*3rd agg. with 4-digit values*	*1st agg. with 2-digit values*
1	-	-	-	-	-	-	-	-	-	-	-
2	-	-	-	-	-	-	-	-	-	-	-
3	-	-	-	-	-	-	-	-	-	-	-
5	**100%**	100%	100%	100%	100%	100%	100%	100%	100%	100%	**100%**
6	**67%**	33%	33%	33%	33%	33%	33%	0%	0%	0%	**0%**
7	**-**	-	-	-	-	-	-	-	-	-	**-**
8	**35%**	20%	20%	20%	20%	20%	18%	0%	0%	0%	**0%**
10	**34%**	22%	22%	22%	21%	20%	19%	0%	0%	0%	**0%**
11	**98%**	22%	20%	20%	20%	20%	20%	3%	3%	3%	**0%**
12	**75%**	75%	75%	75%	75%	38%	38%	38%	38%	38%	**0%**
13	**58%**	48%	47%	47%	38%	38%	38%	0%	0%	0%	**0%**
14	**100%**	42%	42%	42%	42%	42%	42%	2%	2%	2%	**0%**
15	**89%**	43%	43%	43%	43%	43%	43%	8%	8%	8%	**0%**
16	**95%**	24%	24%	23%	23%	23%	21%	1%	1%	0%	**0%**
17	**45%**	31%	31%	31%	31%	31%	31%	1%	1%	0%	**0%**
18	**100%**	93%	93%	93%	93%	93%	87%	87%	20%	15%	**0%**
19	**21%**	15%	15%	15%	15%	15%	15%	6%	6%	6%	**0%**
20	**56%**	29%	29%	29%	28%	28%	28%	0%	0%	0%	**0%**
21	**84%**	35%	35%	35%	35%	35%	35%	2%	2%	0%	**0%**
22	**33%**	12%	12%	12%	12%	12%	12%	1%	1%	0%	**0%**
23	**49%**	26%	26%	26%	22%	22%	20%	1%	1%	1%	**0%**
24	**39%**	34%	34%	34%	34%	33%	33%	0%	0%	0%	**0%**
25	**37%**	16%	16%	16%	16%	16%	15%	2%	0%	0%	**0%**
26	**99%**	19%	19%	19%	19%	19%	18%	2%	2%	2%	**0%**
27	**91%**	19%	19%	19%	19%	19%	19%	0%	0%	0%	**0%**
28	**88%**	16%	16%	16%	16%	16%	16%	0%	0%	0%	**0%**
29	**90%**	32%	32%	32%	32%	32%	26%	1%	1%	0%	**0%**
30	**98%**	56%	56%	56%	54%	52%	43%	6%	6%	6%	**0%**
31	**100%**	12%	12%	12%	12%	12%	11%	1%	1%	0%	**0%**
32	**97%**	32%	32%	31%	31%	31%	31%	0%	0%	0%	**0%**
Total	66%	26%	26%	25%	25%	24%	24%	2%	1%	1%	0%

Fig. D-3 Simplified structure for a level of production data distribution per nine-digit-GP2009 heading

The remaining missing entries for *products of agriculture, hunting and related services* (CPA-01), *products of forestry, logging and related services* (CPA-02), *fish and other fishing products; aquaculture products; support services to fishing* (CPA-03) as well as *coal and lignite* (CPA-05) and *metal ores* (CPA-07) are separately determined. For an output level of CPA-05 and CPA-07 mining statistics for 2012 were published by the German *Federal Ministry of Economics and Technology* (BMWI, 2013, p. 55 ff.). Based on this an estimate for the total national utilisable production tonnage of *hard coal* (GP09-051010300), *brown coal* (GP09-052010000) and *iron ore* (GP09-071010000) can be derived[101].

For CPA-01 to CPA-03 no equivalent sectoral total production is reported. Instead, the totals are derived from multiple statistical reports. For CPA-01 the level of production and prices for sale of the most relevant products are listed in BMEL (2013). The volume of products from hunting activities sold in 2012, as part of CPA-01, is not reported and will be neglected in the following. An overview for CPA-01 is listed in Table G-14.

For CPA-02 the level of production of timber is reported in DESTATIS (2014u). Average market prices per type of timber are taken from LWKN (2014). To apply the latter price declaration, no significant price change from 2012 to 2014 is assumed in accordance to HPE (2014). The conversion rates from 1000m³ per class of timber to tonnes are taken from expert statements. Results of this calculation can be found in Table G-15.

The volume of production of *fish and other fishing products; aquaculture products; support services to fishing* can be determined according to the volume of catches in Germany, reported in EUROSTAT (2014c) and the aquaculture production, reported in DESTATIS (2013c, p. 10). The corresponding monetary value is derivable from *whole-sale market pricings of aquaculture production*, indicated in DESTATIS (2013h). In absence of adequate similar inputs to determine the value of catches, an average price is assumed as a

[101] Iron ore exports of Germany are given by value and quantity in EUROSTAT (2010), indicating a value of 67,60 € per tonne that is added to the production volume from BMWI (2013). For *hard coal* (GP09-051010300) the missing value per weight is taken from reports of the *Federal Bureau of Economics and Export Control* on import prices, indicating a value of 108,26 € per tonne in Germany for 2012 (BAFA, 2014). For *brown coal* (GP09-052010000) an estimated average caloric value factor of 1/3 compared to *hard coal* (GP09-051010300) leads to a value of 36,09 € per tonne in Germany for 2012.

similar price for aquaculture production as well as catches. Results can be found in Table G-16.

An overall derived distribution of the level of production per CPA is given in the following table:

CPA	*Label*	*Thousands of tonnes*	*Millions of €*	*€ per tonne*
1	Products of agriculture, hunting and related services	133,585	46,175	345.7
2	Products of forestry, logging and related services	28,546	2,162	75.7
3	Fish and other fishing products; aquaculture products; support services to fishing	232	784	3,381.5
5	Coal and lignite	196,202	7,858	40.1
7	Metal ores	451	30	67.6
8	Other mining and quarrying products	350,426	3,735	10.7
10	Food products	110,044	127,335	1,157.1
11	Beverages	34,285	16,998	495.8
12	Tobacco products	336	2,215	6,603.0
13	Textiles	1,554	9,532	6,132.0
14	Wearing apparel	28	1,590	56,734.6
15	Leather and related products	57	1,393	24,612.3
16	Wood and products of wood and cork, except furniture; articles of straw and plaiting materials	35,580	16,547	465.1
17	Paper and paper products	36,700	35,099	956.4
18	Printing and recording services	1,819	15,987	8,790.7
19	Coke and refined petroleum products	105,094	39,398	374.9
20	Chemicals and chemical products	104,286	114,120	1,094.3
21	Basic pharmaceutical products and pharmaceutical preparations	1,468	27,683	18,863.3
22	Rubber and plastic products	14,132	63,012	4,458.7
23	Other non-metallic mineral products	170,529	30,869	181.0
24	Basic metals	86,195	91,009	1,055.8
25	Fabricated metal products, except machinery and equipment	22,394	94,278	4,209.9
26	Computer, electronic and optical products	689	51,525	74,804.5
27	Electrical equipment	4,103	70,336	17,143.1
28	Machinery and equipment n.e.c.	14,990	191,408	12,768.8
29	Motor vehicles, trailers and semi-trailers	25,129	269,006	10,705.2
30	Other transport equipment	2,240	28,613	12,771.6
31	Furniture	4,108	16,682	4,061.1
32	Other manufactured goods	644	18,477	28,701.7

Table D-4 Annual national production per producing CPA-heading for 2012 by weight, value and the corresponding ratio

12.4. Firm-specific output allocation

To finalise the *freight generation* module the determined total annual production of each economic sector has to be assigned to each modelled firm. This distribution will be executed according to a proportion of the number of employees per firm in relation to the sectoral total. This processing requires the following assumptions:

- each employee per economic sector features a common level of productivity,
- the size affects the level of productivity in a linear relation, hence the output per firm is proportional to its number of employees and
- a firm's level of productivity is not influenced by its location.

This procedure follows general principles as proposed within the studies of LIEDTKE (2006) and BOCHYNEK ET AL. (2009). Unfortunately, both studies do not publish their results and in consequence, no validation is presented. Nevertheless, an allocation of a sectoral output to its related individual firms may still be an appropriate concept for the presented context. The result in a consolidated format is depicted as follows:

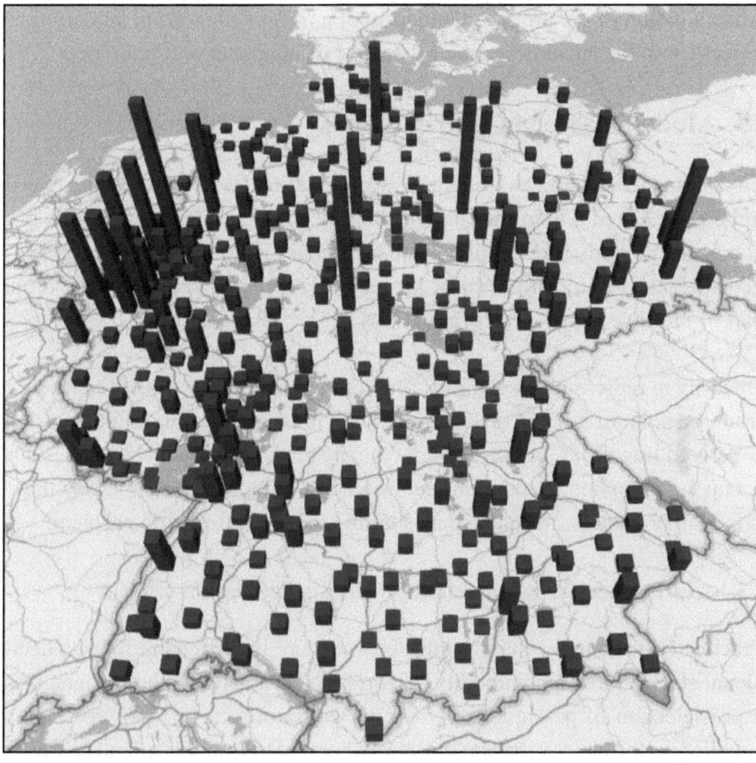

Fig. D-4 Spatial distribution of annual production volumes in German NUTS-3 regions in total for Germany in 2012 [own results, depicted with BING MAPS (2016)]

This illustration depicts the distribution of production capacities in terms of tonnes per annum. The magnitude of a bar on the map represents the production volume per region in comparison to another. Note that the underlying dataset is also specified by a commodity class as well as each firm within the depicted regions as a basis for the subsequent processing[102].

---

[102] The result of this section is to a certain extent comparable to a distribution of domestic freight traffic origins. A more refined distribution that includes also imports will be the result of the following section.

An evaluation of the proposed initiation of a *freight generation* modelling is processed in the ensuing section by the use of import-export references.

13. Modelling freight distribution

To conduct a disaggregate freight flow distribution for Germany, firms need to be represented within their unique supply chain configurations. This is a complex problem since suppliers and customers interact explicitly inhomogeneously to set up partnerships. Firms in the *real-world* practice are often part of multiple supply chains in various superordinate networks.

One way to represent this complex problem as precisely as possible is to start from the point of a final consumption. Each preceding supply chain decision is derived therefrom. In other words, the presented supply chain synthesis is set up as a supplier selection model instead of a likewise possible consumer selection procedure, formulated as a multi-stage, multi-criteria problem.

13.1. Firm-specific supply chain synthesis evaluation criteria

Practically, the supply chain synthesis procedure is carried out for two decisional stages. First, a commodity performance evaluation is implemented as a pre-selection of potential suppliers per consuming firm[103]. Afterwards, a supplier performance criterion will be elaborated to estimate a final selection of suppliers. Thus, the concept follows the question of which product is required and who performs best as a supplier.

Even though most of the individual firm-specific product requirements and supplier performance measures are not made public, practicable indicators can be used for a synthesis of *real-world* supplier choice decisions. Therefore, the supply chain network configuration in Germany is assumed to be the result of the following two effects:

[103] Note assumption on final consumer allocations in section 4.4.2.

- if shorter sourcing distances are of a high relevance, then producers are observed to be closer to its consumers. Vice-versa, if the sourcing distance is less relevant, then supplier and consumer locations are merely scattered in comparison to producer-consumer pairings for other commodities as both do not have to be an immediate adjacent.

It could therefore be assumed that the spatial distribution of production and sourcing capacities is the indicator for an appropriateness of a commodity flow relationship: the supply chain setup. In addition:

- if sourcing units either prefer suppliers of a different or a similar size, then this pattern will be represented within the market configuration, built by the supply and demand side and the related relative establishment sizes. This supposition leads to a preference of larger, smaller or equally sized sources by sinks of a certain size. Pursuing this interpretation, size relations between production and sourcing capacities also point at supply chain partnership likelihoods.

These rather vague interpretations are the result of the consideration of what is at hand. The only information on potential supply and demand interactions is their stand-alone configuration. For one part this is the quantity of suppliers and consumers and for another part it is their average establishment size and the average distance to each other. This is far less than what is necessary to model a versatile changing and complex interlinked market structure – in fact, the entire German economy. Nonetheless, it is a starting point for a structured linking of the unknown market configuration that follows the individual final supplier selection of an establishment in need of inputs or that of a final consumer, respectively.
The representation of this *real-world* process within the presented freight model is based on the following considerations:

- A sourcing unit – be it a producing establishment, a service provider, a wholesale institution or a final consumer – requires a distinct amount of commodity inputs from a selected number of producing establishments.
- These inputs are sourced from the most suitable suppliers in terms of two different evaluation criteria.

A distance-based decision criterion

An illustrative example for the first supply chain configuration-likelihood evaluation parameter – the distance-related decision criterion – will be the production and consumption of *basic pharmaceutical products and pharmaceutical preparations* (CPA-21). In geospatial terms almost everywhere required in various amounts, this commodity class is only supplied by a limited number of sources with large production capacities. These are namely the few production sites within Germany as well as certain imports from other European countries or overseas. Consumption, however, is spread among various other economic sectors (see Fig. D-4), each of which comprises a set of more spatially distributed sinks for potential commodity flow relationships.

For this exemplary commodity class the average distance from sink to source (producer to consumer, supplier to customer) in practice is expected to be outstandingly high. This is contrary to a supply chain distribution for *products of agriculture, hunting and related services* (CPA-01). Here, disperse spread sources will meet disperse sinks. Thus, comparatively small average distances between supply chain partners are assumed to be the result.

In total, supply chain configurations among firms are more likely if they are in a certain distance to each other that reflects the overall average. Products of CPA-21 are sourced in larger distances than products of CPA-01.

A firm-size-related decision criterion

The size-related supply chain potential evaluation is best described by firms that manufacture or produce *chemicals and chemical products* (CPA-20). Compared to its consumers, firms of CPA-20 are significantly larger on average – vice-versa, for the production of *wearing apparel* (CPA-14). Here, the size ratio between producers and potential consumers is small.

Since the average size ratio of producers and consumers of every modelled commodity class is quantifiable, the size-related supply chain potential evaluation allows for the limiting of unlikely configurations. Such a configuration would involve a setting where e.g. the largest firm of CPA-20 is assigned to the smallest of all its potential customers, say, a one-employee firm of *crea-*

tive, arts and entertainment services (CPA-90) instead of a more likely partnership with a larger representative that is found in the segment of *wholesale trade services* (CPA-46).

Overview on selected decision criteria for a firm supply chain synthesis

Within the limits of available data, a supply chain synthesis is performed along with the following supplier evaluation criteria:

- a product performance criterion in terms of appropriateness for a firm's economic activity,
- the cost criterion in terms of trade resistance by average firm-to-firm distance and
- a service performance measure by firm size in terms of number of employees.

The subsequent processing of trade flow partnerships corresponds to a basic scoring model (see section 9.1) with the aim of adequately representing the integration of firms within *real-world* supply chains[104].

13.2. Supplier pre-selection by sectoral economic relationships

An ideal way for an identification of potential trade partners for each modelled firm is the consultation of a comprehensive and at the same time sufficiently disaggregate statistical dataset for firm-to-firm trade interactions. Since no such data is published for Germany, the goal is to strive for the second-best solution.

Implementation of German input-output-statistics

That is why an aggregate – here, sector-specific instead of firm-specific – source of information is explored to determine commodity flows between generated firms. The German input-output-statistics, as documented in DESTATIS (2014t), indicates commodity specific linkages among economic sectors. Based upon these relationships a pre-selection of potential trade partners per firm is arranged in order to identify a class of potential trade partners for each firm.

[104] See introduction of section 9.

A general characteristic of the German input-output-calculation is its focus on monetary flows. Hence, instead of freight volumes in tonnes, the trade volume between economic sectors is indicated in terms of its value. Another characteristic is that input-output-matrices do not focus on institutions but on products. This implies that each establishment of a company is evaluated separately by its own economic activity rather than e.g. for a dominant product or service of an entire company (DESTATIS, 2010, p. 14) – a beneficial feature of the input data in the given context. Another promoting property is the predominant two-digit-CPA equivalent level of detail of German *input-output statistics*[105].

In detail, the input-output-tables are built up for national production as well as for imports separately. Favourably, both indicate the required monetary inputs of a CPA and the corresponding outputs in different dimensions. Furthermore, not only the sectoral sourcing alone but also the final destination of commodities can be derived thereby. The latter encompasses indications for stock variations, manufacturing investments and final consumptions.

Applied to the presented model, each sectoral input is determined on a monetary basis. These inputs are equivalent to a share of another sector's total output. This output by value is equated to the total sectoral output volume in tonnes as determined previously (see section 12.3). In this context, input-output-tables serve as a valuable source of information on linkages between economic sectors in Germany. Incorporating this setting into the presented context implies the following assumption:

- A conversion from sectoral relations in terms of monetary flows into trade relations by volume in tonnes is qualified to represent the *real-world* process of commodity flows.

Thus, the final consumption of a commodity class can be allocated. For the presented model this distribution is considered as an additional supply to *retail trade services* (CPA-47). In consequence, establishments of CPA-47 are assumed to source commodities for their own economic productivity as well

[105] In case of more aggregate headings within the input-output-tables, e.g. when multiple CPA are consolidated, a unique distribution scheme for all incorporated CPA is assumed for a subsequent decomposition. Likewise, for CPA that are more disaggregated than the dominant two-digit level of detail, an average total is built up for an aggregation.

as various commodities that are determined for a final consumption[106]. For indications on stock variations and the manufacturing investments within the German input-output tables, an aggregate is accounted to the total monetary output to ensure the consistency of the corresponding quantitative structure.

In total, the constitution of a supplier pre-selection is interpreted in the context of a sourcing structure. An exemplary interpretation of this processing would be the following:

- Firms per CPA-20 manufacture chemicals and chemical products. Therefore, national inputs of products of agriculture, hunting and related services from firms of CPA-01 are required, as well as products of forestry, logging and related services (CPA-02), coal and lignite (CPA-05) etc.
- German firms of CPA-01 in total distribute inputs of a value of 18 mill. EUR out of 51,872 Mill. EUR[107] to firms in CPA-20. Furthermore, national firms of CPA-02 distribute products by a value of 6 mill. EUR out of 780 mill. EUR to firms in CPA-20. This line is complemented by several others out of the 88 CPA-two-digit-classified economic sectors.
- From international sources of CPA-01, 10 out of 26,964 mill. EUR of their total distributed production is consumed in CPA-20 in Germany. Analogously, international firms of CPA-02 distribute *products of forestry, logging and related services* of a value of 6 mill. EUR out of a total production 780 mill. EUR among German firms of CPA-20 etc.

This procedure is broken down for all national and international supplying sectors of the exemplary CPA-20. Afterwards, the international sectoral sourcing is allocated to particular regions – in this context, selected countries and a representative for the rest of the world. The given example is depicted as follows:

[106] A further development of the presented model may involve e.g. multiregional input-output tables to coordinate commodity flows for all specified commodity classes via related wholesale and/or distribution centres.

[107] The total of 51,872 mill. EUR for further consumption is constituted from an overall total sectoral outcome of 54,807 mill. EUR, whereof a share of 2,935 mill. EUR is used to build up stocks and investments within CPA-02.

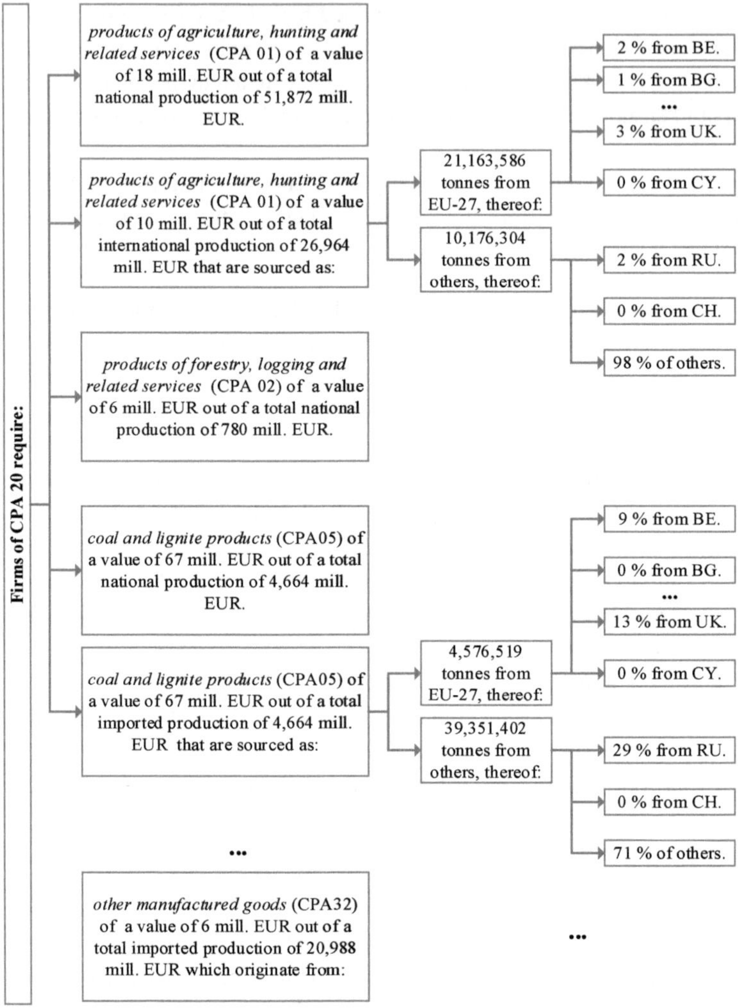

Fig. D-5 Exemplary interpretation of German input-output tables as well as import and export statistics [incl. data from DESTATIS (2014t); DESTATIS (2014j)]

Besides the use of *input-output statistics*, data on German trade relationships complete the supplier pre-selection procedure. This set of information stems from German *import and export statistics* (DESTATIS, 2014j). Therein, German import and export relations with nations of the EU-27 and others are indicated in terms of tonnes on a four-digit level of detail. Based upon this additional set of input data, volumes for imported and exported tonnes per commodity class are transformed to the framework[108].

In total, the indicated input output interactions by value are transferred into trade relations by weight under the assumption of a homogeneous distribution of production and consumption rates per economic sector[109]. This allocation is complemented by import export distributions. Therefore, additional trade network nodes are required to complement the given set of firms.

13.3. Synthetic supply chain elements

After the preceding sections a set of national sinks and sources of produced commodities is at hand for a German freight transport model. Furthermore, an allocation of German imports and exports from or to a country of the EU-27, Switzerland and Russia as well as an aggregate for the rest of the world is established. These additional supply chain elements need to be represented appropriately.

Initially, for each member-state of the EU-27 as well as for Switzerland and Russia a *land-based* transport is modelled. Therefore, a synthetic trade flow network element is located within each country. Furthermore, imports and exports via *seaports* and *airports* are considered for these international supply chain elements and especially for trade relations to the rest of the world. Therefore, the 16 most relevant German freight-handling airports – in terms of turnaround volumes in 2012 – complement the presented model. Additionally, 24 German seaports as well as the ports of *Rotterdam* and *Antwerp* are

[108] Although a significant volume of products for *printing and recording services* (CPA 18) is reported for Germany for the national production, the counterpart listed in the German input output statistics is zero. In contrast, a non-negligible volume of imports and exports for commodities of *publishing services* (CPA 58) is indicated. To meet the input output table's distribution scheme, imports and exports of CPA 58 are accounted to a sourcing of CPA 18.

[109] Cf. introduction of section 12.4.

considered for the German freight transport model. In this context, the following assumption can be made:

- Imported and exported freight flows to or from countries other than the EU-27 member states, Russia and Switzerland are directed via modelled seaports and airports only[110].

German seaports

For Germany, 24 seaports are modelled. The corresponding national as well as international incoming and outgoing volumes are reported with a specification of three-digit NST commodity classes in DESTATIS (2014g). The total sea freight volume handled at German seaports and inland ports are presented in DESTATIS (2014f). As a result of an interpretation of both inputs, it can be stated that the selected 24 seaports comprise more than 98% of the entire outgoing and incoming German transport volume by sea[111]. An aggregate overview is given in the following table:

[110] This specification implies certain false interpretations, for instance for imports from countries other than the EU-27 states, such as Serbia, the Ukraine or Turkey via *land-based* transport modes. In contrast, it allows for a structured representation of international trade relations for the German freight transport system based on national and EU statistics. The focus on seaports is a result of an insignificantly small share of maritime freight traffic arriving on inland ports (cf. Table D-5).

[111] The ports of Bremen and Bremerhaven are only represented as an aggregate. To be more conformed to the model's spatial breakdown, both are disaggregated by approximations. Therefore, the distribution of the overall share for sending and receiving according to DESTATIS (2013g, p. 7) is distributed among the indications for sending and receiving of both ports.

		Sending		Receiving	
		Σt	*share*	Σt	*share*
1	Hamburg	65,086,463	36%	633	41%
2	Wilhelmshaven	26,048,804	15%	9,076	0%
3	Bremen*	12,837,559	7%	293	11%
4	Bremerhaven*	23,447,293	13%	41,050	19%
5	Brunsbüttel	6,129,291	3%	1,041,005	2%
6	Brake	4,273,393	2%	81,212	2%
7	Bützfleth	4,088,410	2%	10,216	1%
8	Emden	2,688,797	2%	1,921	2%
9	Nordenham	2,844,003	2%	296,482	0%
10	Cuxhaven	1,316,859	1%	291,353	1%
11	Leer	43,778	0%	0	0%
12	Papenburg	413,133	0%	112,341	0%
13	Husum	297,729	0%	2,757	0%
14	Rostock	9,425,728	5%	181	6%
15	Lübeck	9,837,695	6%	4,097	6%
16	Puttgarden	1,699,080	1%	106	3%
17	Kiel	2,398,107	1%	633	2%
18	Wismar	2,102,567	1%	9,076	1%
19	Saßnitz	683,378	0%	293	1%
20	Lubmin	57,582	0%	41,050	0%
21	Wolgast	91,906	0%	1,041,005	0%
22	Stralsund	288,132	0%	81,212	1%
23	Flensburg	451,273	0%	10,216	0%
24	Rendsburg	179,495	0%	1,921	0%
	TOTAL	179,125,857	99%	119,666,510	98%

*approximation

Table D-5 German seaports included in the model by reported volume of sending and receiving of loads in total [incl. data from DESTATIS (2014f); DESTATIS (2014g)]

Complementing data on commodity specifications as well as details on origins and destinations of the handled loads by country are retrievable from DESTATIS (2014k)[112]. This sound dataset, however, only indicates shipments by country of origin and/or destination for aggregate regions in Germany – namely, the German Baltic Sea and the German North Sea. To adopt this

[112] Therein, for Germany in 2012 the total incoming volume of 116,274,744 tonnes and total outgoing of 178,813,257 tonnes is reported. Both totals and the related specified distribution are incorporated within the following in favour of variant reports, such as those included in Table D-5.

information to the presented context, a proportional fitting is applied to disaggregate the volumes to each modelled seaport according to Table D-5. The finally compiled table lists the volume of specified incoming and outgoing commodities per seaport. Therein domestic transports are evaluated according to the final German seaport for ingoing and initial national seaport for outgoing shipments according to the following assumption:

- transports by seagoing vessel between German seaports relate to an international transport origin and/or destination.

In other words, a transport from e.g. the port of Hamburg to the port of Rostock relates, at least, to an international origin. Within the model, this transport will be considered as an import via the port of Rostock, regardless of a previous transhipment at the port of Hamburg.

After a conversion from the NST to the CPA classification standard and the subsequent limitation of commodity classes – here, CPA-01 to CPA-32, excluding CPA-06 – a total of 80,782,638 tonnes of incoming and 148,247,609 tonnes of outgoing volumes is finally processed for German seaports. The deviation to reported totals, as indicated in Table D-6, are most notably the result of significant volumes that are originally reported as *unidentifiable goods in containers or swap bodies* (NST-191) and *other unidentifiable goods* (NST-192). In absence of further details on these unidentifiable goods handled at German seaports, the corresponding volumes are not considered in the model. A summarizing overview on volumes transferable from reports to the modelling framework is given in the following table:

	Sending		**Receiving**	
	Σ t	*share*	*Σ t*	*share*
Total reported	116,274,744		178,813,257	
Transferable to CPA	80,782,638	70%	148,247,609	83%

Table D-6 Share of identifiable goods sent and received at German seaports

As a result the volumes considered in the model for German seaports as additional transport network nodes comprise about 70% of the total reported incoming as well as 90% of the total outgoing transports. These distributions require certain further adaptations in accordance with results of section 13.2 and a distribution of ingoing and outgoing volumes for German airports. The subsequent final allocation of ingoing and outgoing transport volumes is given in Table D-8, Table D-9 and Table D-10.

German airports

The annual total volume of goods handled at the 16 most relevant German airports is reported for sending and receiving countries for each one in DESTATIS (2014h) and DESTATIS (2014l). A listing of the modelled airports is given in the following table:

		Sending		**Receiving**	
		Σ t	*share*	*Σ t*	*share*
1	Berlin-Schönefeld	689	0%	407	0%
2	Berlin-Tegel	13,411	1%	12,989	1%
3	Dresden	57	0%	238	0%
4	Düsseldorf	42,921	2%	40,848	2%
5	Frankfurt/Main	1,014,083	49%	947,120	49%
6	Hahn	88,578	4%	86,652	4%
7	Hamburg	12,785	1%	10,581	1%
8	Hannover	2,918	0%	1,943	0%
9	Köln/Bonn	359,095	17%	339,189	17%
10	Leipzig/Halle	385,324	19%	386,419	20%
11	Memmingen	0	0%	0	0%
12	München	135,248	7%	112,025	6%
13	Nürnberg	4,568	0%	2,604	0%
14	Rostock-Laage	50	0%	11	0%
15	Stuttgart	4,858	0%	4,391	0%
16	Zweibrücken	19	0%	9	0%
	Total	2,064,603		1,945,424	

Table D-7 German airports included in the model by volume of sending and receiving of loads in total for the year 2012 [incl. data from DESTATIS (2014h); DESTATIS (2014l)]

In contrast to an allocation of German imports and exports via German seaports, no transits for airfreight cargo are considered in the model. For instance, there will be no outgoing loads from the airport of Leipzig/Halle which originate in an Austrian production establishment. Consequently, the following assumption can be established:

- The reported incoming and outgoing transports via German airports are either destined for or originated in Germany.

A drawback for an inclusion of statistical reported freight rates per airport arises due the fact that the most recent report is available for the year 2010. Since then this particular statistic is no longer updated. This deficit can be dealt with by an approximation through a breakdown of total freight rates from the year 2012 to the airport-specific distribution from 2010 according to DESTATIS (2014e).

Another shortcoming of German airfreight statistics in the context of an implementation to a disaggregate freight model is the lack of specification for the handled commodity classes per airport or at least in total. Since this information is mandatory for the presented modelling framework, the following assumption applies[113]:

- The choice of forwarders and/or transport planners for shipping by mode air corresponds to a value-to-weight ratio of the commodity class in focus.

In other words, the higher the value-to-weight ratio, the higher the assumed probability of choosing air freight transport options. Applied to the model, the air freight volume in tonnes per commodity class is determined for each international trade relation according to the total volumes handled per airport and the value of a selected commodity class in relation to the value of imports and exports, respectively. An example:

- The total reported import of *electrical equipment* (CPA-27) from Finland in 2012 is of a value of about 488 mill. EUR.
- The total value of imports from Finland in 2012 is about 5,730 mill. EUR.

[113] Note that the volume of goods received by airplanes in total is only 0.02% of the volume of all imports. The volume of outgoing goods of all German airports is about 0.03%. Despite the comparatively small overall share, in comparison to basic nodes – here, establishments – German airports are ponderous transport network nodes. The volume handled at other German airports than the reported 16 is neglected.

- Thereafter, imports of CPA-27 comprise about 8.5% of the total import volume by value from Finland.
- In the year 2012, a total volume of about 159 tonnes is received from Finland at Berlin Airport TXL.
- As a result, a volume of about 13.5 tonnes of CPA-27 is assumed to be imported from Finland via Berlin Airport TXL[114].

Following this principle, a set of 24+16 additional national network nodes as synthetic supply chain elements is modelled to represent imports and exports of commodity classes 01 to 32 (excluding CPA-06 and CPA-09) from or to the EU-27, Russia, Switzerland and other countries.

Additional seaports

A significant share of goods produced or consumed in Germany is partly transported as maritime freight via seaports outside Germany. This is mainly relevant for international imports and exports passing through the *Port of Rotterdam* and *Antwerp*. These ports are among the largest ports in Europe in terms of handled volumes[115]. Furthermore, both ports are connected to the German road, rail and IWW network[116]. To determine the role of these two ports in the model, two simplifying assumptions are made:

- The ports of *Rotterdam* and *Antwerp* are the only international ports to handle imports and exports for Germany.
- *Rotterdam* and *Antwerp* exclusively handle German imports and exports from countries other than EU-27 member states or Russia or Switzerland. This refers to a distinct share of all oversea imports and exports[117].

[114] Note that received and sent volumes are assumed to be similar to imports and exports for airfreight statistics. In other words, when commodities are received from an airport in Finland, their origin of production is in Finland only.
[115] See a list of the *top 20 ports* in Europe in EUROSTAT (2014l). According to this ranking, the port of *Rotterdam* is larger in terms of throughput than the aggregate of numbers 2, 3 and 4, namely *Antwerp*, *Hamburg* and *Marseille*. This gives also reason to the limitation to two additional ports considered in the model.
[116] Especially for German regions close to the Dutch and Belgian borderline the ports of *Rotterdam* and *Antwerp* are in a competitive reach compared to German seaports, for instance via the *Betuweroute*.
[117] In this context, seaborne imports from Portugal are only considered for German ports, not via Amsterdam or Antwerp.

For an assignment of oversea imports and exports to *Rotterdam* and *Antwerp*, the required total reported throughput of both ports is given in EUROSTAT (2014l). Additionally, the share of the total volume with German origin and/or destination for *Rotterdam* is indicated in a report of PORT OF ROTTERDAM AUTHORITY (2013). A similar distribution to European origins and or destination is assumed for the throughput of Antwerp in absence of further details.

Further adaptions on ingoing and outgoing loads regarded as imports and exports

Volumes reported for national seaports and airports as well as the two modelled international seaports are compared with German import and export volumes as calculated in section 13.2. This process requires the following assumption:

- Reported incoming and outgoing transports via German seaports are predominantly, but not only, related to German imports and exports.

This assumption is required due to the fact that reports on volumes handled at German seaports are neither specified according to their explicit origin, nor according to their final destination. Consequently, an exemplary outgoing load from the port of Hamburg, originated in an Austrian production establishment, may not be distinguished from German exports. Nevertheless, to make use of the viable seaport handling reports, the above stated assumption is made. To specify import and export volumes accordingly, two complementary procedures apply.

The first process involves a comparison of reported volumes with the total German import and export volume for each commodity class of CPA-01 to CPA-32 as well as every region of imports and exports, the EU-27 countries, Russia and Switzerland and for the aggregate of others. This reveals an excess of certain country specific handling reports when compared to calculated total import volumes, as depicted in the following table[118]:

[118] See Table G-17 for counterpart on outgoing loads.

Total import volume* (in thousands of tonnes)		**Thereof, received via:** *German airports, allocated*	*German seaports, reported*	*German seaports, allocated*	*German seaports, allocated as share of reported*
BELGIUM	24,184	26	1,321	1,318	100
BULGARIA	572	1	57	57	100
DENMARK	5,031	10	1,618	1,488	92
ESTONIA	843	0	1,071	592	55
FINLAND	4,721	8	5,291	4,041	76
FRANCE	24,865	65	1,565	449	29
GREECE	928	4	106	101	95
IRELAND	771	6	87	36	41
ITALY	15,491	57	173	173	100
LATVIA	1,271	0	5,942	608	10
LITHUANIA	1,968	1	1,287	1,131	88
LUXEMBOURG	1,854	1	0	0	100
MALTA	18	2	81	11	14
NETHERLANDS	54,696	10	4,391	3,748	85
AUSTRIA	14,988	14	0	0	100
POLAND	21,036	18	2,516	2,506	100
PORTUGAL	1,529	11	581	496	85
ROMANIA	1,344	2	162	162	100
SWEDEN	11,807	29	4,845	4,530	94
SLOVAKIA	2,962	5	0	0	100
SLOVENIA	1,098	0	0	0	100
SPAIN	8,718	37	1,147	1,061	93
CZECH REPUBLIC	13,067	3	0	0	100
HUNGARY	4,142	4	0	0	100
UNITED KINGDOM	9,954	88	4,615	4,235	92
CYPRUS	31	0	5	5	94
SWITZERLAND	5,918	14	0	0	100
RUSSIA	20,507	72	7,204	6,611	92
OTHERS	127,787	1,492	74,720	64,447	86
Total	382,099	1,979	118,784	97,807	82

* excluding CPA-06

Table D-8 Comparison of reported incomings per country of origin at German seaports with related import assignment

In case of in- or outgoing volumes reported for German seaports exceeding the modelled total oversea imports or exports, the reported volumes are reduced, as follows[119]:

[119] See Table G-18 and Table G-19 for the corresponding export volume allocation.

		Thereof, received via:					
Total import volume from EU-27, CH and RU (in thousands of tonnes)*		*German airports allocated*	*German seaports reported*	*German seaports allocated*	*German seaports allocated as share of reported*	*Port of Rotterdam allocated*	*Port of Antwerpen allocated*
CPA-01	21,369	7	2,174	2,051	94%	0	0
CPA-02	2,510	0	1,579	545	35%	0	0
CPA-03	402	1	34	15	43%	0	0
CPA-05	16,043	5	8,304	4,649	56%	0	0
CPA-06	0	0	0	0	100%	0	0
CPA-07	5,505	1	2,825	891	32%	0	0
CPA-08	18,315	0	3,579	3,461	975	0	0
CPA-09	0	0	0	0	100%	0	0
CPA-10	21,833	20	2,114	1,909	905	0	0
CPA-11	5,935	3	404	266	66%	0	0
CPA-12	90	1	0	0	100%	0	0
CPA-13	728	4	188	95	51%	0	0
CPA-14	228	13	63	32	51%	0	0
CPA-15	85	2	15	5	32%	0	0
CPA-16	11,598	7	808	805	100%	0	0
CPA-17	14,726	22	5,854	5,315	91%	0	0
CPA-18	426	3	12	4	29%	0	0
CPA-19	34,696	28	6,924	5,756	83%	0	0
CPA-20	36,527	43	4,547	4,045	89%	0	0
CPA-21	697	13	0	0	100%	0	0
CPA-22	4,452	19	313	208	67%	0	0
CPA-23	7,933	1	299	255	85%	0	0
CPA-24	27,713	51	1,309	1,214	93%	0	0
CPA-25	4,389	13	329	193	59%	0	0
CPA-26	531	32	102	38	37%	0	0
CPA-27	2,699	38	178	162	91%	0	0
CPA-28	4,647	59	1,054	606	58%	0	0
CPA-29	6,452	69	768	628	82%	0	0
CPA-30	1,782	17	35	28	78%	0	0
CPA-31	1,583	5	146	134	92%	0	0
CPA-32	415	9	104	49	47%	0	0
Total	254,312	487	44,064	33,359	76%	0	0

* excluding CPA-06

Table D-9 Allocation of imports from the EU-27, Switzerland and Russia via German seaports, airports and the ports of *Rotterdam* and *Antwerp* [incl. data from EUROSTAT (2014l)]

This first calibration of reported import and export volumes for German seaports is complemented by a second process: a proportional allocation of the remaining import and export volumes for international relations with countries other than the EU-27, Switzerland and Russia to the ports of *Rotterdam* and *Antwerp*. The corresponding results are depicted for imports in the following table:

Total import volume from EU-27, CH and RU (in thousands of tonnes)*		Thereof, received via: German airports allocated	German seaports reported	German seaports allocated	German seaports allocated as share of reported	Port of Rotterdam allocated	Port of Antwerpen allocated
CPA-01	9,971	16	6,557	6,557	100%	2,038	1,360
CPA-02	247	0	812	247	30%	0	0
CPA-03	91	1	442	89	20%	0	0
CPA-05	27,885	5	5,281	5,281	100%	13,551	9,048
CPA-06	0	0	0	0	100%	0	0
CPA-07	38,957	12	13,964	13,964	100%	14,980	10,002
CPA-08	5,619	0	7,120	5,618	79%	0	0
CPA-09	0	0	0	0	100%	0	0
CPA-10	7,631	40	8,287	7,591	92%	0	0
CPA-11	431	1	557	430	77%	0	0
CPA-12	4	0	0	0	100%	2	1
CPA-13	719	20	1,382	698	51%	0	0
CPA-14	977	254	1,439	722	50%	0	0
CPA-15	256	29	260	227	87%	0	0
CPA-16	1,545	3	1,320	1,320	100%	133	89
CPA-17	2,539	11	1,477	1,477	100%	630	421
CPA-18	86	3	177	82	46%	0	0
CPA-19	4,707	8	2,382	2,382	100%	1,389	928
CPA-20	5,467	27	3,963	3,963	100%	886	591
CPA-21	208	18	0	0	100%	114	76
CPA-22	1,370	28	1,358	1,342	99%	0	0
CPA-23	2,662	2	2,057	2,057	100%	361	241
CPA-24	4,701	26	1,704	1,704	100%	1,781	1,189
CPA-25	1,530	30	1,786	1,500	84%	0	0
CPA-26	1,108	380	1,177	727	62%	1	0
CPA-27	1,429	112	1,802	1,317	73%	0	0
CPA-28	1,983	116	1,958	1,866	95%	0	0
CPA-29	1,803	89	2,536	1,714	68%	0	0
CPA-30	2,540	149	355	355	100%	1,221	815
CPA-31	576	11	2,179	565	26%	0	0
CPA-32	746	98	2,387	648	27%	0	0
Total	127,787	1,492	74,720	64,444	86%	37,088	24,763

* excluding CPA-06

Table D-10 Allocation of imports from other countries than within the EU-27, Switzerland and Russia via German seaports, airports and the ports of *Rotterdam* and *Antwerp* [incl. data from EUROSTAT (2014l); PORT OF ROTTERDAM AUTHORITY (2013)]

The final allocation of imports and exports to the ports of *Rotterdam* and *Antwerp* deviates from the reported total handling volumes. This divergence may be interpreted as the result of a missing exclusion of *crude petroleum and natural gas* (CPA-06) within the reported total imports and exports volumes.

Finally, a set of 37,381 synthetic supply chain elements is added to the model. This set comprises 24 national and 2 European seaports as well as 16 national airports that represent potential import and export relations with 27 European trade partners, complemented by Russia, Switzerland and an aggregate for the rest of the world. Each of the aforementioned set is a potential synthetic supply chain element for 1 out of 29 commodity classes from CPA-01 to CPA-32[120]. This configuration is complemented by one *land-based* transport option for 29 countries that likewise represent trade options for 1 out of 29 commodity classes.

13.4. Designation of synthetic supply chain elements to network nodes

Part of the subsequent modelling procedures are:

- 228,786 firms as national sources,
- 2,062,282 firms as national sinks of production,
- 37,381 additional national and international sources and
- 37,381 additional national and international sinks.

National sinks and sources represent firms, seaports and airports. They are located within the centroid of a German NUTS-3 region. International sinks and sources within the EU-27, Russia and Switzerland are either represented by national aggregates of establishments within the corresponding NUTS-0 centroid or by seaports and/or airports as stated above. As a result of this separation, only the *land-based* section of international trade relations remains unassigned for a subsequent *combined modal split and assignment* module. An excerpt of the overall distribution is given in the following figure:

[120] Cf. section 264.4 for details on exclusion of headings CPA 06 and CPA 09.

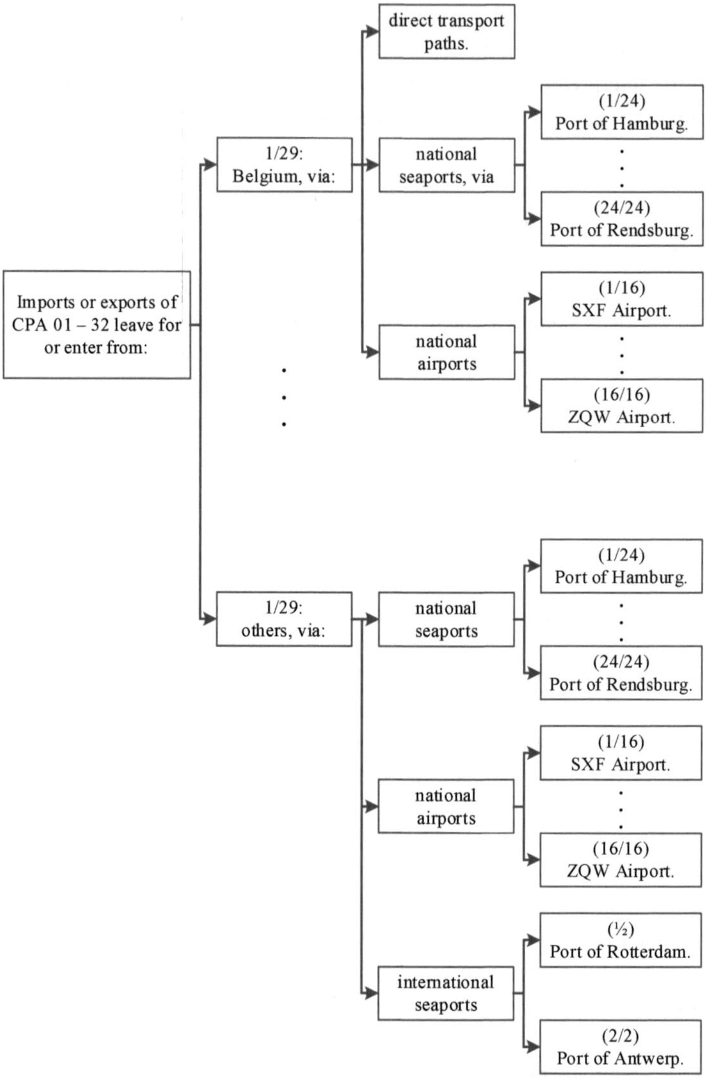

Fig. D-6 Excerpt of modelled international trade flow options

An exemplary result of the presented data processing for a supplier pre-selection is depicted in the following figure[121].

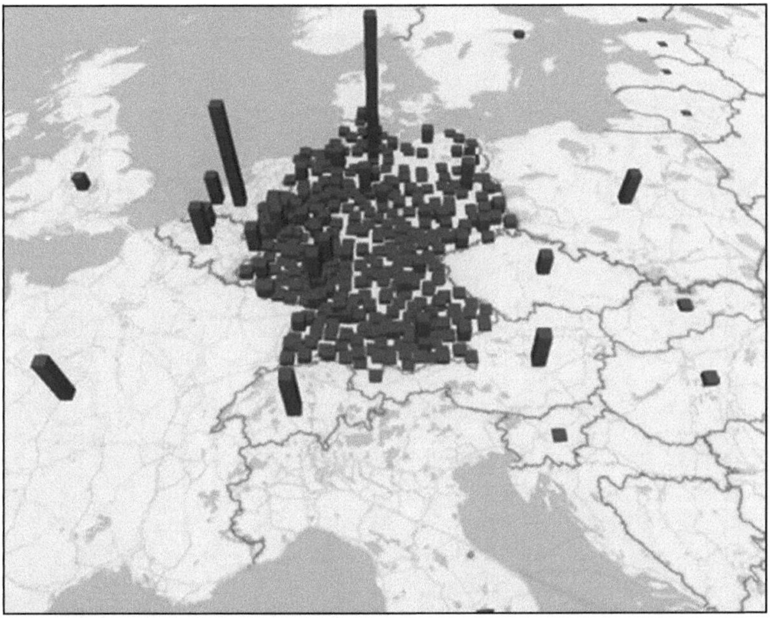

Fig. D-7 Spatial distribution of annual consumption volumes related to the German economy for national NUTS-3 and international NUTS-0 regions in total for 2012 [own results, depicted with BING MAPS (2016)]

13.5. Validation of allocated sectoral input and output volumes

The result of the previous calculations is an allocation of input and output volumes to national and international sourcing and distribution units in the form of a supplier pre-selection. Since this configuration is determined as an input for a subsequent trade flow assignment, a primary validation is reasonable. Therefore, aggregate freight volumes per mode of Table A-3 are compared with the totals of the calculated distribution of freight flows, as depicted in Table D-11 and Table D-21.

[121] Within this depiction, a complement to Fig. D-5 is given displaying the spatial distribution of annual production volumes in total for Germany in 2012. Additionally, Fig. D-7 displays a distribution by annual import volumes by corresponding consumption units.

		Transport volume in thousands of tonnes for NST-01 to NST-13	Transport volume in thousands of tonnes for CPA-01 to CPA-32	*Deviation*
CPA	*NST*	*total reported*	*total determined*	
1 2 3	1	200,172	183,670 33,572 864	9%
5 6	2	79,165	246,806 0	212%
7 8	3	1,029,354	46,074 407,532	-56%
10 11 12	4	340,128	172,176 50,518 530	-34%
13 14 15	5	12,072	4,133 1,696 547	-47%
16 17 18	6	152,850	59,117 70,413 2,573	-14%
19	7	170,419	176,084	3%
20 21 22	8	219,564	206,590 3,605 27,774	8%
23	9	355,335	236,327	-33%
24 25	10	227,897	146,102 36,319	-20%
26 27 28	11	65,039	3,594 12,125 35,910	-21%
29 30	12	101,898	49,634 10,831	-41%
31 32	13	19,771	9,123 2,570	-41%
TOTAL		2,973,664	2,236,807	-25%
Weighted absolute deviation				*39%*
Weighted absolute deviation for limited scope a)				*34%*
Weighted absolute deviation for limited scope b)				*22%*

a) excluding CPA-05 and CPA-06 that relate to NST-02
b) excluding CPA-05 and CPA-06 that relate to NST-02 as well as CPA-07 and CPA-08 that relate to NST-03

Table D-11 Comparison of NST two-digit-specific transport volumes reported to a calculated counterpart of CPA two-digit headings for Germany 2012 in thousands of tonnes [incl. data from Table A-3; Table G-10]

In total, a transport volume of 2,236,807 thousand tonnes is modelled for trade and resulting freight flows within further modelling stages. This corresponds to a share of around 75% of the statistically reported transport volume of produced goods in Germany.

Furthermore, not all commodity classes are represented within the model in the same quality. A validation in detail reveals e.g. large shares of *coal and lignite* (CPA-05) as well as of *metal ores* (CPA-07) and *other mining and quarrying products* (CPA-08) that are not adequately represented within the model, when compared to its statistical counterpart. Thus, although the dominant share of the total reported transport volume – the road freight transport – is only based on projections of a survey, the comparison may indicate shortcomings within the determined freight volumes. Potential causes for the most significant deviations are:

- the NST-02 related headings within the model include large volumes of *coal and lignite* (CPA-05) that are produced and consumed within an immediate distance, wherefore no transports are carried out by road, rail or IWW
- the NST-03 heading also includes e.g. *sand* and *gravel* that not only correspond to economic activities of CPA-08[122]. For instance, various constructional projects for buildings, roads etc. may yield into large amounts of transport volumes that are not related to firms of CPA-08. As a consequence, parts of the reported NST-03 transport volume are out of scope of the model in analogy with e.g. *secondary raw materials; municipal wastes and other wastes* (NST-14) or *goods moved in the course of household and office removals* (NST-17).

This is why the modelled transport volume is assumed to be within an acceptable range for further modelling stages without recalibration. If the goal is to focus explicitly on one of the outlying commodity classes, additional measures are required, as e.g. discussed in section 16.1.

[122] When commodities of CPA-07 and CPA-08 as well as the related NST-03 volumes are excluded from the comparison, the overall deviation from modelled to reported total freight volumes diminishes to about 8%.

13.6. Final supplier selection by performance evaluation

Apart from a general commodity performance evaluation within the format of a pre-selection of potential suppliers, a specific allocation of firm-to-firm pairings needs to be identified. As this modelling stage refers to a final supplier selection among potential suppliers in practice, as briefly discussed in 9.1, individual-specific information per sourcing firm are processed in line with section 13.1. The representation of this *real-world* process within the presented freight model is consequently based on the following considerations:

- A sourcing unit – be it a producing establishment, a service provider, a wholesale institution or a final consumer – requires a distinct amount of commodity inputs from a pre-selected number of producing establishments.
- These inputs are sourced from the most suitable suppliers in terms of different evaluation criteria.

Both the allocation of the required inputs among a distinct number of suppliers, as well as the evaluation of their eligibility, are unique decisions for each sourcing unit. These decisions may follow distinct reasons or not – they may vary over time or may be influenced by various individual-specific factors etc. – making it excessively challenging to represent them in a model. Nevertheless, a segregation of the aforementioned two decision procedures as the result of certain underlying individual decisions will be helpful to at least *avoid the estimation of an unlikely German commodity trade flow configuration.*

The subsequent modelling section therefore incorporates the estimation of a number of suppliers per sourcing unit – here, the number of sources per sink – as well as a supplier evaluation procedure[123].

[123] Since the model is set up to represent international suppliers and sourcing units only as aggregates, the terms sources and sinks are used within a description of the model's execution.

Number of sources per sink

An initial factor to estimate the number of sources per sink is the *average number of sources for a given commodity class*. This initial measure is derived from the ratio of the *total number of sinks for a given commodity* and the *total number of sources for a given commodity*. For instance, when there are three sources and two sinks of a commodity class α, the average number of sources for a sink of α is two, with a ratio being rounded to the nearest non-zero integer.

This ratio is adopted as an average per sink of α. Thus, sinks with a higher demand for α will be related to multiple sources and vice versa. This procedure allows one to represent the observation that sinks of different economic sectors and/or size may have a different number of sources for the same commodity that in total lead to the observed overall average distribution.

As a result, the commodity specific ratio of sources per sink is used as an initial solution for the subsequent modelling stages in an iterative framework. This implies a scaling – when necessary – to meet an overall transport demand measured in total tonne-kilometres per commodity class in accordance with the *combined modal split and network assignment module*.

Final supply chain network configuration

After designating an average number of sources per sink, an evaluation procedure for potential partnerships of each eligible source is developed. Therefore, a cross-sectional two-criterion comparison between the distinct economic segment of a source with multipple segments of potential sinks is processed. This interpretation of the supply chain network configuration follows the subsequent structure:

- Sinks shall be connected to their best-ranked producing counterparts. Each source that is an eligible trade-partner according to the pre-selection evaluation is evaluated individually. An exception is the exclusion of connecting a sink as a source to itself as well as the prevention of pairings between international sources and sinks[124].

[124] The first effect potentially occurs since certain CPA require products from the same CPA as an input. For instance, for an output of *Electrical equipment* (CPA-27) an input from the same economic segment but another firm is required. The second exclusion is required to eliminate allocations of direct trade relations that do not contribute to the German economy and hence are not necessarily relevant for the German freight transport system.

- The size and distance evaluation criteria are initially equally rated for the final ranking. Nevertheless, there will be an option for future adjustments, e.g. when additional input data is available.
- When multiple sinks per commodity type have to be sourced according to the initial average number of sources per sink, not all sourcing requirements are fulfilled by the best-ranked candidate. Instead, the sourcing is spread among multiple well-ranked sources in descending order. According to this, the goal of meeting the overall observed average number of sources per sink is prioritised in direct comparison to the aim of identifying the best-ranked candidate.
- When the sourcing requirements will not be met by the remaining alternative sources, then the priority is to fulfil sourcing requirements first, independently of the number of sources per sink.
- A first-come, first-served order gives a priority among sourcing establishments. This creates opportunities to either prioritise e.g. larger establishments as a result of their market dominance for a search of the best-ranked producing counterparts or to avoid any prioritisation within the model along with a random order. In absence of any indication of such priorities at present, a random order is processed within the following context.

On the one hand, this order represents another input to the final supplier selection in the context of individual specific conditions in practice. On the other hand, it assures a complete and compensated modelling of freight flows between sinks and sources within the model's computation.

The goal is to give an approximate answer to questions, such as: Is a production firm more likely to be sourced by a sink from economic sector *a* or from *b*? Or: Is source *x* or source *y* a more likely candidate to be related with a given sink? A more tangible example for an interpretation of these assumptions could be outlined for the sourcing of an establishment that produces *chemicals and chemical products* (CPA-20):

- For its own output an input of *products of forestry, logging and related services* (CPA-02) as well as *machinery and equipment* (CPA-28) is required according to the pre-selection process.

- Sources and sinks for commodities of CPA-02 are assumed to be located away from each other whereas sources and sinks of CPA-28 are geographically more proximate.
- For this setting, the firm that produces *chemicals and chemical products*, sources its inputs from CPA-02 preferably rather locally. In contrast, the sourcing of CPA-28 takes place among more distant establishments.
- The example is continued when, for the latter case, a setting is assumed, whereupon sources of CPA-28 are predominantly smaller sized compared to their sinks. Then, an establishment of CPA-28 that is not only sufficiently distant but also adequately smaller than the one in focus is a favourable candidate for a final firm-to-firm pairing.

The relevant two criteria are evaluated for each potential supplier-customer combination of:

- 266,167 national and international sources for trade flows and
- a set of 2,099,663 national and international modelled sinks.

Therefore, reference values are determined according to the following scheme:

$$\overline{d^p} = \frac{\sum\sum d^p_{so,si} * PRO^p_{so} * CON^p_{si}}{\sum\sum PRO^p_{so} * CON^p_{si}} \tag{13}$$

and:

$$\overline{\left(\frac{NEP^p}{NEC^p}\right)} = \frac{\sum\sum \frac{NEP^p_{so}}{NEP^p_{si}} * PRO^p_{so} * CON^p_{si}}{\sum\sum PRO^p_{so} * CON^p_{si}} \tag{14}$$

where:

$\overline{d^p}$	average sourcing distance of product $p = 1\ (CPA01), 2\ (CPA02), \dots, 31\ (CPA32)$
$d^p_{si,so}$	distance from source $so = 1, 2, \dots$ to sink $si = 1, 2, \dots$ for product p
PRO^p_{so}	level of production of product p in source so
CON^p_{si}	level of consumption of product p in sink si
$\overline{\left(\frac{NEP^p}{NEC^p}\right)}$	average establishment size ratio for sources and sinks of product p
NEP^p_{so}	number of employees for a production of product p in source so
NEP^p_{si}	number of employees for a consumption of product p in sink si

For each evaluated source the average distance to all its potential sinks $d^p_{si,so}$ is calculated as an *as the crow flies* distance – the shortest distance between two points on a spherical surface – according to the given geographic coordinates for the national NUTS-2 and international NUTS-0 regions and equatorial radius $R = 6378.137$ in line with:

$$d_{ij} = acos\left(sin(lat_i) * sin(lat_j) + cos(lat_i) * cos(lat_j) * cos(long_j - long_i)\right) * R \tag{15}$$

The average size relation $\overline{\left(\frac{NEP^p}{NEC^p}\right)}$ is calculated based on a comparison of each source with each eligible sink in line with the supplier pre-selection configuration. An initial average number of sources per sink is derived from the basic ratio of the number of counts of potentially related sources and sinks. Since this ratio is determined by a significantly higher number of sinks compared to sources for all commodities, the initial integer average number of sources per sink is one. The final rating of potential supplier-customer relationships is subsequently processed according to the following equation:

$$SCO^p_{so,si} = \frac{\alpha}{2} * \left| \frac{d^p_{so,si} - \overline{d^p}}{\overline{d^p}} \right| + \frac{(1-\alpha)}{2} * \left| \frac{\frac{NEP^p_{so}}{NEP^p_{si}} - \frac{\overline{NEP^p}}{\overline{NEC^p}}}{\frac{\overline{NEP^p}}{\overline{NEC^p}}} \right| \quad (16)$$

where:

$SCO^p_{so,si}$ relationship's score according to an evaluation of source so of product p in a comparison to sink si

α weighting parameter

Afterwards, a set of best ranked candidates is assembled individually for each modelled sink and each of its required inputs. Thereof a number of sources per sink is assigned accordingly. Within this scope and despite the low level of information on the supply chain network configurations in Germany, the model is set up as an evaluation of a two-criteria sourcing decision[125]. A simplified format of this procedure is depicted within the following figure[126]:

[125] Hence, it allows for a future extension to a more advanced sourcing decision modelling in case of additional information. The chosen aggregate evaluation criteria are implemented to compare the almost incomparable, or even more idiomatic, to compare apples and oranges as parts of one group for *products of agriculture, hunting and related services* (CPA 01).

[126] Therein, an eligible source per sink is predefined according to the supplier pre-selection process. Furthermore, pairings of international sources and sinks are ruled out at this stage. Subsequently, a descending order of ranked sources per sink is processed in order to assign pairings apart from best matches only if the number of best ranked firms is either lower than the target number of sources, or their total remaining output level is too low for the given setting. In case an average volume sourced per sink is assigned to a firm-to-firm pairing, it results from a ratio of the sink input level and the average number of sources per sink. In this particular case the sink input level is equal to the sink input level left and so is the average number of sources per sink and the target number of sources per sink.

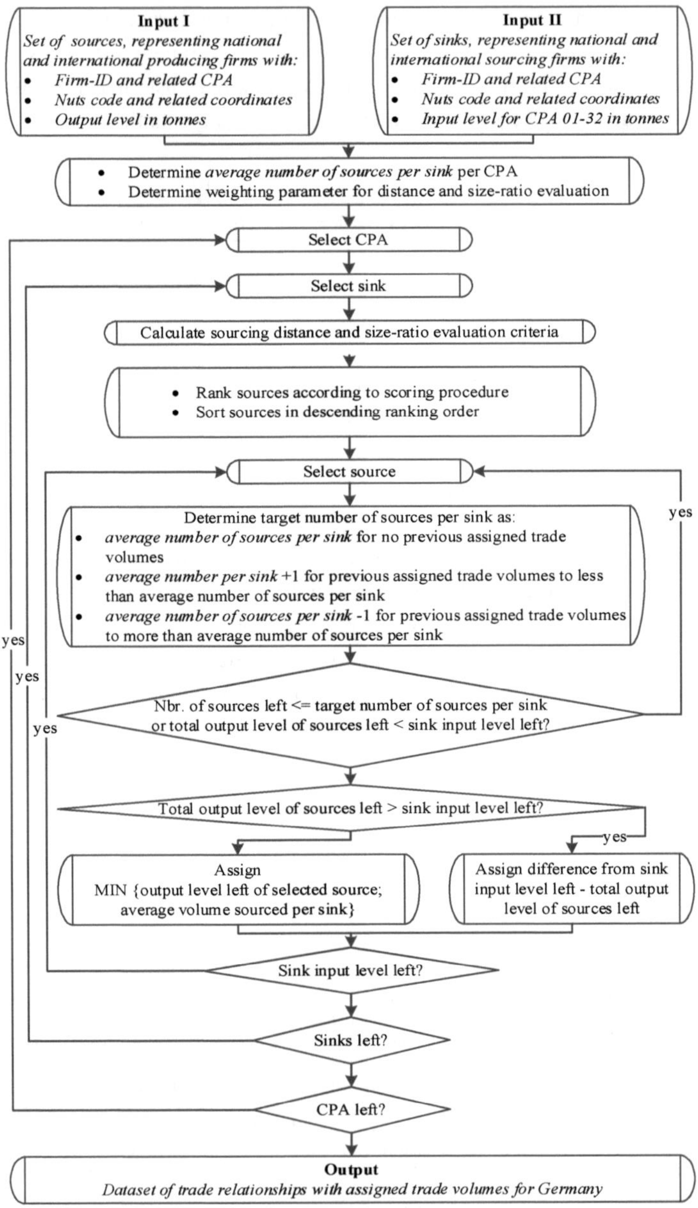

Fig. D-8 Simplified supplier evaluation processing

The output of this procedure describes trade interactions from an initial point of production – here, shippers – to a point of consumption – here, receivers – each of which is further specified by a commodity class. Subsequently, the result is also referred to as multi-commodity production/consumption (PC) matrix. Therein the exchanged freight volume is determined on an annual basis. Batches for each shipment are determined as shipment sizes hereinafter. The overall result will accordingly be a *flow coordination* including distinct average shipment sizes.

13.7. Completion by a shipment sizing

An additional breakdown of an annual trade flow volume to transport shipment batches is performed for the demand side of the modelled transport market. It is the result of an interpretation of firm specific preferences and evaluations of a transport supply configuration.

Implementing an EOQ Model

This decision is usually analysed in the context of the inventory theory, a field of *operations research.* In that case the choice of sizing of shipments is interpreted as an optimisation of a given objective function. This function consists e.g. of cost components for inventory holding, for transport and handling or for the achievement of a certain service level for stock management. Although it is a rather normative perspective on a *real-world* determination of shipment sizes, the concept is assumed to be transferable to the aspired descriptive framework of the presented model of the German freight transport market[127].

This approach is in line with results from a comparison of the *basic inventory theoretic optimal shipment size determination model* with survey results from French shippers by COMBES (2012a). This proceeding can also be stated as:

[127] Similar to one of the most advanced freight model systems, the ADA freight model approach of BEN-AKIVA AND JONG (2013, p. 82 ff.), the presented freight model for Germany relies on a normative shipment sizing module in absence of more individual specific descriptive input data.

- The analytical concept of an EOQ model is assumed to be an adequate description of the shipment size dimensioning for commodity transports in practice for Germany in 2012[128].

Apart from the proven empirical relevance of the model in general, further exemplary hypotheses are stated to favour the basic EOQ format within the proposed framework, whereupon:

- additional warehousing options between shippers and receivers are set out of scope,
- a consolidation of shipments is not considered,
- each shipper-receiver-relation is individually evaluated for a single commodity type,
- the demand of a receiver is given as fixed and assumed to be equally distributed per annum,
- the presented model excludes shipment sizes of less than a truck load and
- a routing between shippers and receivers is separately analysed.

A basic EOQ model, as introduced by HARRIS (1913) and formalised e.g. by WILSON (1934), includes the following variables:

C	total cost of inventory,
C_i	storage or inventory holding cost,
C_o	ordering or transport operation cost,
Q	annual trade flow of a constant rate,
q	standard shipment size,
b	fixed cost per transport order as costs that arise regardless of the volume ordered,
a	commodity value of time cost factor that is e.g. the result of capital and insurance cost.

[128] In the given context, an order and shipment quantity are equilibrated to each other. Thus, the choice of words only depends on the perspective on transport market actors.

The total costs C for transport or ordering are then expressed as:

$$C = C_i + C_o = \frac{q}{2} * a + \frac{Q}{q} * b \tag{17}$$

Consequently, the optimal lot size q^* is calculated by minimizing this cost function through its first derivative as:

$$q^* = \sqrt{\frac{2 * Q * b}{a}} \tag{18}$$

As a result, the EOQ-based shipment size determination only relies on information about an annual trade quantity, a value of time variable per commodity type and a transport ordering cost variable.

Applying the EOQ model to derive shipments from an annual trade generation

To apply the basic EOQ model to the presented context three inputs are required:

- an annual trade flow rate,
- fixed order costs per transport and
- a value of time factor.

The first input variable set for an *annual trade flow rate* is given in the format of the PC matrix as the result of the previous section 13.6.
One way to identify *fixed order cost per transport* is a specific transport cost evaluation for each considered shipper-receiver. Thus, the required factor is approximated by the lowest per tonne cost rate of accessible transport paths and corresponding modes[129].

[129] Note that this approach implies an evaluation of *expected* fixed cost per transport operation that is not necessarily realised according to the model's final *combined modal split and network assignment* module. A recursive modelling – hence, a recalculation of the shipment size in accordance to the finally assigned transport path and mode selection is not part of the present state of the model due to an absence of appropriate calibration measures. However, this option may be considered for future refinements, when the corresponding input data is at hand.

The potentially available alternatives for each modelled shipper-receiver pairing are depicted exemplarily in Fig. D-21. The related cost analysis to determine mode specific per tonne transport costs is presented in section 14. According to COMBES AND TAVASSZY (2016, p. 44), an adequate approximation to the *value of time cost factor* within the EOQ modelling context is the average value density of a shipment, hence the value per weight unit in € per tonne. This input is at hand due to the elaborated output generation of producing establishments in Germany (see Table D-4).
An optimal shipment size with regard to transport costs and a value of time per commodity for each shipper-receiver pairing is processed according to following setup:

$$q_{ij}^{p} = \sqrt{\frac{2 * Q_{ij}^{p} * b_{ij}^{p}}{a^{p}}} \tag{19}$$

with:

$$b_{ij}^{p} = min\{C(1)_{ij,m}^{p}\} \tag{20}$$

and:

$C(1)_{ij,m}^{p}$ fixed transport order costs for commodity class p from an origin i to a destination j by mode $m = road, rail, IWW$ or combined rail transport paths$c1, c2, c3, c4$ or combined IWW transport paths $c5, c6, c7, c8$ in [€][130]

Q_{ij}^{p} annual transport volume of commodity class p from establishment i to a destination j in $[t]$

a^{p} value per tonne of commodity class p according to Table D-4, in $\left[\frac{€}{t}\right]$

[130] According to the modelled cost structure, the minimum fixed order costs will be predominantly determined by mode road since it usually offers the lowest per order cost. This assumption, however, is not required since all modelled mode and related transport path alternatives are compared at this stage. For combined modes, the relevant costs are set up as a sum of the pre-, main- and post-haul transport network link costs – hence, designated total transport path costs. Equation (20) excludes this additional dimension for each indexed leg in favour of an improved readability, similar to (21) and (22). See also Table D-12 and Table G-21 or an exemplary application.

Until now, the shipment size will not reflect step functional specific threshold values within the selected costing scheme[131].

These values reflect the assumption that for the given scheme an optimal shipment size smaller than a full truck load will not be reasonable – here, in terms of commodity and transport relation specific load configurations as further discussed in section 14.

This is why a second optimal shipment size value is determined – a processing that is comparable to e.g. an application of the EOQ model within the context of global shipment discounts for *operations research.* This re-adjusted optimal shipment size is determined as:

$$q_{ij}^{p*} = \left\lceil \frac{q_{ij}^{p}}{NLF_{ij,m^*}^{p}} \right\rceil * NLF_{ij,m^*}^{p} \tag{21}$$

with:

NLF_{ij,m^*}^{p} net load factor for a transport of commodity class p from origin i to a destination j by mode m^*, chosen in line with $C(1)_{ij,m}^{p}$ according to equation (20)

Mode specific total annual transport costs for each shipper-receiver pairing are then evaluated according to volume:

$$x_{ij,m}^{p} = \left\lceil \frac{Q_{ij}^{p}}{q_{ij}^{p*}} * \left\lceil \frac{q_{ij}^{p*}}{NLF_{ij,m}^{p}} \right\rceil \right\rceil \tag{22}$$

with:

$NLF_{ij,m}^{p}$ net load factor for a transport of commodity class p from origin i to a destination j by mode $m = road, rail, IWW$ or combined rail transport paths $c1, c2, c3, c4$ or combined IWW transport paths $c5, c6, c7, c8$

[131] See e.g. Fig. D-10, Fig. D-12 or Fig. D-15.

This formalisation is transferred to an example of a given firm-to-firm paring – say firm a and firm b – with a in *Ludwigshafen am Rhein* (NUTS-ID: 07314) and b in *Bottrop* (NUTS-ID: 05512) with an annual trade flow volume of $Q_{ab}^{20} = 15{,}156.68$ tonnes of *chemicals and chemical products* (CPA-20) and minimum fixed order costs of $b_{ab}^{20} = 638.11$ € according to equation (20) for mode $m = road$ as well as with a given value of time factor for CPA-20 of $a^{20} = 1{,}094.30\ €/t$. In consequence, an initial optimal shipment size $q_{ab}^{20^*}$ of about 132.95 tonnes for this particular setting is calculated in line with equation (19).

The determined shipment size is thereupon comparable with a next larger size configuration for the selected mode with similar costs. That $q_{ij}^{p\ *re}$ is 147.09 tonnes, reflecting 7 potential transports per order by mode road with an average observed load configuration[132]. For this re-adjusted optimal shipment size the amount of annual road freight transports would by 722. Alternative modes and their total transport costs, in line with the modelled costing scheme, are depicted within the subsequent table:

[132] Cf. section 14.

	Road	**Rail**	**IWW**	**Combined-rail path 1**			**Combined-rail path 2**		
				pre-haul	*main-haul*	*post-haul*	*pre-haul*	*main-haul*	*post-haul*
EOQ in t					132.95				
Net load in t	21.01	1,966.39	1,058.09	22.87	1,075.08	22.87	22.87	1,075.08	22.87
EOQ_re in t					147.09				
Number of trips per order	7	1	1	7	1	7	7	1	7
Costs per order in EUR	638	8,860	14,943	0	9,727 9,847	119	83	12,506 12,709	119
Number of trips in total	722	104	104	722	104	722	722	104	722
Total annual costs in EUR	460,715	921,395	1,554,038		1,097,814			1,447,007	

Table D-12 Exemplary application of shipment size evaluation for CPA-20

As a result, the lowest overall annual transport costs among all accessible mode configurations for the exemplary setting is realised by mode road. For an exemplary setting where everything is similar apart from the commodity class and its respective value of time factor, the EOQ outcome will be different[133]. For instance for a transport of *products of forestry, logging and related services* (CPA-02) the average modelled shipment size is larger than in the previous example, due to a significantly lower value of time factor.
The results of the annual transport costs modelled for such a setting are depicted in Table G-21. For this case, the order quantity evaluation suggests a different mode choice. Instead of mode road, as favoured for CPA-20, a transport by mode rail would lead to the lowest overall transport costs for the exemplary firm-to-firm paring within a final modal selection – a choice that is modelled within the subsequent *combined modal split and network assignment* module.

Prior to this, however, potential freight transport network paths and corresponding link combinations need to be evaluated. This, in turn, is preparatory work to identify link specific transport costs for each modelled mode and

[133] Note that this is an exemplary setting to point out the effect of a trade flow bundling in the format of a trade-off between inventory holding costs and mode specific order costs.
In the *real-world* practice, as well as within the presented model, costs per mode would be different for a transport of commodities from CPA heading 02 and 20.

commodity class, as they are another essential input to the *combined modal split and network assignment* module (cf. section 15.1).

14. Modelling the freight network

The modelled transport network comprises a set of national and international locations of production and consumption that represent an origin and/or a destination node. These nodes comprise:

- 402 national Nuts-3-regions,
- 24 national seaports,
- 16 national airports,
- 28 international Nuts-0-regions and
- 2 international seaports.

In total, a number of around 266,000 origins of production and 2,100,000 destinations of consumption that result according to the synthetic supply chain configuration of section 13 in a set of nearly 43,000,000 firm-to-firm pairs are modelled.

Road network links

Each of these 43 million pairings is conected, at least, by one road freight link that is equal to a direct transport path. For this set of transport network paths minor adaptions apply to those that contain a node at either end which, in turn, is not directly connected to the European continental road infrastructure, or at least only with extensive detours. These particular nodes represent the countries of Ireland, Malta and Cyprus as well as Finland. Their Nuts-0-centroid is changed to a synthetic network node that is processed within the model's transport distance-related evaluations. These nodes are:

- Finland – routing by ferry via Stockholm, Sweden
- Ireland – routing by ferry via Liverpool, United Kingdom
- Malta – routing by ferry via Palermo, Italy
- Cyprus – routing by ferry via Athens, Greece

In total, a set of more than 220,000 direct road freight paths is evaluated for multiple commodity specifications within the model.

Rail network paths

A unimodal transport by mode rail and likewise IWW is not feasible between all modelled trade flow pairings since a specific access to the corresponding network is required. In case an access is modelled, the relevant nodes represent private (or equivalent) rail network and/or IWW access points. Otherwise, a rail and IWW network infrastructure access is only feasible via public intermodal transfer nodes.

For Germany a set of 2,374 private rail access nodes is reported for the year 2012 according to DB NETZ (2013, p. 5). However, neither information about the related operator nor any indication of accessibility restrictions is at hand. It would be favourable to announce a distinct set of operating establishments for each (inland-) port or freight yard. In consequence, two assumptions apply, whereupon:

- predominantly highly productive establishments maintain private rail access nodes and
- transport bundling of multiple trade flows are not explicitly modelled

In line with the first assumption, the number of private rail access points is distributed in descending order among the 2,374 most productive firms in Germany. Since each of them is located in one out of 402 national NUTS-3 regions, fewer corresponding rail freight yards are finally processed. To determine the total number of modelled private rail network nodes, a nearest-neighbour search is processed to identify the closest freight yard per NUTS-3-centroid according in line with a request to DB CARGO (2015a)[134]. This process leads to a number of 389 different national regions and national seaports with at least one located firm operating a private freight yard (cf. Table G-61 to Table G-71).
The second initial assumption ensures, first of all, a limited access to direct transport options per rail. Only large production and/or consumption firms in the modelled network possess direct rail transport options. Otherwise, small-sized establishments without a direct network access would favour options within the model that are not feasible in practice. Besides, it ensures an ade-

[134] The request is based upon the supplementary assumption that the rail network configuration is not subject to relevant changes on routing options from year 2012 to 2015.

quate *real-world* representation of intermodal transport paths without considering timetables for IWW and rail mainhaul link alternatives. Although the consolidation of multiple trade flows along a similar link is an elementary component of the competitiveness of intermodal transport paths, it is the result of various conditions[135] that are out of scope of this study, due to an absence of adequate input data. As a result, intermodal transport paths are only evaluated for shipment sizes of each firm-to-firm pair individually.

A separate test within on the impact of this simplification reveals that a breakdown of the overall intermodal rail or IWW mainhaul costs to single TEU – thus, assuming a complete vehicle utilisation as the result of a sufficient consolidation potential and unrestricted transport schedules to match each individual shipment frequency – would lead to highly unrealistic transport cost advantages that are not reflected within the related *real-world* link flow observations. Instead, adequate modelling results are obtained for a breakdown of costs to complete containerised train or vessel loads for each individual firm-to-firm trade flow[136].
One interpretation might be that, even though not only complete loads but also consolidations are observed in practice, intermodal transports are only favoured for sufficiently large average shipment sizes, whereof one part of the annual total trade flow volume is e.g. shipped along intermodal transport paths, whereas another part is directly transported in a multi-mode transport strategy.

The ports of *Rotterdam* and *Antwerp* are modelled as two additional international direct IWW network access nodes, together with a set of 26 rail network nodes for international Nuts-0-regions (cf. Table G-71). These nodes are subsequently allocated to transports with a German origin and/or destination as indicated in DESTATIS (2013b). In total, a set of around 150,000 direct rail transport paths is part of the model.

---

[135] Exemplary conditions are the frequency of scheduled intermodal freight trains and IWW vessels as well as their link specific occupancy rate. The latter condition not only influences the transport costs per unit but likewise the remaining load capacity in a nearly opposite direction. As a result, the exemplary second condition relates to the first one.

[136] For a detailed transport cost evaluation, such factors need to be identified and individually apportioned. See section 14.3 and 14.4 for details on the final implementation within the model, as well as Table D-12 for an example.

Inland waterway paths

The number of private inland ports is significantly lower than its rail network counterpart. According to PLANCO (2013, p. 44), 11 ports in Germany are entirely private. A complementary listing of the 60 busiest inland ports that are surrounded by one or multiple firms within direct reach is given in DESTATIS (2014o, p. 27 ff.). Eligible ports from this listing are adopted to the model. This includes a similar process as that for private rail access nodes, along with the following resembling assumptions:

- predominantly highly productive firms either maintain private ports or are directly located around public inland and
- transport bundling of multiple trade flows are not explicitly modelled.

This approach allows for the consideration of inland ports within the model that are accompanied by at least one firm in an immediate proximity – e.g. connected via private railways or comparable in-company transport options – and likewise offer public access.
A prerequisite is their identification within the IWW routing engine of WSV (2015) for a request on merchant navy navigation[137]. Additionally, 14 eligible national seaports are represented within the model's IWW transport network. Double counts of ports are excluded. As a result, 75 national IWW network nodes are included in the presented model (cf. Table G-58 and Table G-59).

The required assignment of ports to German NUTS-3 regions is processed according to EUROSTAT (2015c). Additionally, a set of 14 international IWW network access nodes is modelled, including the ports of *Rotterdam* and *Antwerp* (cf. Table G-59 and Table G-60). This selection is derived from a request on reported in- and/or outgoing loads within the link count data from DESTATIS (2013a), similar to rail freight operations[138]. In total, the set of possible direct IWW transport paths comprises about 9,000 alternatives that are modelled for different commodity class specifications.

[137] The IWW network configuration is not subject to relevant changes on routing option from the years 2012 to 2015, as available from WSV (2015).
[138] Apart from the modelled international sea shipping transports, there is also a reported annual IWW short-sea transport volume with particular vessels for sea and inland port access capabilities for Denmark, Finland, Russia, Sweden, United Kingdom, Latvia, Ireland and Lithuania that correspond to a German inland port origin and/or destination. In total, however, these specific transports

Intermodal transport paths, national

For intermodal transport options specific freight yards and inland ports are included. A list of rail freight yards with handling capabilities for at least 20-foot ISO containers in Germany is given in DB CARGO (2012). Therein, 44 intermodal rail freight yards are listed that service intermodal containers, semi-trailers, swap bodies and assimilated devices. Besides, a list of intermodal terminals in Germany and Europe is also given e.g. in AGORA (2014). The latter database comprises a sound listing of European intermodal terminals, including their location and the modes served, respectively. The 104 listed terminals for Germany are initially considered for the presented model.

If applicable, 24 national seaports of the presented framework are included as intermodal rail and IWW network nodes, whereas intermodal transports to or from airports are not considered. Similar to private rail freight yards and equivalent private inland ports, a successful request for routing options to or from seaports within the data bases of either DB CARGO (2015a) or WSV (2015) is therefore required[139]. Within this total of inputs – double entries excluded[140] – a nearest-neighbour search as well as a search for a second-nearest neighbour is performed to identify potential intermodal transport paths for each modelled OD-pairing. This results in a selection of two main-haul links by rail and two links by IWW, respectively.

This initial set of intermodal transfer nodes is compared to recorded operation of containerised or equivalent transports in 2012 according to DESTATIS (2013b) and DESTATIS (2013a). This procedure that results from the following assumption:

- Intermodal transports only take place in distinct transport units (see Fig. A-3).

comprise only 0.29% of the overall annual IWW transport volume. In absence of a generic – hence, for all freight vessels usable – IWW routing path, these countries are not considered within the model for an IWW transport option.

[139] Therefore, the database is assumed to be without relevant changes for operating terminals from year 2012 to the time of query in 2015.

[140] The method applied is the exclusion of entries in a linear distance of less than 3 kilometres to each other in case of multiple entries per NUTS-3-region. For instance, instead of multiple ports for the German city *Duisburg* that are listed in the input datasets, only one representative is modelled.

The resulting number of intermodal rail freight and IWW network nodes is reduced for subsequent calculations to one unique terminal per NUTS-3-region for rail and/or IWW line-hauls. Finally, a total of 90 intermodal rail freight access nodes and 44 national intermodal IWW nodes is processed within the following (cf. Table G-72 and Table G-73 as well as Table G-75 to Table G-77).

Intermodal transport paths, international

For a set of intermodal network nodes, representing the modelled international trade partners – here, the EU-27 as well as Russia and Switzerland – the following simplification applies:

- For each international origin and/or destination only a single – here, the nearest identified terminal in terms of the shortest straight-line distance – intermodal terminal is modelled.

First, a nearest-neighbour search within the set of international AGORA (2014) terminals is processed. Afterwards, the corresponding intermodal pre- and/or post-haul distances out of Germany are estimated, in line with the following simplifying assumption:

- Intermodal terminals in European countries are nearly equally distributed – hence, the distance to each other is close to uniform.

Consequently, the mean distance between a country's centroid and its three closest intermodal terminals – as listed in AGORA (2014) – is assumed to represent an average pre- or post-haul distance out of Germany of the related intermodal transport chain[141].
In line with this procedure, the nearest of the trimodal terminals of an international centroid is modelled for the main haul distance evaluation of intermodal transport chains. Additionally, the international seaports of *Rotterdam* and *Antwerp* are considered as international intermodal network nodes. Both nodes represent potential rail and IWW access nodes.

[141] Here, bi- and tri-modal terminals are equally considered. In case of multiple listings of the same port – e.g. Constanta Port and Constanta South Container Terminal in Romania – the list is limited to only one entry per port in line with a distance criterion of less than 3 km from port to port *as the crow flies.*

In total, 60 synthetic international intermodal rail freight and/or inland waterway network nodes are modelled. The corresponding international rail freight distance relations are taken from DB CARGO (2015a), inland waterway navigation distances are adopted from WSV (2015) and BINNENREEDEREI (2014) along with data from Table G-74.

Consolidation of a set of intermodal network nodes

The modelled set of about 43,000,000 firm-to-firm pairs can be reduced to around 220,000 different regional combinations. An assignment of, at maximum, the two nearest intermodal rail freight and the two nearest intermodal IWW network nodes with a reported annual transport volume to each origin and destination allows to reduce the overall number of combinatorial options for intermodal transports significantly. As a result, only around 400,000 distinct commodity specific network paths are modelled to represent a variety of real-world intermodal transport alternatives. In general, a pre- and post-haul by mode road is involved within this paths. An exemption is made for seaport-origins and seaport-destinations within the model (cf. Fig. D-21)[142].

14.1. Network link transport cost evaluation principles

Apart from designating the set of potential transport network paths for each modelled firm-to-firm pair, estimating the corresponding link transport costs is a prerequisite for an application of the *combined modal split and network assignment* model – hence, for obtaining the overall model's result.

Concept for a modelling of transport costs

Transports may be performed in unimodal or multimodal configurations. A question in practice – and consequently within the presented modelling approach – is whether one particular mode or a combination of multiple modes performs best for a given transport demand configuration.

The basic performance measure within this context is derived from an evaluation of pecuniary and time costs reflected in transport market prices. Since

[142] It may occur that a containerised load is directly transported from a private rail yard or inland port to a seaport or vice versa. Since this phenomena is not separated within the intermodal transport statistics for Germany, it is likewise not modelled.

no adequately exploitable information on profit margins within freight transport market prices is at hand, the following assumption is employed:

- The German freight transport market is a market under almost perfect economic competitive conditions. Subsequently, a transport service is supplied at a market price equal to its corresponding marginal costs.

Although the level of competition among road freight forwarders in Germany is high – and so is the intermodal competition – profits are feasible. That is why this simplification is required to still enable a modal choice modelling based upon cost structures of the market supply side. In contrast to profit margins, cost structures are derivable for alternative modes and their combinations.

This gives reason to the application of the selected scheme for a performance comparison of competing modes within the presented framework. It is generally based upon the idea that the most relevant performance measures in the contemporary German and European freight market result from an evaluation of:

- pecuniary cost rates per transport and distance unit as well as
- hourly cost rates that are factorised with travel time units.

Following the notion principles of HANSSEN, MATHISEN AND JØRGENSEN (2012, p. 192), complemented by a volume-related factor, shippers as purchasers of freight transport services seek to minimize the total costs of a transport *C*, that are defined as:

$$C(x,d,p) = \left\lceil \frac{x}{NLF} \right\rceil * P(d,p) + H(p) * T(d) \tag{23}$$

Thereafter, total costs *C* are determined by:

- a transport volume *x*,
- a transport vehicle specific net load factor *NLF*,
- pecuniary costs *P* for transport services,
- time costs as a product of costs per time unit *H* and time *T*,
    - a transport distance *d*,
    - a commodity specification *p*.

This global format for a transport cost analysis allows one to consider effects of economies of scale by distance – i.e. the cost per distance unit decreases with an increasing transport distance – as well as effects of economies of scale by volume – thus, a decrease of costs per transport unit with an increase of the overall volume in transport[143].

A comparable cost concept is e.g. applied within a study of JANIC (2007). Therein, average cost per door-to-door distance for 20-foot container for road and rail transports are quantified based on the results of a survey on cost structures in the freight transport market in Europe, the European RECORDIT-project from 2000 to 2002 (EC, 2003b).
A more recent empirical study on transport costs for the year 2010 in Germany can be found in BMVI (2014a). This study mainly contributes to a modal split evaluation within the German *Federal Transport Infrastructure Plan* for 2015 (cf. section 7.2). Valuable aspects for the modelling of costs for German road, rail and freight transport on IWW are contained therein. The presented cost evaluation framework, however, lacks a standardisation of the alternative modes. Beyond that, the study is inconsistently executed in detail for relevant parts, e.g. the breakdown of road freight transport costs.
As a result, transport cost evaluations from BMVI (2014a) are not applicable to the presented context. Nevertheless, the sketched framework will be a major impetus for the following section.
Additional relevant breeding ground for transport cost modelling concepts in detail is the study of WITTENBRINK (2010) as well as the report of ZEITZEN (2012). Therein valuable indicators, especially for the road freight segment, are presented.

Commodity class handling specifications

To evaluate commodity specific transport costs per mode, the corresponding classification scheme is evaluated in terms of the most probable category of cargo handling. This is the result of the fact that different commodity groups

[143] A survey among German freight transport actors for the German *Federal Transport Infrastructure Plan* for 2015 reveals a perceived reliability for transports on IWW of 99%, for road of 97%, followed by conventional rail (91%) and combined rail transports (95%) (BMVI, 2014a, p. 123). As a result, a separate evaluation of the level of volatility of a determined time in transit will not be part of this study.

of the CPA classification scheme require comparable transport vehicles and related handling activities[144].

For road freight transports, loads of dry bulk (1), liquid bulk (2), palletised cargo (3) or break bulk cargo (4) are differentiated. For rail and IWW freight transports, as well as intermodal main-haul links, a different vehicle specific categorisation of transported commodity groups is applied.
This concept is an attempt to deal with still limited information on commodity class specifications on a level of two-digits. Due to this fact, for certain CPA-coded commodity classes an unambiguous allocation to the selected forms of material handling is not feasible. For *coke and refined petroleum products* (CPA-19) for example either dry bulk or liquid bulk handlings and related transport units are required. In the presented model, a liquid bulk handling is assumed to be the most representative cargo handling category to evaluate the transport costs of this particular commodity class. The results of the mode specific commodity group classifications are depicted in Table G-22, Table G-29, Table G-37 and Table G-44.

14.2. Road freight transport link costs

To determine transport and time costs for road freight transports, the study of the BMVI (2014a, p. 126 ff.) proposes an evaluation of the following three exemplary classes of lorries:

- standard tractor-trailer combination with a maximum permissible gross laden weight of 40 tonnes,
- tractor-trailer combination with combined transport optionality with a maximum permissible gross laden weight of 40 tonnes and
- tank tractor-trailer combination with a maximum permissible gross laden weight of 40 tonnes.

A review on the number of trips and the related length per road freight vehicle size class in Germany, as given in the review on a trucking survey evaluation from the German *Federal Motor Transport Authority* in KBA (2013a), reveals that a similar classification is adequate for the presented disaggregate freight transport model for Germany.

[144] Thereafter, the computational burden of the transport cost related *combined modal split and network assignment* module is significantly lowered without relevant restrictions.

The classification stems from the observation that the selected tractor-trailer combination classes cover a dominant share of all transport activities by road. Although lorries with or without trailers outperform tractor-trailer combinations in terms of a number of trips in Germany – 44% of all trips are conducted by lorries and 38% by tractor-trailer combinations – the contrary holds true for the distance travelled – hence the average trip length. The overall distance travelled is performed in 57% of all cases by tractor-trailer combinations and only 19% are carried out by lorry. As a result, the average trip length of lorries is only one third of the trip length for tractor-trailer combinations. Since the average load capacity for regional and long-distance transports in Germany is above 22 tonnes, as indicated in KBA (2013b), these three 40-tonne tractor-trailer combinations are assumed to be eligible representatives for road freight transport units within the presented model[145].
The three selected types are consequently assigned to the modelled commodity classes of the CPA system in line with dominant handling specifications, as depicted in Table G-22.

14.2.1. Reference road freight cost calculation scheme

To estimate the related road freight transport costs then, the following cost calculation data input is required:

[145] For an own weight of a standard tractor-trailer combination within an average range of 13 to 15 tonnes, the related maximum permissible gross laden weight of 40 tonnes is nearly met.

	Input, according to:			*Tractor-trailer combination 1 (standard)*		*Tractor-trailer combination 2 (container)*		*Tractor-trailer combination 3 (tanker)*	
				MB Actros 1848 LS Euro 6	Three-axle curtainsider	MB Actros 1848 LS Euro 6	Three-axle 40 ft. container carrier	MB Actros 1848 LS Euro 6	Three-axle tanker
				Basic cost-related input factors					
1	*a)/ f)/ g)*	Own weight	t	6.60	6.40	6.60	5.08	6.60	6.70
2	40 (44) t -Input 1	Payload	t	27.00		[29.16]		26.70	
3	*a)/ f)*	Net purchase price	€	98,500	27,000	98,500	23,200	98,500	110,000
4	*a)*	Fuel consumption	l/km	0.33	-	0.33	-	0.33	-
5	*b)*	Operating lifetime	a	6	6	6	6	6	6
6	*b)*	Annual mileage	km/a	135,000	135,000	135,000	135,000	135,000	135,000
7	*c)*	Annual capital interest rate	%/a	3.38	3.38	3.38	3.38	3.38	3.38
8	*b)*	Residual value for resale	%	30	30	30	40	30	30
9	*d)*	Replacement purchasing price	%	108.2	108.0	108.2	108.0	108.2	108
10	*a)*	Price for tires	€	3,564	3,306	3,564	3,306	3,564	3,306
11	*a)*	Max. mileage tires	km	147,000	173,000	147,000	150,000	147,000	150,000
12	Input 3 * Input 8	Residual value for resale	€	29,550	8,100	29,550	9,280	29,550	33,000
13	Input 3 * Input 9	Residual value for replacement	€	106,577	29,160	106,577	25,056	106,577	118,800
14	(Input 3 - Input 12) * 0,5 *e)*	Average operating capital	€	34,475	9,450	34,475	6,960	34,475	38,500
				Fixed cost components/costs per operating time unit					
15	Input 7 * Input 14	Annual costs for capital	€/a	1,164	319	1,164	235	1,164	1,300
16	(Input 13 - Input 12) / Input 5 * 0,5	Depreciation per time unit (50%)	€/a	6,419	1,755	6,419	1,315	6,419	7,150
17	*a)*	Liability and comprehensive vehicle insurance	€/a	6,315	480	6,315	480	6,315	1,159
18	*a)*	Vehicle tax	€/a	665	671	665	671	665	671
19	*a)*	Vehicle depot costs	€/a	1,104	1,104	1,104	1,104	1,104	1,104
20	*a)*	Vehicle depot management costs	€/a	8,134	8,013	8,134	8,013	8,134	8,013
21	Σ (Input 14,...,19) / 240 d / 9.6 h	**Σ of hourly fixed costs** (without wages)	**€/h**	**10.33**	**5.36**	**10.33**	**5.13**	**10.33**	**8.42**
22		**Hourly lorry driver wages**	**€/h**	**18.22**	-	**18.22**	-	**18.22**	-

Table D-13 Cost calculation data input for modelled road freight vehicles, p. 1

	Input, according to:			*Tractor-trailer combination 1*		*Tractor-trailer combination 2*		*Tractor-trailer combination 3*	
				Variable cost components/costs per distance travelled					
23	(Input 13 - Input 12)/ Input 5 * 0,5 / Input 6	Depreciation per distance travelled (50%)	€/km	0.048	0.013	0.048	0.010	0.048	0.053
24	Input 10 / Input 11	Wear of tires	€/km	0.024	0.019	0.024	0.022	0.024	0.022
25	*a)*	Maintenance and servicing	€/km	0.097	0.018	0.097	0.018	0.097	0.037
26	*a)/e)*	Diesel exhaust fluid additives and lubricants	€/km	0.015	-	0.015	-	0.015	-
27		Σ of variable costs (without fuel)	€/km	0.18	0.05	0.18	0.050	0.18	0.11
28	Input 4 * (1,181 € / litre)	Variable fuel costs	€/km	0.39	-	0.39	-	0.39	-

Table D-14 Cost calculation data input for modelled road freight vehicles, p. 2

Sources of information in Table D-13 and Table D-14 are referenced as[146]:

a) ZEITZEN (2012)
b) BMVI (2014a)
c) BUNDESBANK (2014)
d) DESTATIS (2014p)
e) WITTENBRINK (2010)
f) FLIEGL (2014)
g) FELDBINDER (2014)

General details

As a representative for road tractor specific cost factors, the exemplary evaluated vehicle type is a Mercedes Actros 1848 LS Streamspace Euro 6[147]. For

[146] Therein, both the residual value for resale of input 8 as well as the replacement purchasing price for an equivalent vehicle of input 9 are reflected in a forward projected purchasing price related to the calculated lifetime. Input 16 is the result of the sum of inputs 14 tthrough 19, divided by 240 vehicle operation days and further divided by 9.6 hours of operating time per day for driver and vehicle. For details on payloads for tractor-trailer combination 2 see discussions on *Containerised road freight transports* in section 14.2.2.

[147] According to KBA (2014), the selected manufacturer Daimler represents more than 25% of all new registrations in the segment of lorries and road tractors in 2012, followed by Volkswagen (i.e. MAN and Scania) and others.

this vehicle, valuable references are given in ZEITZEN (2012, p. 289). Specific cost factors for the corresponding semitrailers are likewise primarily adopted from the valuable cost calculation of ZEITZEN (2012)[148].
For capital relevant calculations, an interest rate of 3.38% per year is assumed. This value is derived from a 10-year average of yields on outstanding debt securities according to the BUNDESBANK (2014). Thereafter, a depreciation rate is calculated ensuing from an annually recalculated replacement value, considering specific repurchase price changes for road tractors and semitrailers. In the study of BMVI (2014a, p. 133), the output price index for *tractor trailers* (GP09-291043) and *coachwork bodies for motor vehicles; trailers and semi-trailers* (GP09-292) are proposed as an indicator in this context. For a 6-year period from 2006 to 2012, this results in price increases of 8,2% for road tractors (based on GP09-291043) and 8,0% for semitrailers (based on the GP09-292 output price index) (DESTATIS, 2014p, 196 f.).

Upon this setting, a fixed – in terms of an independence from transport activity level – as well as a variable – hence, distance dependent – depreciation cost component of equal importance are calculated in accordance with BMVI (2014a, p. 133 ff.). The former component includes costs for insurance, taxes as well as a vehicle fleet administration and a maintenance factor which are adopted from ZEITZEN (2012, p. 289 ff.). These costs components are broken down to 250 business days as a national average for regular working days in 2012 – hence, no transports on weekends are modelled. Finally, 240 vehicle operation days is processed taking into account 10 days for maintenance and servicing.

Personnel cost details

Apart from vehicle specific costs per operating time unit, one of the most significant fixed cost components – the personnel deployment of a transport service provider – is evaluated. Therefore, a report based on a survey of the German labour union on the actual wages paid – including estimates for surcharges for 48 working hours per week – for 2010 by region is consulted (BERGRATH, 2010, p. 19). The study reveals variations of more than 100%

[148] For an estimate of the average purchasing price of a three-axle 40 ft. container carrier and the trailers' own weight indications, exemplary manufacturers are consulted for missing values, here FLIEGL (2014) and FELDBINDER (2014).

within Germany for the regional specific wages paid. In consequence, a national weighted average by employment rates in the sector of *land transport and transport via pipelines* (CPA-49) is calculated according to Table G-23. As a result, an annual income of 26,788 € for lorry drivers in 2012 for Germany is assumed for the subsequent evaluations accordingly.

As proposed in the concept of BMVI (2014a, p. 134), associated wages, social security contributions and taxes are taken into account as well as additional surcharges on a lorry driver's income. The sum of all these surcharges for the year 2012 is estimated at 23% of the actual wages paid (LOHNDIREKT, 2012). Additionally, business expenses of 15 € per working day are considered as a further premium in accordance to WITTENBRINK (2010, p. 26). This likewise reasonable value for 2012 is within the range of the usual 12 €, in case of 8 to 24 hours, and 24 €, in case of more than 24 hours of absence from the driver's home location – in many cases the trucking company's base location.
Furthermore, a value of 20 days per annum is assumed in the following context to cover absences from work by drivers due to illness, mandatory driver training etc. Another 24 days per year are considered to account for the number of paid holidays[149]. In consequence, for a national average of 250 regular business days the number of days to account for personal deployment costs is set to 206. For 32,950 € – as the total lorry driver costs – and 15 € per day for expense allowances, total personnel costs of 174.95 € per day result. For a weekly working time of 48 hours[150] the processed hourly driver costs are 18.22 €, as incorporated into Table D-13 and Table D-14.

Fuel cost details

Fuel prices within the road freight transport cost evaluation for the year 2012 are set to an annual average value of 1,181 € per litre for large-scale consumers, based on DESTATIS (2014r, p. 59). Additionally, a consumption of specific fuel additives and lubricants is considered. This approach follows the

[149] Within this scheme, additional personnel factors per lorry are not required. For instance, a factor of 1.2 drivers per lorry, as proposed in the work of WITTENBRINK (2010, p. 26), when the absences of work of a lorry driver is not covered otherwise, as e.g. stated above.
[150] Multiple drivers per vehicle deployed are not considered in the presented context. German law allows driving time peaks but limits the total to 48 hours per week.

suggestions of BMVI (2014a, p. 135), based on assumptions as stated in WITTENBRINK (2010, 40 f.). Thereafter, a road freight tractor of the modelled type requires exhaust fluid additives with a value of 0.4 € per litre in 2010 to the extent of 5% of the vehicle's fuel consumption (ibid.). An identical consumption rate as well as an updated price of 0.42 € per litre[151] are considered in the following framework. Finally, additional costs for lubricants, distance specific maintenance and servicing as well as tire abrasion are included. The corresponding values per kilometre are obtained from ZEITZEN (2012, p. 289 ff.)[152].

14.2.2. Additional road freight cost factors

For the year 2012, the share of national transports performed by international road freight forwarders – the cabotage penetration rate – is about 3% (BAG, 2013a, p. 54). In contrast, a significant share of all road freight transports outgoing from and incoming to Germany is performed by international forwarders. In accordance with BMVI (2014a, 136 f.) different cost factors are considered for these transports that are primarily the result of variant personnel costs and fuel cost differences, but not from vehicle-related costs, granted through e.g. less advanced technical equipment in Germany's neighbouring countries.

Subsequently, employment costs as well as fuel price index tables from various sources are consulted. Thereof, potential price differences to the national reference cost factors of Table D-13 and Table D-14 are identified. These price levels are then processed in line with the share of foreign forwarders for international transport relations according to Table G-3. Therefore the following assumption applies:

- For each modelled international OD pairing, no third-country freight forwarders are involved, e.g. for a transport from France to Germany, either transport costs for German or French freight forwarders are evaluated.

---

[151] Here, a price index for diesel fuel price changes from the year 2010 to 2012 according to a report of the EC (2014) is incorporated.

[152] Originally, the given factors are set up for an annual mileage of 150,000 km for tractors and 125,000 for trailers respectively. However, in lack of further insight into the calculation basis, these factors are applied for an annual mileage of 135,000 km without further adaptions.

International road freight driver wages

Personnel cost differences within the EU-27 can be determined from results of a labour cost survey on behalf of the European Commission (EUROSTAT, 2014r)[153]. Therefrom, the regional variety of annual total labour costs in the section of *land transport and transport via pipelines* (CPA-49) among firms of 10 and more employees is adopted for the presented context. The outcome is presented in Table G-24.

International road freight fuel costs

A second additional cost component for international road freight forwarders comprises a fuel price index. Following BMVI (2014a, p. 139), market observations for prices of petroleum products in the EU by the European Commission on Energy are consulted for a determination of diesel fuel price differences among the EU-27 states in 2012, as presented in EC (2014). In analogy with the processing of labour costs, reference to the German fuel price level is set up in Table G-25[154].

Motorway charges

In addition to fuel and personnel specific additional cost components, infrastructure charges are considered for road freight transport costs. For this intent the study of BMVI (2014a, p. 141 ff.) proposes a uniform infrastructure charge for all roads in Germany of 0,155 € per vehicle kilometre[155] – a value that is adopted in the presented context. For other European countries with mileage-dependent and mileage-independent tolls likewise uniform approximations apply, as indicated in Table G-26.

Empty road freight vehicle shares

To accomplish a commodity transfer, road freight vehicles often perform empty trips before and/or after loading and unloading. In order to consider

153 Except for Belgium, Greece, Italy and Malta observations for the year 2012 are available. The exceptions are calculated analogously with 2008 data from EUROSTAT (2014s)

154 It is assumed that both German and foreign freight forwarders equally make use of the lowest fuel price along a border-crossing transport link, either at the origin or the destination.

155 In practice, the German road freight infrastructure toll is charged for motorways only. It is furthermore dependent on the vehicle's exhaust emission class and the number of axles of a lorry and/or tractor-trailer combination.

this observation within the costing framework, the study of BMVI (2014a, p. 146) proposes consulting the survey results of the German *Federal Motor Transport Authority*. Thus, empty trip data for the year 2012 are interpreted from the report of KBA (2013a, p. 60).

Thereafter, the share of empty trips decreases with the length of distance travelled for a commodity transfer[156]. As a result, a road travel distance (d_{ij}^{road}) dependent share of empty trips (SET) for a transport from origin i to destination j can be incorporated into the road freight cost function for the year 2012 in the format of a two-termed exponential function, as follows[157]:

$$SET\left(d_{ij}^{road}\right) = w^{xd_{ij}} + y^{zd_{ij}} \tag{24}$$

with:

$w = 0{,}4976$, $x = -0{,}005217$, y= $0{,}01842$ and $z = 0{,}0003759$.

Road freight travel time and turnaround time

To calculate a travel time for road freight transports, the distance for each out of more than 220,000 considered OD transport links is determined first. In practice, each pairing is requested for a motorised *shortest path* routing from OSMF (2014) or VIAMICHELIN (2013)[158].

The travel time is calculated according to the transport distance and an average estimated vehicle velocity for the considered tractor-trailer combinations as well as a time buffer. The latter component is modelled in line with the study of BMVI (2014a, p. 145), wherein a time-buffer of 20% per transport is proposed as a travel time supplement observed for road freight schedules to maintain contracted delivery time-frames with shippers and/or receivers. An additional surcharge to account for national regulations on driving times and rest periods in the context of a travel time estimation is not considered[159].

---

[156] An interpretation might be that long-distance transport loads on the return trip are easier to acquire for road freight forwarders than for local distances.

[157] A depiction of the fitting procedure result is given in Fig. G-2.

[158] In order to make use of a free road network routing software, no potential restrictions – e.g. due to weight limitations – for road freight lorries are included. Compared to a routing of inland vessels, however, this impact is assumed to be negligible.

[159] Neither mandatory interruptions after at most four-and-a-half hours of driving, nor the daily driving time limit of 9 hours or – twice a week – 10 hours, nor the weekly limit of 56 hours of driving

An average vehicle speed for the modelled tractor-trailer combinations can be derived from the work of KAISER AND ZADEK (2015). Their research examines the underlying database of the *Handbook Emission Factors for Road Transport* (HABEFA) for Germany from INFRAS (2004)[160]. As a result, an average speed of 68.56 km/h is indicated to best represent the average speed for the entirety of the German road freight network – including an average level of congestion and an annual mean value for the occurrence of different delivery tour types as a weighting of the HABEFA-specific driving cycles – here, for vehicles of a maximum permissible gross laden weight of 40 tonnes in Germany. Since the HABEFA database covers transports within Germany as well as to or from Germany, this value is assumed to be applicable to all road freight links within the model without further spatial specifications.

For an estimation of average turnaround times, e.g. at industrial warehouses – here, the time span from the point of registration before loading or unloading at the ramp to the subsequent departure – a survey in Germany among forwarders as well as shippers and receivers was conducted in 2012 by HWH (2013, 27 f.). For a loading and unloading procedure of commodities transported by standard tractor-trailer combinations, an average value of 85 minutes is adopted therefrom to the presented context. An equal duration is assumed for turnarounds of tank tractor-trailer combination. Only for containerised commodity transfers a deviant time of 0.75 hours is assumed for the model in accordance with suggestions from BMVI (2014a, p. 144) that, in turn, are the result of expert interviews. These values, in total, are adopted to the model for all national as well as international origins and destinations of road freight transports.

Road freight transport unit load factors

In practice, road freight lorries are often not loaded up to their maximum payload. This is due to manifold, partly overlapping reasons. A first im-

---

are appropriately assignable to unique transport relations within the presented context. However, it shall be noted that within the regular driving time, the average German road freight transport distance of 106 km according to KBA (2013b, p. 7) is often feasible without these travel time breaks.

[160] The *Handbook Emission Factors for Road Transport* is a tool to calculate e.g. emissions for heavy duty vehicles based upon a sound database of observations on various vehicles and routings. The database comprises specific data for Germany, Austria, Switzerland, Sweden and Norway based upon a set of various driving cycles for heavy duty vehicles.

portant influence is the volume-to-weight ratio of products of a certain commodity class as well as the corresponding cargo handling form. Another major impact on the level of capacity utilisation of road freight vehicles is the result of individual settings within the shipper-receiver relationship (cf. Fig. A-5). Within this understanding, tractor-trailer load factors are, among other factors, likewise the result of reciprocal effects from a shipment sizing, as e.g. discussed in 13.7.

A valuable statistical report on average loading factors for Germany in 2012 can be obtained from KBA (2013a, p. 78). Within this publication a load factor of about 59% of the maximum permissible gross laden weight for all aggregate transport links is indicated (ibid.)[161]. In contrast, significant variations of the loading coefficient for road freight vehicles are reported for different transported commodity types and cargo handling categories. For instance, for commodities handled as dry bulk, a load factor of 85% is reported in KBA (2013a, p. 76 ff.). For commodity classes that are predominantly handled in a palletised form, the utilised share of the total truck payload is only about 42% (ibid.).
In consequence, a tractor-trailer combination related load factor per cargo handling category TLF_p according to Table G-22 is modelled.

Containerised road freight transports

Within Table D-13 and Table D-14 a maximum permissible gross laden weight of 44 tonnes for containerised loads of combined transports is considered, instead of the regular 40 tonnes. The respective payload within Table D-13 and Table D-14 is set to 29.16 tonnes, in order to comply with:

- an estimated average own weight of 2.33 tonnes for a 20-foot ISO containers[162],
- an average payload of 10.54 that is uniform with the respective load configuration of combined transport sections by mode rail and IWW according to Table G-32 and
- the vehicle load factor indication in Table G-22.

---

[161] Likewise, for aggregate transport links of domestic transports and international transports, load factors within a small range of deviation from 59% and 55% are indicated (KBA, 2013a, p. 54).
[162] The estimate is the result of expert consultations and the related reference in Table D-20.

The composition of additional applying time dependent *container provision cost factors* of the subsequent rail costing scheme will be discussed in section 14.3.3.

14.2.3. Overall road freight cost calculation scheme

Total link costs for a road transport of a commodity class with an assigned handling specification h, along a route from origin i to destination j, are calculated in line with the general format of equation (23) as:

$$C_{h,ij}^{road}(x) = \left\lceil \frac{x}{TLF_h * TPL_h} \right\rceil * \left(P_{h,ij}^{road} + H_{h,ij}^{road} * \left(T_{ij}^{road} + t_h^{turn}\right)\right) * \left(1 + SET\left(d_{ij}^{road}\right)\right) \tag{25}$$

with:

$$P_{h,ij}^{road} = d_{ij}^{road} * \left(c_h^{\Sigma var} + c^{vfc} * min(FPI^v)\right) + \sum_{V=1}^{29} d_{ij}^{road,v} * ICI^v * c^{ic} \tag{26}$$

and:

$$H_{h,ij}^{road} = c_h^{\Sigma hfc} + c_h^{fcc} + c^{tdw} * \left(\left(1 - SFF_{ij}\right) + SFF_{ij} * LCI_{ij}^{road}\right) \tag{27}$$

and:

$$T_{ij}^{road} = 1.2 \frac{d_{ij}^{road}}{v^{road}} \tag{28}$$

where:

$C_{h,ij}^{road}(x)$ total road freight transport costs for a volume x in tonnes of the modelled commodity classes p, assigned to a handling specification h according to Table G-22, from origin i to destination j, in: [€]

TLF_h load factor of a tractor-trailer combination for commodity handling h according to Table G-22, in: $[1/t]$

TPL_h payload of a tractor-trailer combination for commodity handling h as indicated in Table D-13 and Table D-14, in: $[t]$

$P_{h,ij}^{road}$ pecuniary costs for a freight transport with commodity handling specification h from origin i to destination j by road, in: $[€]$

$H_{h,ij}^{road}$ cost per time unit for a freight transport with commodity handling specification h from origin i to destination j by road, in: $[€/h]$

T_{ij}^{road} travel time from origin i to destination j by road, including a time-buffer of 20%, in: $[h]$

t_h^{turn} turnaround time of 2.83 hours for $h = 1, \dots, 4$ and 1.5 hours for $h = 5$

$SET(d_{ij}^{road})$ empty trip share according to equation (24)

d_{ij}^{road} *shortest path* road distance from origin i to destination j as determined from routing procedure, in: $[km]$

and:

$c_h^{\Sigma var}$ sum of variable costs without fuel of a tractor-trailer combination for commodity handling h as indicated in Table D-13 and Table D-14, in: $[€/km]$

c^{vfc} variable diesel fuel costs in Germany of 0.39 $€/km$

FPI^v fuel price index according to Table G-25 if a foreign country v is involved when travelling from origin i to destination j for $v = i$ or $v = j$ and 1, otherwise

c^{ic} road infrastructure charges for Germany of 0.155 $€/km$

$d_{ij}^{road,v}$ *shortest path* road distance in foreign country v from origin i to destination j as determined from routing procedure, in: $[km]$

ICI^v road infrastructure charge index for foreign country v according to Table G-26

and:

$c_h^{\Sigma hfc}$ sum of hourly fixed costs without wages of a tractor-trailer combination for commodity handling h as indicated in Table D-13 and Table D-14, in: $[€/h]$

c^{tdw} hourly German truck driver personnel costs of $18.22\ €/h$

c_h^{fcc} hourly container provision costs of $0.10\ €/h$ if a transport of commodities with handling specification $h = 5$ is involved[163] and $0\ €/h$, otherwise

SFF_{ij} share of international road freight forwarders that is involved when travelling from origin i to destination j according to Table G-3

LCI_{ij}^{road} labour cost index for driver wages paid by foreign forwarders that is involved when travelling from origin i to destination j according to Table G-24

and:

v^{road} average road freight vehicle velocity of $68.56\ km/h$.

In total, the road freight cost calculation is processed for each modelled OD transport link as follows:

[163] See section 14.3.3 for details.

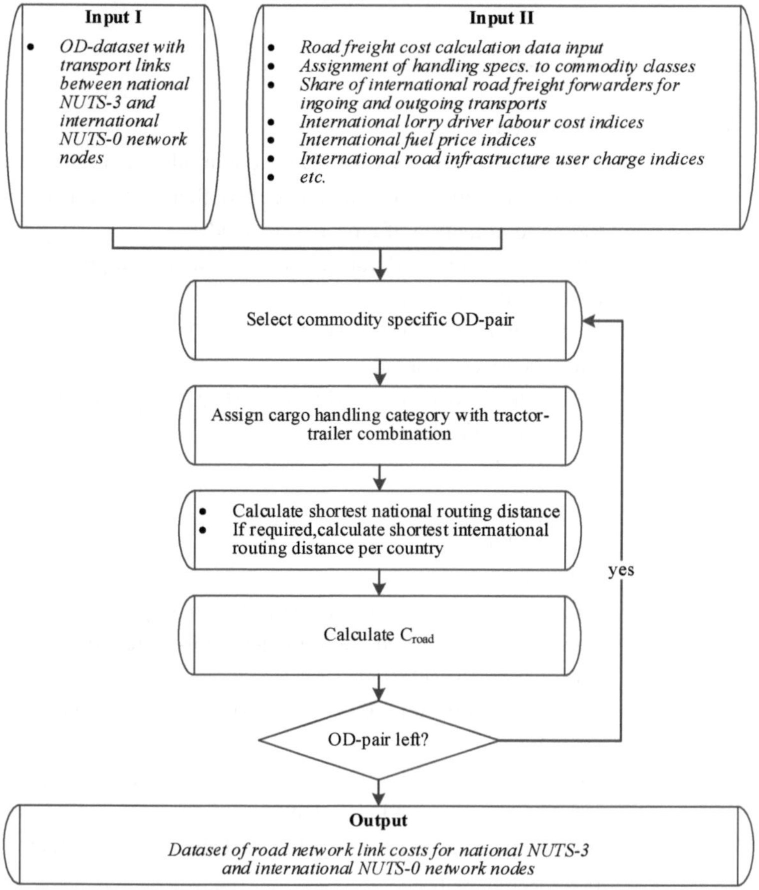

Fig. D-9 Simplified processing scheme to determine road freight transport network link costs

Exemplary application

An exemplary link cost calculation for a direct road freight transport of *food products* (CPA-10) from an origin in the German city *Ludwigshafen am Rhein* (ID: 07314) in Germany and a destination in the city *Flensburg*, Germany (ID: 01001) is constituted as follows:

d_{ij}^{road}	= 731 km
T_{ij}^{road}	= 12.79 h

c^{tdw}	= 18.22 €/h
$c_h^{\Sigma hfc}$	= 15.69 €/h, since for CPA-10 the commodity handling category 3 (palletised goods) is modelled, therefore a tractor-trailer combination of type 1 is selected
c_h^{fcc}	for non-containerised transports no transport unit provision costs are considered
SFF_{ij}	for transports from i = 07314 to j = 01001 no foreign forwarders are considered
LCI_{ij}^{road}	for transports from i = 07314 to j = 01001 no foreign driver wages are considered
$H_{h,ij}^{road}$	= 33.91 €/h

c^{ic}	= 0.16 €/km
c^{vfc}	= 0.39 €/km
$c_h^{\Sigma var}$	= 0.23 €/km, since for CPA-10 the commodity handling category 3 is modelled, therefore a tractor-trailer combination of type 1 is selected
$d_{ij}^{road,v}$	for transports from i = 07314 to j = 01001 no foreign country is involved
ICI^v	for transports from i = 07314 to j = 01001 no foreign road infrastructure charge is involved
FPI_{ij}	for transports from i = 07314 to j = 01001 no foreign fuel price is considered
$P_{h,ij}^{road}$	= 569.33 €/h

t_h^{turn}	= 2.83 h, since for CPA-10 the commodity handling category 3 is modelled, therefore a tractor-trailer combination of type 1 is selected
$SET\left(d_{ij}^{road}\right)$	= 0.04, according to equation (24)
TPL_h	= 27,00 t, since for CPA-10 the commodity handling category 3 is modelled, therefore a tractor-trailer combination of type 1 is selected

TLF_h = 0,50*1/t, since for CPA-10 the commodity handling category 3 is modelled

$C_{h,ij}^{road}(1)$ = 1,137.83 € for one exemplary tonne of CPA-10

For an increasing transport volume within the exemplary setting, the course of the road freight transport cost function is depicted in the following figure:

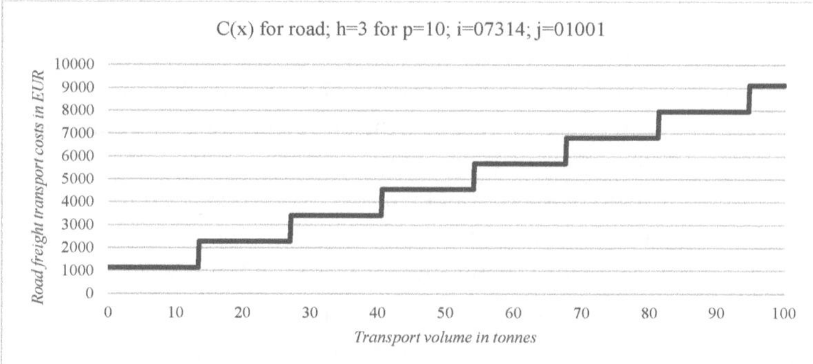

Fig. D-10 Exemplary road freight transport cost curve for transports of *food products* from *Ludwigshafen am Rhein* to *Flensburg* within Germany

For a second exemplary cost calculation, an origin in Germany and a destination in Belgium are considered. The setting comprises a road freight transport of *electrical equipment* (CPA-27) from the city *Wolfsburg* (NUTS-ID: 03103) to *Belgium* (NUTS-ID 20004) that leads to the following road freight transport costs:

d_{ij}^{road} = 571 km

T_{ij}^{road} = 9.99 h

c^{tdw} = 18.22 €/h

$c_h^{\Sigma hfc}$ = 15.69 €/h, since for CPA-27 the commodity handling category 3 (palletised goods) is modelled and the tractor-trailer combination of type 1 is selected

c_h^{fcc} for non-containerised transports no transport unit provision costs are considered

SFF_{ij}	= 0.55, since for transports from i = 03103 to j = 20004 about 54.80% of all transports is performed by Belgian forwarders
LCI_{ij}^{road}	= 1.40, since for transports from i = 03103 to j = 20004 to the share of foreign lorry drivers a different wage is paid
$H_{h,ij}^{road}$	= 37.86 €/h
c^{ic}	= 0.16 €/km
c^{vfc}	= 0.39 €/km
$c_h^{\Sigma var}$	= 0.23 €/km, since for CPA-27 the commodity handling category 3 is modelled, therefore a tractor-trailer combination of type 1 is selected
$d_{ij}^{road,v}$	for a transport from i = 03103 to j = 20004 Germany and two foreign countries are involved, a transit via the Netherlands ($v = 16$) and the destination in Belgium ($v = 03$) that results in: $d_{ij}^{road,16}$ = 118 km; $d_{ij}^{road,3}$ = 69
ICI^v	for a transport from i = 03103 to j = 20004 Germany and two foreign countries are involved, but no foreign road infrastructure charges apply and as a result, ICI^{03} = 0; ICI^{16} = 0
FPI_{ij}	= 0.97, since for a transport from i = 03103 to j = 20004 different fuel prices for Belgium are considered
$P_{h,ij}^{road}$	= 407.64 €/h
t_h^{turn}	= 2.83 h, since for CPA-27 the commodity handling category 3 is modelled, therefore a tractor-trailer combination of type 1 is selected
$SET(d_{ij}^{road})$	= 0.05
TPL_h	= 27,00 t, since for CPA-27 the commodity handling category 3 is modelled, therefore a tractor-trailer combination of type 1 is selected
TLF_h	= 0,50* 1/t, since for CPA-27 the commodity handling category 3 is modelled
$C_{h,ij}^{road}(1)$	= 936.13 € for one exemplary tonne of CPA-27

14.2.4. Validation

Representative road freight rates are difficult to identify. In consequence, only a few exemplary rates were found. Two freight rates – or more precisely: market price ranges – are published in a report of BAG (2012, p. 19) for unspecified transports in the year 2010 from the *south of Germany* to *Bulgaria* and to *Romania.* These two references are also incorporated into the study of BMVI (2014a, p. 147 ff.), complemented by two more unspecified and undated freight rates for containerised transports between the *Port of Rotterdam* and the German cities of *Heidelberg* and *Dormagen* as well as one unspecified bulk freight fare for a transport from the cities of *Frankfurt am Main* and *Aachen.* Even though these freight rates are limited in terms of their representativeness, they serve at least as indicators for the validity of the corresponding results from the presented study. A comparison is depicted in the following table:

Origin	*Destination*	*commodity handling*	**Reported transport costs in EUR**	**Calculated transport costs in EUR**	**Deviation**
South of Germany *(München)*	*Bulgaria* *(Sofia)*	*unspecified* *(tractor-trailer combination 1)*	1,600 to 1,700 (1,650)	1615.47	-2%
Bulgaria *(Sofia)*	*South of Germany* *(München)*	*unspecified* *(tractor-trailer combination 1)*	1,000 to 1,400 (1,200)	1615.47	35%
South of Germany *(München)*	*Romania* *(Bucharest)*	*unspecified* *(tractor-trailer combination 1)*	1,600 to 1,700 (1,650)	1430.40	-13%
Romania *(Bucharest)*	*South of Germany* *(München)*	*unspecified* *(tractor-trailer combination 1)*	1,000 to 1,400 (1,200)	1430.40	19%
Port of Rotterdam	*Heidelberg*	*containerised* *(tractor-trailer combination 2)*	665.0	754.60	13%
Port of Rotterdam	*Dormagen*	*containerised* *(tractor-trailer combination 2)*	451.0	384.71	-15%
Frankfurt am Main	*Aachen*	*dry bulk* *(tractor-trailer combination 1)*	385.3	464.46	21%

Table D-15 Comparison of calculated road freight costs with reference freight fares [incl. data from BMVI (2014a, 147 ff.); BAG (2012, p. 19)]

Road freight prices for transports between the *south of Germany* and *Bulgaria* or *Romania* are compared with calculations for the German city of *München* and *Sofia* as well as *Bucharest*, respectively. Unfortunately, no specification further is given for the data in BAG (2013a).
As a result the transport distance may vary significantly, when, for instance, a routing between the centroid of *Romania* or *Bulgaria* is evaluated instead. Another drawback for a transparent validation is the missing weighting of in- and outgoing transports. Since the share of German freight forwarders for road freight transports to or from these two countries is not reported within EUROSTAT (2014f), the share of foreign forwarders is assumed to be about 100%. In consequence, the calculated road freight costs are equivalent for ingoing and outgoing transports.
Additionally, freight rates for *Bulgaria* in 2010 consider foreign diesel fuel prices of about 1.22 € at the pump (*Romania*: 1.26 €/litre). In contrast, for the year 2012, fuel prices of 1.27 € in *Bulgaria* and 1.32 € per litre diesel fuel in *Romania* are processed (cf. Table G-25). Nevertheless, the calculated prices for these two transport relations are assumed to be within an exactable range of deviation from these published freight fares.

Reference freight fares for national transports are available, too. They cover different types of road freight vehicles that, in turn, relate to different commodities. However, these references are undated. In consequence, reasons for a deviation of around 21% for the calculated dry bulk cost rate for a transport between Frankfurt and Aachen to its reported counterpart in BMVI (2014a, p. 149) are difficult to identify – especially, when compared to the deviation of calculated and observed freight fares for transports from *Rotterdam* to *Dormagen*.

A major reason for small numbers of published freight fares might be their low degree of representativeness. As shown within the development of the road freight cost evaluation scheme, the number of individual factors for the overall result is high. In practice, this may lead to a large bandwidth of prices for road freight transport activities, especially when profit margins come into play. The elaborated scheme is not able to deal with this freight fare heterogeneity in further detail. However, it is assumed to suffice the requirements of an intermodal competition analysis.

14.3. Rail freight transport link costs

For a cost evaluation of rail freight transport network links two factors – in analogy to road freight cost calculations – for mileage dependent and time dependent costs are examined. Likewise, similar to road freight network costs, only national rail freight transports as well as incoming and outgoing loads are considered within the model. This is the result of findings derived from DESTATIS (2013b). Hereafter, the cabotage rate for the year 2012 only is about 4.42% of all rail freight activities in Germany. Furthermore, rail freight costs are evaluated based on the following assumption:

- The German national and international rail freight network is adequately represented by modelling block train compositions only. This applies to direct rail freight transports as well as to intermodal transport legs by rail.

This simplification is in line with the initial simplification for a consolidation of multiple shipments and likewise the result of observations of the share of block trains compared to single wagon compositions. Weighted by the distance travelled, only about 22% of all German rail freight transports are realised by trains that are composed and/or decomposed at multiple stops (cf. Table G-28). Although this type of transport has an impact on the overall rail freight cost evaluation, it is excluded therefrom due to a lack of data. Neither the German *Federal Bureau of Statistics* nor the expert consultations provide detailed information on general single wagon transport procedures.
In accordance with the assumptions in BMVI (2014a, p. 150), the share of non-electric locomotives is comparatively small for regular transport activities – without minor exceptions, diesel fuel locomotives are deployed for shunting movements only. In consequence, exclusively electric locomotives will be considered for the subsequent cost evaluations.

Time- and distance-related costs

Time dependent costs of rail freight activities mainly comprise costs for a provision of locomotives and wagons. Therefore, purchase prices as well as indicators for operating times and corresponding costs are required. These factors are likewise relevant for modelling distance-related costs that will be furthermore complemented by an energy consumption rate.

14.3.1. Reference rail freight cost calculation scheme

Based on the scheme and respective inputs of BMVI (2014a, p. 151 ff.), the following reference costs for locomotives on national and international transport relations is set up as[164]:

[164] In contrast to cost factors as given in BMVI (2014a, p. 153), an approximated 10%-surcharge is added to fixed vehicle costs in order to update cost evaluations more easily for further developments. Furthermore, an increase of purchasing prices for *rail vehicles* (GP09-302) from 2010 to 2012 of about 5,4% is included, according to DESTATIS (2014p, p. 199).

	Input, according to:			*Standard electrified locomotive* (Bombardier TRAXX F140)	*Multi-system electrified locomotive* (Bombardier Multi-system TRAXX F140)
				Basic cost-related input factors	
1	*a)*	Net purchase price for 2010	€	3,000,000	3,400,000
2	*b)*	Rate of price increases 2010 to 2012	%	5.40	5.40
3	Input 1 * Input 2	Net purchase price for 2012	€	3,162,000	3,583,600
4	*d)*	Operating lifetime	a	25	25
5	*a)*	Total mileage	km	5,000,000	5,000,000
6	Input 5 / Input 4	Annual mileage	km/a	200,000	200,000
7	*a)*	Operating days	d	280	280
8	*a)*	Operating time per day	h	12	12
9	*c)*	Annual capital interest rate	%/a	3.38	3.38
10	*a)*	Residual value for resale	€	0	0
11	*a)*	Replacement purchasing price	%	130.0	130.0
12	Input 3 * Input 11	Residual value for replacement	€	4,110,600	4,658,680
13	(Input 3 - Input 10) * 0,5 *a)*	Average operating capital	€	1,581,000	1,791,800
				Fixed cost components/costs per operating time unit	
14	Input 9 * Input 13	Annual costs for capital	€/a	53,383	60,500
15	(Input 12 - Input 10) / Input 4 * 0,5	Depreciation per time unit (50%)	€/a	82,212	93,174
16	*a)*	Annual general inspection	€/a	20,000	20,000
	d)	Liability and comprehensive vehicle insurance in relation to purchase price in 2012	%/a	1.50	2
17	*a)*	Fleet provision and administration surcharge on fixed costs	%/a	10	10
18	Σ (Input 14,…,17) / Input 7 / Input 8	**Σ of hourly fixed costs without wages**	**€/h**	**66.47**	**74.46**
19		**Hourly lorry driver wages**	**€/h**	**18.48**	**18.48**
				Variable cost components/costs per distance travelled	
19	(Input 12 - Input 10) / Input 4 * 0,5 / Input 6	Depreciation per miles travelled (50%)	€/km	0.4111	0.4659
20	*a)*	Maintenance and servicing	€/km	0.60	0.60
21		**Σ of variable costs without cost for energy**	**€/km**	**1.01**	**1.07**

Table D-16 Cost calculation data input for modelled rail freight transport locomotives

Sources of information in Table D-16 are indicated as[165]:

a) BMVI (2014a)
b) DESTATIS (2014p)
c) BUNDESBANK (2014)
d) WSDO (2007)

Additionally, a cost calculation scheme for rail freight wagons is required. A list of potential exemplary wagon types for rail freight transports in Germany for different commodities is given in BMVI (2014a, p. 154). The given listing also comprises valuable information on purchasing prices and annual operating hours (ibid.). They are projected from 2010 to 2012 according to a vehicle allocation as given in Table G-27, complemented by relevant technical data. The final cost calculation data input for rail freight wagons is given in the subsequent formation:

[165] Both the residual value for resale of input 10 as well as the replacement purchasing price for an equivalent vehicle of input 11 are reflected to a forward projected purchasing price in accordance with the calculated lifetime.

Wagon type	*1*	*2*	*3*	*4*	*5*	*6*	*7*	*8*	*9*	*10*
Net purchase price for 2010 in 1000 €	19.2	27.0	28.8	28.8	29.6	44.0	46.6	58.2	58.2	40.7
Annual operating time in 1000 hours	6.4	4.5	5.5	4.5	6.3	5.1	6.1	5.7	5.1	6.0
Purchase price increase 2010 to 2012 in %	5.4									
Net purchase price for 2012 in 1000 €	20.2	28.5	30.4	30.4	31.2	46.4	49.1	61.3	61.3	42.9
Operating lifetime in years	15									
Annual capital interest rate in %	3.38									
Residual value for resale in €	0									
Replacement purchasing price in relation to 2012 in %	141									
Residual value for replacement in 1000 €	28.5	40.1	42.8	42.8	44.0	65.4	69.3	86.5	86.5	60.5
Average operating capital in 1000 €	10.1	14.2	15.2	15.2	15.6	23.2	24.6	30.7	30.7	21.4
Annual costs for capital in €/a	342	480	512	512	527	783	829	1,036	1,036	724
Depreciation per time unit (50%) in €/a	951	1,338	1,427	1,427	1,466	2,180	2,308	2,883	2,883	2,016
Liability and comprehensive vehicle insurance in %	1.3									
General inspection costs in relation to purchase price for 2012 in %	5.5									
Hourly fixed costs in €/h	0.42	0.83	0.73	0.89	0.65	1.2	1.06	1.42	1.59	0.94

Table D-17 Cost calculation data input for modelled rail freight transport wagons [incl. data from BMVI (2014a, p. 154)]

General details

Within Table D-17, purchasing prices for 2010 are updated to 2012 by a factor of 1.054, derived from a price increase of *rail vehicles* (GP09-302) from DESTATIS (2014p, p. 199). Furthermore – equivalent to locomotives – fixed depreciation costs as hourly fixed costs (without wages) c_{hfc}^{wagon} are determined for the presented wagons. This rate includes the following components:

- an assumed value for resale of zero € in absence of further indications,
- an annual capital interest rate of 3.38%, according to the BUNDESBANK (2014) data analysis in analogy with the road freight transport cost calculation scheme in section 14.2.1,
- a depreciation time of approximately 15 years as well as costs for liability and comprehensive vehicle insurance of 1.3% and general inspection costs of 5.5% of the purchase price, according to the suggestions of WSDO (2007, p. 215),
- a replacement purchasing price after the depreciation for equivalent wagons of 141% of the original purchase price due to a price increase of 16.4% for *rail vehicles* (GP09-302) within 10 years, according to DESTATIS (2014p, p. 199) and
- cost estimates for liability and comprehensive vehicle insurances as well as general inspections that correspond to 2012 net purchase prices.

In contrast to the road freight costing scheme, a reasonable factor for overhead costs on transport activities cannot be individually assigned to transport vehicles. Instead, a general surcharge of 15% on the overall costs is assumed to cover the rail specific efforts to perform freight transports[166]. This surcharge comprises management costs for a train composition, network registration efforts, depot and depot management costs, insurance costs etc. A similar approach can also be found in BMVI (2014a, p. 167) and DEUTSCH (2013, p. 396).

An assignment of commodity classes of the CPA classification scheme to rail freight wagons can be found in Table G-29.

[166] The magnitude as a '*rule of thumb*' is the outcome of expert consultations on this specific topic.

14.3.2. Additional rail freight cost factors

Freight train energy consumption rates

Unlike for road freight transport, where fuel consumption is assumed to be fixed for an average load factor, an energy consumption calculation for freight trains needs to consider the gross laden weight of a specific train configuration. For electric trains with weights between 600 and 1800 gross tonnes, a functional form – derived from several different European railway observations – to determine the energy consumption is given in EWI (2014, p. 55) as:

$$EC_w = 1{,}2 * {GW_w}^{-0{,}62} \tag{29}$$

Therefrom, an energy cost calculation scheme for electric trains is calculated as:

$$c_w = EC_w * GW_w * c_{kWh} = 1{,}2 * {GW_w}^{0.38} * c_{kWh} \tag{30}$$

where:

EC_w	energy consumption in kilowatt hours per gross tonne kilometre for a freight train of configuration w, in: $[kWh/Gtkm]$
GW_w	average gross tonne weight of a composed block train of configuration w in: $[Gt]$
c_w	energy costs per freight train kilometre for a transport by block train of configuration w in: $[€/km]$
c_{kWh}	average energy cost per kilowatt hour in: $[€/kWh]$

In absence of an equally generalised functional form for freight trains with a weight of more than 1800 gross tonnes – e.g. transports of *metal ores* (CPA-08) – the given energy consumption calculation is equally adopted for all modelled freight train configurations.

National and international electricity costs

To determine c_{kWh}, the price publication of the German network infrastructure operator is recommended in BMVI (2014a, p. 156). For the year 2012, a dominant[167] hourly energy cost rate is indicated as 0,125 $€/kWh$ (DB NETZ, 2012). For international transports, a price index in relation to Germany for international energy costs is determined, as depicted in Table G-30. Subsequently, this concept is comparable to a processing of costs for diesel fuel in Europe for road freight transports in section 14.2.2.

Personnel costs

For the year 2012, the average monthly income as wages paid, including estimates for surcharges of up to 12 hours of working per day on public holidays, for locomotive drivers of German Railways (Deutsche Bahn) is 2,764 € (SZ, 2012)[168].
Equivalent to the road freight lorry driver calculation scheme, the sum of all surcharges of an employer is estimated at 23% of the actual wages paid (LOHNDIREKT, 2012). Likewise, an estimate of 250 for the national average number of regular working days as well as an average of 24 days to account for the number of paid holidays and 20 days of absence due to illness, driver training etc. are considered for personal deployment costs. The resulting daily labour cost factor is about 198.04 €. For a regular working time of 8 hours per day for locomotive drivers, the hourly cost rate is determined as 24.76 €.
To account for significant rail-driver-specific additional working time, e.g. for travel time between home and the place of deployment, the waiting and transfer times as well as pre and post examinations of composed trains, the study of BMVI (2014a, p. 158) proposes a surcharge of 70% on the total hourly personnel cost rate. As a result, an hourly labour cost component of 42.08 € is further processed.

In addition to pecuniary costs, a time-related personnel cost component is proposed in BMVI (2014a, p. 158). Thereafter, a premium on the determined

---

[167] The presented value of 0,125 $€/kWh$ is applicable from 05:30 a.m. to 10:00 p.m. every day. Only for the remainder, a different value of 0,106 $€/kWh$ is charged according to DB NETZ (2012).
[168] Therein, the initial salary for newly recruited locomotive drivers and the maximum income with 25 years of professional experience as well as the collective wage agreement negotiated in 2012 are incorporated.

travel time of 20 minutes by rail is proposed to account for mandatory driver changes every 4 hours of driving on average (ibid.).
In analogy to the road freight personnel cost calculation scheme, total labour costs for locomotive drivers within the EU-27 are calculated based on a price relation to their German counterpart. Again, annual total labour costs in the section of *land transport and transport via pipelines* (CPA-49) among firms of 10 and more employees from EUROSTAT (2014r) and EUROSTAT (2014s) are consulted and depicted in Table G-24[169].
Deviant from road freight and IWW transports, only national drivers are considered for each country. In other words, a change of drivers is effectuated within the presented model when trains cross borders[170].

Train composition costs

For block trains, costs of 32.70 € per wagon are indicated in BMVI (2014a, p. 158) to calculate the costs for train composition, including the corresponding locomotive costs. Although this value is determined for 2010, it is directly incorporated in the presented context for base year 2012 in absence of adequate insight into its composition for possible cost updates. An additional time-dependent cost surcharge for block train compositions is not considered.

Railway network user charges

For Germany, as well as for other European countries, charges tied to the use of the rail network infrastructure apply. The German train path price for rail freight services includes a charge based on the classification of the network section as well as the gross weight of the composed train, if higher than 3,000 tonnes. Rail freight infrastructure charges in other European countries are set up from different factors. Therefore, as proposed in BMVI (2014a), the study of OECD (2008) provides an overview on average infrastructure costs per freight train-km travelled for most European countries. The BMVI (2014a) study further proposes a simplified update on European infrastructure charges for the year 2007 by an indicator for the German infrastructure usage cost development as an approximation that will be equally applied in the presented study. In this regard, the report of BNETZA (2013, p. 27) indicates an increase of 13% for rail freight transport path prices. The resulting estimates

[169] The mandatory second locomotive driver in Italy is calculated with labour costs of factor two.
[170] A similar assumption can be found within the rail freight cost estimation of BMVI (2014a).

for different infrastructure charges within the EU-27 for 2012 can be found in Table G-31. Thereafter, international rail network user charges are referenced to a German average toll of 2.83 € per kilometre.

Empty rail freight vehicle shares

In contrast to the presented link cost calculation scheme for road freight transport, encompassing empty trip estimates are not appropriate for the rail freight cost setting. For each composed wagon of a freight train return loads are comparatively more difficult to dispatch than for road freight operations. This procedure takes into account a typical specification of rail freight wagons. Apart from wagon *type 7* used for containerised transports (Sgns 691) as well as *type 5* for pallet transports (Habbis 345) a wagon's specification is mostly unsuitable for return loads[171]. In consequence, for any other wagon type than those for containerised loads and pallets, the empty return cost surcharge is set to 100%.
For the transports of containers or pallets, a wagon-type-specific percentage of empty returns is incorporated into the costing.
Therefore, an empty return share is determined upon a relation's total demand for transports of all commodities that require the use of container or pallet wagons on either end of each modelled rail freight network link, as presented in the following example:

Origin ID (NUTS-2)	Destination ID (NUTS-2)	Total annual transport volume related to wagon type 7 in t	Share of empty return loads
Austria (AT)	*Mittelfranken (DE25)*	44,939	0,98
Mittelfranken (DE25)	*Austria (AT)*	997	0,00

Table D-18 Exemplary calculation of empty return surcharges for rail freight transport links

The related processing incorporates the dataset of DESTATIS (2013b). To incorporate this empty return share evaluation scheme, it is assumed that d_{ij} =

[171] See Table D-17and Table G-29 for more details. In certain cases it may be possible to take a return load of a sufficiently similar specification compared to the first trip's load and to prepare the related wagons accordingly, e.g. cleaning, inspection etc. This, however, is considered to be an exception.

d_{ji} for the modelled railway transport network. Hence, the distance from an origin to a destination of a rail freight link is equivalent to its return.

Length of rail freight trains and train composition assumptions

Differing from the modelled loading configurations of road freight vehicles, variable energy costs per kilometre are notably correlated to a train's gross weight. Likewise, train composition costs are determined by the type of train setup – namely, block train or single wagon composition. For both cost components the elementary factor is the length per train and its gross weight, respectively.

Unlike for road freight vehicles, there is no survey data for German rail freight operators that yield to load-specific capacity utilisation factors. Moreover, the *Federal Bureau of Statistics* neither collects any information about the number of realised trips for a given transport volume per time unit, nor the number of wagons nor trains. This status is challenging for a specified rail freight transport cost analysis.
A viable option is the appraisal of average train lengths and load weights in Germany per commodity class and the handling type, respectively. A supporting indication is the maximum rail freight train length of 740 metres in Germany for the year 2012, as given in DB NETZ (2014). Until year 2010, the train length was limited to 670 m due to rail network infrastructure restrictions. Even though the network constraints are homogenised to more than 670 metres in the year 2012, private rail access nodes often are not equally adapted to such train lengths – e.g. along loading and unloading facilities[172].
Another indication is the average gross laden weight of 1560 tonnes for freight trains in Europe, given in JANIC (2007, p. 41). This quantity is also within the range for average European train gross weights that are indicated in EWI (2014, p. 53 ff.). Valuable information on average freight train lengths and weights is also given in WSDO (2007, p. 36). These values, however, are based on observations of the German freight train network setting

[172] This is the outcome of expert consultations on this specific topic. Apart from technical restraints, the average train length is likewise the result of other drivers resulting from the demand for transport services – partly, equivalent to road freight vehicle load factors. This effect, however, is less present for rail freight activities that are more frequently used for bulk freight cargo, than for their road freight counterpart.

before the year 2010. In absence of any further reference, the few values given in WSDO (2007) are incorporated into the presented context as a form of a validation – indicated by [b] – for the train length dimensions of [a], obtained from the expert consultations:

Wagon type	*1*	*2*	*3*	*4*	*5*	*6*	*7*	*8*	*9*	*10*
Average length of train in m	(580)	(580)	(450)	(600)	(700)	(580)	(580)	(450)	(600)	(450)
Number of locomotives per train	(1)	(1)	(1)	(1)	(1)	(2)	(1)	(1)	(1)	(1)
Number of wagons per train*	37	44	27	29	34	35	23	26	22	50
Load capacity per train in 1000 t	2.1	2.6	1.3	1.5	2.4	3.5	1.1	1.5	0.4	1.3
Load capacity per wagon in t	56.5	59	48	53	70	100	47	56	17	25
Average load per wagon	54	60	49	54	32	99	0	58	16	49
Number of observations in 1000	303	393	23	35	0	156	0	814	366	135
Net load per train in 1000 t	2.0	2.6	1.3	1.5	1.1	3.5	0.0	1.5	0.4	1.3
Net load per train in t	2.0	2.5	1.3	1.5	1.1	3.5	1.0	1.3	0.4	1.3
Average train load capacity in 1000 t	2.1	2.6	1.0	0.0	0.8	3.5	0.0	1.5	0.0	0.0
Gross weight per loaded train* in 1000 gross tonnes	2.9	3.6	2.0	2.4	2.1	4.8	1.8	2.1	1.0	1.9
Gross weight per unloaded train* in gross tonnes	950	1,180	725	863	760	1,385	765	881	615	620
EC, loaded in kWh/tkm	24.9	27.05	21.64	23.1	21.92	30.15	20.68	22.08	16.4	21.01
EC, unloaded in kWh/tkm	16.2	17.64	14.66	15.66	14.92	18.75	14.96	15.78	13.77	13.81
Maximum load capacity in 1000 t for740 m train length	2.7	3.4	2.2	1.9	2.3	4.6	1.4	2.4	0.5	2.1
Assumed net load in % of maximum capacity	73	73	60	81	48	75	76	52	78	60

* locomitve length of 20 metres

Table D-19 Estimated average load capacity of modelled block trains per wagon type

The freight train transport weight capacity estimation is based upon fully loaded wagons for a train composition[173]. For instance, the average load capacity of *type 8* wagons for motor vehicle transports is assumed to be 20 tonnes. According to STATISTA (2013), the average weight of German passenger cars is about 1.53 tonnes. That is why every additionally coupled wagon per train is assumed to be loaded by 13 vehicles on average. Likewise, for the remaining commodity classes and corresponding wagons, an average train length and rail freight transport capacity is modelled, based on complete loads per wagon[174].

An exemption to full wagon loads is made for containerised rail freight transports in Germany, for which an average load factor per TEU of 10.54 tonnes and an estimated average loading unit own weight of 2.33 tonnes are considered[175]. As a result, instead of a maximum of 70 tonnes in total loaded to a wagon of *type 7* (Sgns 691) – with a load capability of 3 twenty-foot equivalent units – only $3 * (10.54 + 2.33) = 38.61$ tonnes are modelled as an average load per wagon with a net load weight of $3 * 10.54 = 31.62$ tonnes.

Rail freight travel time

To calculate a travel time for rail freight transports, the distance for each of the modelled direct and combined rail freight transport paths is determined first[176]. In practice, each modelled OD pairing is requested for a *shortest path* routing from DB CARGO (2015a). A significant drawback for step two, the determination of a travel time in relation to a distance travelled, is the absence of a profound analysis on *real-world* rail freight travel times within the contemporary literature. The question is, whether there is a distance-related average travel time and a delay factor. Further considerations would involve regional varieties or commodity specifications. Since neither of these input

---

[173] The maximum load capacity in tonnes per train does not consider any practical technical restraints for a train length of 740 metres, e.g. a wagon brake performance limit for a heavy duty transport of *metal ores* (CPA06). It serves instead as sizeable value for the finally assumed net load.

[174] The weight-to-volume ratio from the results of section 12.3 can be used to cross-check possible rail freight wagon loading capabilities. The freight train capacity transport estimation also includes the length and own weight of an exemplary Bombardier TRAXX F140 locomotive, as obtained from MRCE (2014).

[175] See Table G-32 and section 14.2.2 for details.

[176] See introduction of section 14 for details. Timetables and frequencies of freight trains are not considered.

values is at hand, a second-best solution is aimed for. Therefore, the study of JANIC (2007, p. 41) proposes an average train speed for Europe of $v^{rail} = 40km/h$ and an average anticipated delay – here, a time-buffer similar to road freight activities – of $0.5h$[177]. In absence of more specified data, these suggestions are incorporated into the following context for German and international freight train configurations.

14.3.3. Additional costs factors for containerised transports

A combined transport is generally performed with containers, swap bodies and road lorries[178]. For these transport units an additional provision cost component needs to be included into the overall intermodal cost structure and performance evaluation. In case of vehicles as a transport unit, additional wages for transports accompanied by road freight drivers may be relevant as well. This configuration, however, is comparatively rare. The long haul section by rail of national intermodal transports is realised by vehicles in less than 4% of all transport cases (DESTATIS, 2014q, p. 12 ff.). Moreover, the share of accompanied transports is only about 0.66% for national transports (ibid.). Even though the share of vehicles as loading units for intermodal rail freight transports in international transport relations is about 30%, the overall share of accompanied transports is still less than 1%, as depicted in Table G-33[179]. Since the share of accompanied vehicle transports is comparatively small, it will be neglected in the following considerations[180].

Unfortunately, there is no statistical reference for the number of entire tractor-trailer combinations compared to the number of trailer-only transports – nor is there any indication on the specification of containers and swap bodies

---

[177] These values are within the range of what was discussed by experts on this topic. Unanimously, the variability of these estimates was emphasized as well. Representatives of the largest German rail freight operator as well as from smaller-sized (private) operators were questioned.

[178] See e.g. Fig. A-3.

[179] For intermodal transports on IWW no accompanied vehicle transports can be observed for 2012 in Germany, according to DESTATIS (2013a).

[180] The choice of using trailers or entire tractor-trailer combinations as loading units for combined transports with a main transport leg by rail is rather motivated by other individual-specific factors than those included in a basic cost and performance structure of rail freight services. For instance, laws and regulations applicable for a crossing of the Alps via Switzerland are out of scope of this model.

that might have at least different hourly provision costs. For trailers and tractor-trailer combinations, time dependent costs are determinable according to Table D-13 and Table D-14. For containers and swap bodies, different cost components for the time in transit may be set up. However, in the absence of further processing opportunities, a mutual cost calculation has to be assumed for both transport units.

Therefore, the study of BMVI (2014a, 143 f.) provides a valuable cost processing scheme and a proposition of input data for 20- as well as 40-foot ISO containers[181]. Within this study, valuable price indications for containers from a report of BUSS CAPITAL (2011, p. 10) for the year 2010 are referenced (ibid.). They are incorporated into the presented context, in line with price updates of containers and swap bodies of about 3.7% from 2010 to 2012 – based on DESTATIS (2014p) price index data for *tanks, reservoirs and containers of iron, steel and alloy* (GP09-2529). If the operating lifetime is set to 15 years, as proposed e.g. in BMVI (2014a), a replacement price estimate of 131% should be incorporated into 2012 cost considerations. For 340 calculated days of operation and 12 operating hours each day as well as estimates for maintenance and administrative costs, the following cost calculation for containers results:

[181] For rail 30-foot ISO container transports are also reported (see Table G-33) that are assumed to have no specific price other than the subsequently calculated average.

	Input, according to:			20 foot ISO container	40 foot ISO container	20 foot ISO tank container
				Basic cost-related input factors		
1	a)	Net purchase price for 2010	€	2125	3,400	-
2	c)	Net purchase price for 2013	€	-	-	3,400.00
3	b)	Load capacity in tonnes	t	28.23	26.70	26.29
4	b)	Own weight in tonnes	t	2.25	3.78	4.19
5	c)	Rate for price differences 2010 to 2012	%	3.70	3.70	
6	c)	Rate for price differences 2013 to 2012	%	-	-	0.40
7	Input 1 * Input 5	Net purchase price for 2012	€	2,204	3,526	3,414
8	b)	Operating lifetime	a	15	15	15
9	estimate (equivalent to barge)	Operating days	d	340	340	340
10	estimate	Operating time per day	h	12	12	12
11	a)	Annual capital interest rate	%	3.38	3.38	3.38
12	estimate	Residual value for resale	€	0	0	0
13	separate calculation	Replacement purchasing price	%	131.0	131.0	131.0
14	= Input 7 * Input 13	Residual value for replacement	€	2,887	4,619	4,472
15	= (Input 7 - Input 12) * 0,5	Average operating capital	€	1,102	1,763	1,707
				Fixed cost components/costs per operating time unit		
12	= Input 11 * Input 15	Annual costs for capital	€/a	37	60	58
13	= (Input 14 - Input 12) / Input 8	Depreciation per time unit (100% for containers, 50% for vehicles)	€/a	192	308	298
14	estimate (equivalent to barge)	Annual maintenance costs in relation to replacement purchasing price	%	2.50	2.50	2.50
15	estimate (equivalent to barge)	Annual administrative costs in relation to replacement purchasing price	%	2.50	2.50	2.50
16	costs per operating time unit for road freight vehicles	Liability and comprehensive vehicle insurance, vehicle tax, depot management costs	€/a			
		Σ overall for fixed costs	**€/h**	0.08	0.13	0.13

Table D-20 Cost calculation data input for modelled containers

The most relevant sources of information are referenced as:

a) BMVI (2014a)
b) GDV (2014)
c) BUSS CAPITAL (2013)
d) BUNDESBANK (2014)

The annual transport volume of intermodal transports by rail (see Table G-33) – and likewise for IWW, according to DESTATIS (2014q, p. 29) – is almost equally distributed among 20- and 40-foot ISO standard containers. Thus, the average cost per operating hour for containers is set to $0.08 * 2/3 + 0.13 * 1/3 = 0.10$ € for a 20-foot ISO container, hence, a twenty-foot equivalent unit.

In line with this setting, intermodal transports within a national freight transport model, based on a comparative cost-performance-analysis of different transport modes, are represented by an analysis of 20-foot containers as transport units only.

Transhipment costs

Compared to unimodal freight transports, where door-to-door (ramp-to-ramp) transports are viable, intermodal freight transports may require additional transhipment activities. To account for these additional transhipment activities, the study of BMVI (2014a, p. 160) conducted a cost evaluation survey among operators of intermodal nodes, railway companies as well as among inland and seaport operators. A factor of 20 € per container handling as well as an average handling capability of 20 units per hour for average transhipments activities are indicated as a result (ibid.).

The proposed handling capacity is adopted as an average time of 0.05 hours per TEU for all transfer nodes within the presented model. The cost factor is incorporated into the following processing as $20 * 2/3 + 10 * 1/3 = 16.67$ € per TEU handling. The average modelled transhipment time is set to $0.05 * 2/3 + 0.025 * 1/3 = 0.04$ hours per TEU accordingly.

The occurrence of additional transhipment activities within intermodal – hence, containerised – transport chains is considered as follows:

- 0 times between two seaports,
- 1 time between a seaport and any non-seaport link origin or destination,
- 2 times for transports from a non-seaport origin to a non-seaport destination.

Seaports are identified according to their NUTS-3-ID within the model. This procedure requires the following assumptions:

- All containerised rail freight transports to or from a region, within wich at least one seaport is located, are handled at seaports only.

For instance, a reported containerised transport by rail, outgoing from the city *Hamburg* (NUTS-3-ID: 02000), is originated at the *port of Hamburg* and no additional pre-haul activities by mode road are considered. German and international seaports within the model are listed in
Table D-5 and e.g.
Table D-10.

14.3.4. Overall rail freight cost calculation scheme

Total link costs for rail freight transports are calculated according to an empty return surcharge and the assumptions of the previous section, in line with the general format of equation (23), as:

$$C_{w,ij}^{rail}(x) = \left\lceil \frac{x}{NLT_w} \right\rceil * ACS * \left(P_{w,ij}^{rail} + H_{p,ij}^{rail} * T_{ij}^{rail} + S_{w,ij}^{rail}\right) \quad (31)$$

with:

$$S_{w,ij}^{rail} = NBT_{w,ij} * NBW_w * \left(c_w^{trs} + t_w^{trs} * NBW_w * \left(c_w^{fcw} + c_w^{fcc}\right)\right) \quad (32)$$

and:

$$P_{w,ij}^{rail} = \left(1 + SER_{w,ij}\right) * \left(d_{ij}^{rail} * c_{w,ij}^{\Sigma\text{var}} + c^{ric} * \sum_{V=1}^{29} d_{ij}^{rail,v} * RCI^{v}\right) + \left(c_{w}^{ecl} + SER_{w,ij} * c_{w}^{ecu}\right) * \sum_{V=1}^{29} d_{ij}^{rail,v} * NEC^{v} + NBW_{w} * c^{tcc} \qquad (33)$$

and:

$$H_{w,ij}^{rail} = NBL_{w} * c_{w}^{fcl} + NBW_{w} * \left(c_{w}^{fcw} + c_{w}^{fcc}\right) + \frac{c^{ldw}}{d_{ij}^{rail}} * \sum_{V=1}^{29} d_{ij}^{rail,v} * LDC^{v} \qquad (34)$$

and:

$$T_{ij}^{rail} = \left(1 + SER_{w,ij}\right) * \left(\frac{d_{ij}^{rail}}{v^{rail}} + 0.83h\right) \qquad (35)$$

where:

$C_{w,ij}^{rail}(x)$ total rail freight transport costs for a volume x in tonnes of the modelled commodity classes p, assigned to a freight train of configuration w from origin i to destination j, in: [€]

NLT_{w} net load for a freight train of configuration $\boldsymbol{w}$ according to Table D-19, in $[\boldsymbol{t}]$

ACS global rail freight administration cost surcharge of 15%

$P_{w,ij}^{rail}$ pecuniary costs for a freight train transport of configuration w from origin i to destination j, in: [€]

$H_{w,ij}^{rail}$ costs per time unit for a freight train transport of configuration w from origin i to destination j, in: [€/h]

T_{ij}^{rail} travel time from origin i to destination j by rail, including an additional time-buffer of 0.5 hours and a supplement for mandatory locomotive driver changes of 0.33 hours, in: $[h]$

$S_{w,ij}^{rail}$ transhipment costs for a freight train transport of configuration $w = 5$ from origin i to destination j, in: [€]

$NBT_{w,ij}$ number of transhipments for a freight train transport of configuration $w = 5$ from origin i to destination j, with:

- 0, for $i \wedge j \neq ID25, ID26, ID2000, \ldots, ID1058$
- 1, for $i = ID25, ID26, ID2000, \ldots, ID1058 \wedge j \neq ID25, ID26, ID2000, \ldots, ID1058 \vee i \neq ID25, ID26, ID2000, \ldots, ID1058 \wedge j = ID25, ID26, ID2000, \ldots, ID1058$
- 2 for $i = ID25, ID26, ID2000, \ldots, ID1058 \wedge j = ID25, ID26, ID2000, \ldots, ID1058$

NBW_w number of wagons for a freight train transport of configuration w according to Table D-19

c_w^{trs} transhipment costs per wagon for a freight train w of configuration $w = 5$ of $3 * 16.67$ €

c_w^{fcw} hourly fixed costs per wagon for a freight train transport of configuration w according to Table D-19, in $[€/h]$

c_w^{fcc} container provision costs for a freight train transport of configuration $w = 5$ of $3 * 0.10$ $€/h$ according to Table D-20

t_w^{trs} transhipment time for a freight train transport of configuration $w = 5$ of $3 * 0.04$ hours per wagon

$SER_{w,ij}$ share of empty returns for transports by freight train of configuration w from origin i to destination j according to separate calculations in line with Table D-18

c^{tcc} composition costs per wagon for a freight train transport of configuration w of 32.70 €

d_{ij}^{rail} *shortest path* rail freight distance from origin i to destination j as determined from routing procedure, in $[km]$

$c_{w,ij}^{\Sigma var}$ sum of variable costs without electricity for a freight train transport of configuration w, with 1.01 $€/km$ for $i \wedge j \leq ID2000$ – hence, an origin i and destination j within Germany – and otherwise, 1.07 $€/km$ as indicated in Table D-16, Table D-17 and Table D-20

$d_{ij}^{rail,v}$ *shortest path* rail freight distance in foreign country v from origin i to destination j as determined from routing procedure, in $[km]$

RCI^{v} rail network infrastructure charge index for foreign country v according to Table G-31

c^{ric} rail network infrastructure charges for Germany of 2.83 $\frac{€}{km}$

NEC^{v} electricity cost index for foreign country v according to Table G-30

c_{w}^{ecl} kilometre specific electricity costs for a loaded freight train of configuration w according to equation (30) and Table D-19 in: $[€/km]$

c_{w}^{ecu} kilometre specific (vehicle) electricity costs for an unloaded block train of configuration w according to equation (30) and Table D-19, in: $[€/km]$

NBL_w number of locomotives for freight train of configuration w according to Table D-19

c_w^{fcl} hourly fixed costs without wages for a locomotive, deployable for freight train of configuration w according Table D-16, Table D-17 and Table D-20 with $66.47\ €/h$ for $i \wedge j \leq ID2000$ – hence, an origin i and destination j within Germany – and $74.46\ €/h$, otherwise

LDC^v locomotive driver labour cost for foreign country v according to Table G-24

c^{ldw} hourly locomotive driver labour cost for Germany of 42.08 $\frac{€}{h}$

v^{rail} average rail freight train velocity of 40.00 $\frac{km}{h}$

The presented rail freight cost calculation is processed for each modelled *direct* as well as for each *combined* rail freight network link as follows:

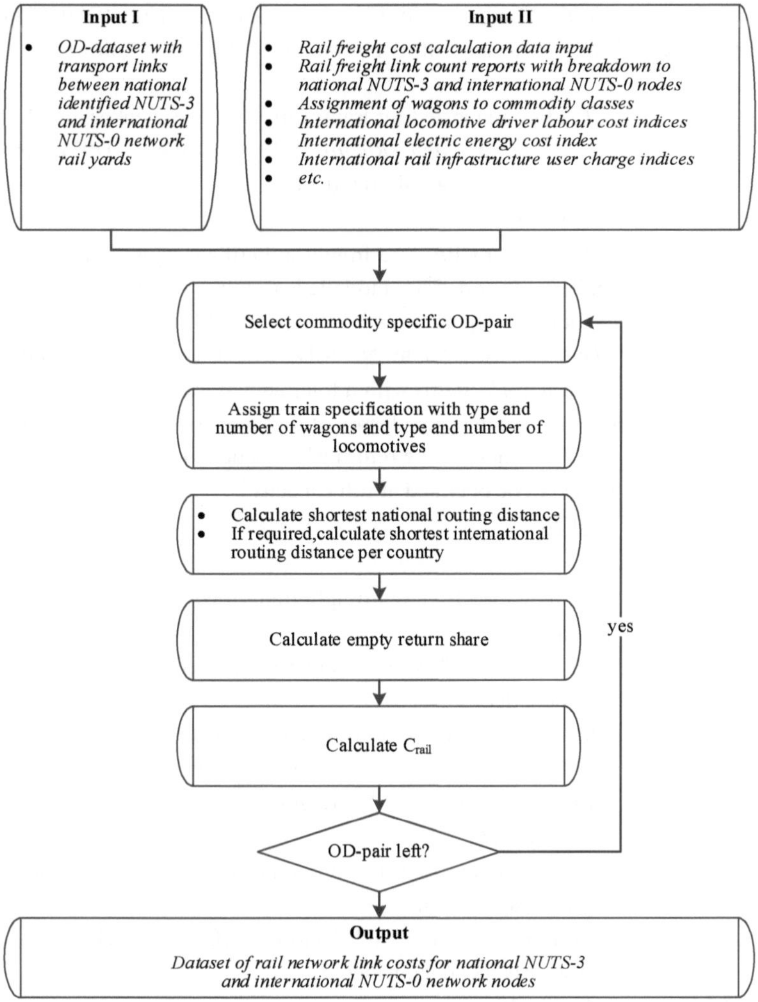

Fig. D-11 Simplified processing scheme to determine rail freight transport network link costs

Exemplary application

An example link cost calculation in analogy with the road freight example is given for an origin in the city *Ludwigshafen am Rhein* (NUTS-ID: 07314) in

Germany and a destination in the city *Flensburg*, Germany (Nuts-ID: 01001) for a direct rail freight transport of *food products* (CPA-10) as follows:

d_{ij}^{rail}	= 778 km, requested for transports via DB-Schenker Freight yards *Ludwigshafen (Rhein) BASF* (DB-ID: 190793) and *Flensburg Grenze* (DB-ID: 0406)
v^{rail}	= 40.00 km/h
$SER_{w,ij}$	= 0, since for CPA-10 a freight train of configuration $w = 7$ (palletised goods[182]) is modelled, an empty return share is relevant for transports from NUTS-ID: 07314 (related to *DEB3: Rheinhessen-Pfalz*) to Nuts-ID: 01001 (related to DEF0: *Schleswig-Holstein*) with a total annual transport volume of relevant commodity classes of 0 tonnes for the outward and 13,753 tonnes for the return trip – hence, return trips are assumed to be operated at full capacity
T_{ij}^{rail}	= 20.28 h
NBL_w	= 1, since for CPA-10 a freight train of configuration $w = 7$ is modelled
NBW_w	= 23, since for CPA-10 a freight train of configuration $w = 7$ is modelled
c_w^{fcl}	= 66.47 €/h, since for rail freight transports from i = 07314 to j = 01001 a standard electrified locomotive is applicable
c_w^{fcw}	= 1.06, since for CPA-10 a freight train of configuration $w = 7$ is modelled
c_w^{fcc}	for non-containerised transports no transport unit provision costs are considered
LDC^v	for transports from i = 07314 to j = 01001 no foreign locomotive drivers are considered
c^{ldw}	= 42.08 €/h
$H_{w,ij}^{rail}$	= 132.97 €/h

[182] For a rail freight transport of *food products* (CPA-10) a handling of pallets is assumed. Hence, the modelled rail wagon type is a *Habbis 345* ($w = 7$). In line with Table G-29 trains of a similar configuration are related to e.g. *beverages* (CPA-11), *tobacco products* (CPA-12), *textiles* (CPA-13) etc.

c^{tcc} = 32.70 €/wagon

$c_{w,ij}^{\Sigma var}$ = 1.01 €/km, since for transports from i = 07314 to j = 01001 a standard electrified locomotive is applicable

$d_{ij}^{rail,v}$ for transports from i = 07314 to j = 01001 no foreign country is involved

RCI^{v} for transports from i = 07314 to j = 01001 no foreign infrastructure charge is considered

c^{ric} = 2.83 €/km

NEC^{v} for transports from i = 07314 to j = 01001 no foreign electricity use is considered

c_{w}^{ecl} = 2.59 €/km

c_{w}^{ecu} = 1.87 €/km

$P_{w,ij}^{rail}$ = 5,748.03 €

$NBT_{w,ij}$ for non-containerised transports no transhipments are considered

c_{w}^{trs} for non-containerised transports no transhipment costs are considered

t_{w}^{trs} for non-containerised transports no transhipment time surcharges are considered

$S_{w,ij}^{rail}$ = 0 € for non-containerised transports no transhipments are considered

NLT_{w} = 1,029.20, since for a freight train of configuration $w = 7$ is modelled

$C_{h,ij}^{road}(1)$ = 9,711.46 € for one exemplary tonne of CPA-10

For an increasing transport volume within the exemplary setting, the course of the rail freight transport cost function is depicted in the following figure:

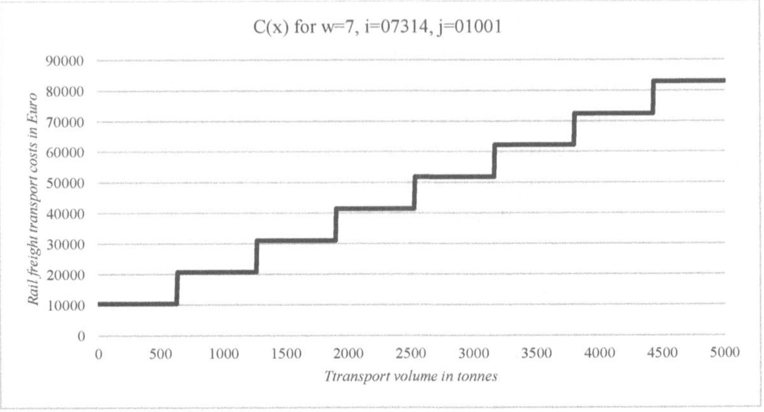

Fig. D-12 Exemplary rail freight transport cost curve for transports of *food products* from *Ludwigshafen am Rhein* to *Flensburg* within Germany

For a second exemplary cost calculation, an origin in Germany and a destination in Belgium are considered. The setting describes a rail freight main haul section of an intermodal transport of containerised commodities train from the city *Wolfsburg* (NUTS-ID: 03103) to *Belgium* (NUTS-ID 20004) and leads to the following results:

d_{ij}^{rail} = 557 km, requested for transports via DB-Schenker Freight yards *Wolfsburg Ubf* (DB-ID: 373100) and *Schaerbeek/Schaarbeek* (DB-ID: 110072)

v^{rail} = 40.00 km/h

$SER_{w,ij}$ = 0.68, since for containerised transports a freight train of configuration $w = 5$ is modelled, an empty return share is relevant for transports from NUTS-ID: 03103 (related to *DE91: Braunschweig*) to Nuts-ID: 20004 (related to BE: *Belgium*) with a total annual containerised transport volume of 611 tonnes for the outward and 194 tonnes for the return trip – hence, 68% of the return trips are assumed to be empty

T_{ij}^{rail} = 24.83 h

NBL_w	= 1, since for containerised commodities a freight train of configuration $w = 5$ is modelled
NBW_w	= 34, since for containerised commodities a freight train of configuration $w = 5$ is modelled
c_w^{fcl}	= 74.46 €/h, since for transports from i = 03103 to j = 20004 a multi-system electrified locomotive is applicable
c_w^{fcw}	= 0.65 €/h, since for containerised commodities a freight train of configuration $w = 5$ is modelled
c_w^{fcc}	= 0.30 €/h, since for containerised commodities a freight train of configuration $w = 5$ is modelled
LDC^v	for a transport from i = 03103 to j = 20004 German and two foreign locomotive driver cost factors are involved, a transit via the Netherlands ($v = 16$) and the destination in Belgium ($v = 03$) that result in: $LDC_{ij}^{03} = 139.57$; $LDC_{ij}^{16} = 137.41$
c^{ldw}	= 42.08 €/h
$H_{w,ij}^{rail}$	<u>= 154.58 €/h</u>
c^{tcc}	= 32.70 €/wagon
$c_{w,ij}^{\Sigma var}$	= 1.07 €/km, since for transports from i = 03103 to j = 20004 a multi-system electrified locomotive is applicable
$d_{ij}^{rail,v}$	for transports from i = 03103 to j = 20004 Germany and two foreign countries are involved that result in: $d_{ij}^{rail,03} = 110$; $d_{ij}^{rail,16} = 62$ km
RCI^v	for transports from i = 03103 to j = 20004 German and two foreign infrastructure charges are involved that result in: $RCI^{03} = 66.04$ and $RCI^{16} = 156.64$
c^{ric}	= 2.83 €/km
NEC^v	for transports from i = 03103 to j = 20004 Germany and two foreign electricity costs factors are involved that result in: $NEC^{03} = 84.81$ and $NEC^{16} = 75.10$
c_w^{ecl}	= 2.74 €/km
c_w^{ecu}	= 1.87 €/km
$P_{w,ij}^{rail}$	<u>= 6,979.79 €</u>

$NBT_{w,ij}$ = 2, since for containerised transports from i = 03103 to j = 20004 no ports are involved

c_w^{trs} = 50.01 €/wagon, since for containerised commodities a freight train of configuration $w = 5$ is modelled

t_w^{trs} = 0.12 h, since for containerised commodities a freight train of configuration $w = 5$ is modelled

$S_{w,ij}^{rail}$ = 3665.11 €, since for containerised commodities a freight train of configuration $w = 5$ is modelled

NLT_w = 1,075.08, since for containerised commodities a freight train of configuration $w = 5$ is modelled

$C_{h,ij}^{road}(1)$ = 16,594.63 for one exemplary containerised tonne

14.3.5. Validation

Publications of transport costs or market prices for different rail freight services in practice are rare. One exception is the freight transport price publication for the year 2012 of DB SCHENKER (2012i). Therein, base prices for wagon-load transports for national and, in part, for international rail freight transports are published. Base prices for wagon load rail freight transports probably only represent prices with limited relevance for larger transport volumes of block trains. In consequence, these price tables are not qualified as an indicator for the validity of the presented framework.
Similar to publications from practice, generic transport cost evaluations of German or European rail freight activities are only seldom discussed in the corresponding literature. Especially specific elements – such as the consideration of additional costs for return trips, individual assumptions on average train configurations or individual data inputs – over-complicate a comparison of different rail freight cost evaluations. However, two studies can be identified that deal with explicit rail freight transport cost evaluations – the study of BMVI (2014a) and the work of DEUTSCH (2013).

The framework presented by the BMVI (2014a) is a major impetus for the presented context and some of the propositions and assumptions stated therein are incorporated analogously. On the other hand, relevant inputs to their overall rail freight cost evaluation are not transparent. Moreover, the presented framework is not entirely consistent as it contains significant miscalculations. Still, the results from BMVI (2014a, p. 150 ff.) comprise a large share of comparable inputs and cost factor assumptions – e.g. train configuration costs per wagon – being set up for the year 2010 in favour of a comparison with the presented model's results.

The study of DEUTSCH (2013, p. 385 ff.) is set up from different inputs that are in most cases only briefly discussed and often involve extensive assumptions. Nevertheless, the study comprises an analysis of multiple national and international transport relations and the corresponding cost factors in a certain depth.

An overview on explicitly evaluated cost factors within both reference studies is given in the following figure:

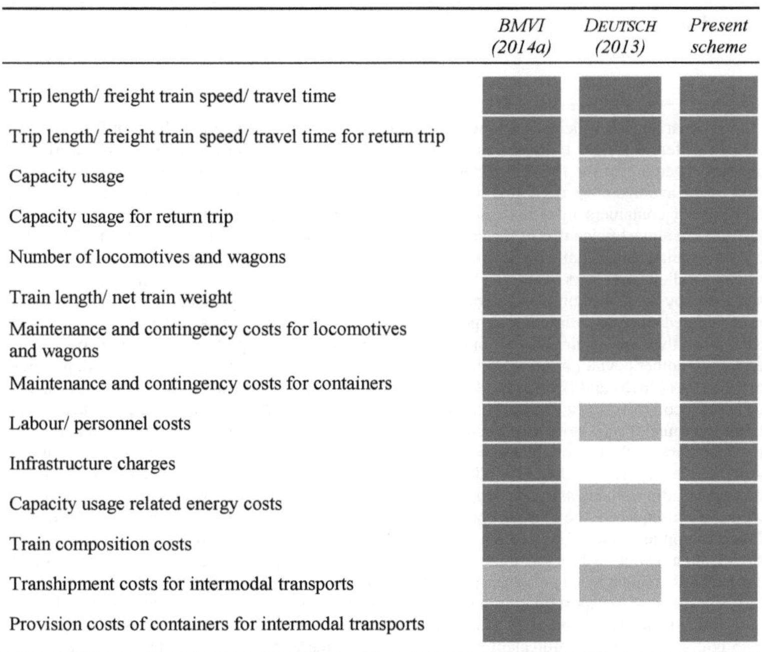

Fig. D-13 Overview on elementary factors involved for a rail network link transport cost evaluation in comparison to reference studies

Both referenced studies do not differentiate costs per tonne for x_{ij} and x_{ji}. Since the demand for e.g. containerised transports is in general not the same on either end of an OD pair, different levels of empty returns and different prices for each direction occur. The study of BMVI (2014a), at least, considers different transport demand levels within its exemplary calculations. In the end, however, only an average cost factor of both transport directions is indicated.

The study of DEUTSCH (2013) assumes a compensated transport demand for a rail freight round trip within the presented exemplary settings. For instance, for a transport scenario from the *Port of Rotterdam* to the German city *Duisburg*, an equilibrated transport demand is assumed for both directions, though it is not reflected within the reports of DESTATIS (2013b).

Although of limited practical relevance for a validation within the presented context, both calculation schemes are helpful indicators. A comparison of results from BMVI (2014a) and DEUTSCH (2013) with results of the presented scheme are depicted in Table G-34 and Table G-35[183].

---

[183] A comparison is explicitly not processed per tonne-kilometre – the preferable unit for a comparison of different freight transport costs – to ensure transparency on the inclusion of return load charges. Therein, for the first rail freight link cost example significant deviations on total costs as well as for average costs per tonne occur in comparison to the referenced studies. The costs for a transport of containers from the *Port of Rotterdam* to *Duisburg* calculated according to BMVI (2014a) are almost twice that of the calculation results of DEUTSCH (2013). Apart from administrative costs being significantly higher in the first reference, transhipment costs for the assumed number of 158 handled containers result in a large spread of both outcomes. Within the calculation scheme suggested by the author only one extra transhipment compared to a unimodal road freight transport is considered since the origin is a seaport. Therefore, no surcharges on handling activities are modelled since they are required for all modes in a comparable kind. Apart from handling cost considerations, another deviation stems from the different time dependent cost evaluations. Both the study of DEUTSCH (2013) and BMVI (2014a) charge costs for the contingency as well as for the maintenance of locomotives, wagons and containers more than twice as high as within the presented model. Charging round trip costs on this specific transport relation neglects the fact that more containerised transports are performed on the route from *Duisburg* to *Rotterdam* than on the way back within the year 2012, as reported in DESTATIS (2013b). In this regard, a return load for containerised rail freight transports from *Rotterdam* to *Duisburg* is assumed to be likely feasible within the presented framework. In consequence, the transport price on this OD pairing will not include return trip surcharges. This also applies for other cost components, such as personnel costs. The results of DEUTSCH (2013) stem from a labour cost factorisation of entire round trips and the related total travel time – that is consequently almost twice as high as the cost result of the given analysis. Within the study of BMVI (2014b), the total transport time for a round trip from the *Port of Rotterdam* to *Duisburg* and return is calculated with a duration of 23 hours. The traction itself on either leg, however, is calculated with 3.45 hours only – that is equivalent to a travel speed of more than 72 km per hour, still neglecting additional time surcharges for this routing. In consequence, personnel costs for a round trip are close to the calculation results presented by the author for only a one-way factorisation with a travel time of about 7.21 hours, an average travel speed of 40 km/h and additional time buffers and average supplements. Comparable discrepancies can be observed for the remaining rail freight cost examples.

In total, a tangible validation of the obtained results on rail freight transport costs in the transport practice for the entirety of potential national and international transport relations is not feasible. However, the presented calculation appears to be within a limited range of deviation on costs calculated in practice – according to expert consultations on selected transport relations – and likewise comparable to cost evaluations in the corresponding literature.

14.4. IWW freight transport link cost

Inland waterway freight transports to or from, as well as within Germany, are enrolled by national and international freight forwarders. An overview on the share of IWW vessel operating under the German flag compared to other countries for international transports can be found in Table G-7[184]. Similar to road freight transports, costs for foreign transport services need to be considered in a separate comprehensive costing analysis. But unlike for the road freight segment, the share of foreign forwarders for cabotage operations is an additional element of the following considerations, since already the share of Dutch-flagged barges serving German inland transport relations is about 22% in terms of tonne-kilometre. An overview on the cabotage level per EU-27-country for Germany is depicted in Table G-36.
Apart from the nationality, different types of operating IWW vessels can also be determined from a market segment overview as follows:

IWW transport by type of freight vessel in millions of tkm for 2012 in Germany						
Type of Vessel	Self-propelled barge	Not self-propelled barge	Self-propelled tanker barge	Not self-propelled tanker barge	Other goods carrying vessel	Total
Transport volume	38,763	7,058	10,881	247	1,539	**58,488**
Share	**66%**	12%	**19%**	0%	3%	

Table D-21 IWW transport volume for Germany by type of freight vessel [incl. data from EUROSTAT (2014o)]

[184] This setting might be interpreted similar to an assumption made for the road freight segment, according to which no third-country freight forwarders are involved – e.g. for a transport from France to Germany, where either German or French freight vessel crews are considered only.

Thereafter about 85% of all national and international IWW transports are executed by either a general self-propelled barge or a self-propelled tanker barge. These types are considered as representative for a commodity class specific IWW costing scheme within the following (cf. Table G-37).
For both freight vessel types, an average load capacity of 1826 tonnes and 1864 tonnes is determined in line with Table G-38. Furthermore, the load capacity in terms of space for a general self-propelled barge is set to a maximum of 143 units of 20-foot standard containers[185]. The own weight of these reference IWW vessel types is estimated according to the ships hydrostatic pressure at 467 and 860 tonnes, respectively[186].

Time-related costs

Time- and mileage-related cost components are identified for IWW transport costs, equivalent to the presented road and rail freight cost calculation process. Therefore, the first factor of time dependent costs incorporates the purchase prices for barges. Similar to the rail freight cost setting, these prices are more difficult to estimate than for road freight vehicles. However, for general self-propelled barges, a purchase price of 3.5 million Euro can be derived from UNICONSULT (2009, p. 51). Acquisition costs for self-propelled tanker barges of approximately 4.5 million Euro are indicated in DOMMER (2013). Since both prices are only vague estimates with potentially large individual deviations in practice, no further price updates will be considered. They will be directly incorporated into the following processing for the year 2012, instead.

According to WSDO (2007, p. 217 ff.) barge acquisition costs are a compound of costs for the hull – with a depreciation time of 25 years – and costs for the engine, navigation equipment, pumps etc. – calculated with an amortisation after approximately 10 years. The study further proposes to split the acquisition cost total into 70% for the hull of general self-propelled barges

---

[185] This value is derived as an average from the *European barge* ship class with a capacity of 1300 tonnes or 90 TEU and the *Large IWW motor vessel* class with a capacity of 2300 tonnes or 208 TEU. See also estimated size measures, indicated in Table G-39.

[186] For the *European Vessel* with a size in length and width of 85 ∗ 9.5 metres, a draught of 2.5 metres is processed. For the *Large motor vessel*, sized with a length of 110 metres and a width of 11.4 metres, a draught of 3.5 metres is calculated. With an assumed block coefficient of 0.9 for a representable hull, an own weight of 467 tonnes for the *European barge* and 860 tonnes for the *Large IWW motor vessel* are finally determined.

and 65% for the hull of self-propelled tank barges (ibid.). To determine an annual replacement value – similar to that for road freight and rail freight vehicles – German price index tables are consulted. The hull is updated with a price index of *other vehicles* (GP09-30), whereas the engine and the remaining technical equipment are processed with a price update for *engines for the maritime sector, railed vehicles and other industrial purposes* (GP09-28111)[187] from DESTATIS (2014p). In absence of appropriate indicators, the residual value for resale after amortisation is assumed to be equal to zero. For the subsequent calculations, an interest rate of 3.38% per year as for previous road and rail freight costing is assumed.

Unlike for road and rail freight vehicles, no mileage dependent – hence, variable – costs for maintenance and servicing as well as vehicle abrasion are calculated. As proposed in WSDO (2007, p. 218), maintenance and equivalent costs are calculated in relation to an annual replacement value only. The resulting cost calculation scheme will be:

[187] Backwards directed data from 2002 to 2012 indicate a price increase for GP09-28111 of 19.5% and, from the earliest listing in 2000 to 2013 for GP09-30, an increase of 16.7% that is set to 32.1% for 25 years.

	Input, according to:			*Self-propelled barges for bulk carries or containers*		*Self-propelled tank barges*	
				Basic cost-related input factors			
1		Net purchase price for 2012	€	3,500,000		4,500,000	
2	*a)*	Purchase price hull/ technical equipment	€	2,450,000	1,050,000	3,150,000	1,350,000
3	*a)*	Operating lifetime hull/ technical equipment	a	25	10	25	10
4	*b)*	operating days	d	340		340	
5	*b)*	operating time per day	h	16		16	
6	*c)*	Annual capital interest rate	%	3.38		3.38	
7		Residual value for resale for hull/ technical equipment	€	0	0	0	0
8		Replacement purchasing price of hull/ technical equipment	%	132.1	119.5	132.1	119.5
9	Input 2 * Input 8	Residual value for replacement of hull/ technical equipment	€	3,236,450	1,254,750	4,161,150	1,613,250
10	(Input 2 - Input 7) * 0,5	Average operating capital	€	1,750,000		2,250,000	
				Fixed cost components/costs per operating time unit			
11	Input 6 * Input 10	Annual costs for capital	€/a	59,089		75,972	
12	(Input 9 - Input 7) / Input 3	Depreciation per time unit (100%) for hull/ technical equipment	€/a	129,458	125,475	166,446	161,325
13	*a)*	Annual maintenance costs in relation to replacement purchasing price	%	2.50		2.50	
14	*a)*	Annual operating material consumption costs in relation to replacement purchasing price	%	0.40		0.40	
15	*a)*	Annual insurance costs in relation to replacement purchasing price	%	1.20		1.20	
16	*a)*	Annual overhead costs in relation to replacement purchasing price	%	0.25		0.25	
18	Input 9 * Σ (Input 13,…,16)	Sum of annual costs for maintenance, operating material, insurance, overhead and administration	€/a	140,786	54,582	181,010	70,176
19	(Input11 + Input 12 + Input 18) / Input 4 / Input 5	**Σ of hourly fixed costs without wages for hull/ technical equipment**	**€/h**	**60.54**	**43.96**	**77.84**	**56.52**
				Variable cost components/costs per distance travelled			
20		**Variable energy costs**	**€/tkm**	**11.08**		**11.31**	

Table D-22 Cost calculation data input for modelled IWW barges

Relevant sources of information in table Table D-22 are indicated as:

a) WSDO (2007)
b) BMVI (2014a)
c) BUNDESBANK (2014)

14.4.1. Additional IWW freight cost factors

National and international IWW crew wages

Labour costs for German crews can be obtained from wage tables for German inland navigation, published by BSV (2012). A distinction of general IWW navigation and the continuous IWW navigation can be found therein. According to BMVI (2014a, 177 f.), the preferential operating mode for both selected representative freight vessel classes is non-continuous and can be estimated at 16 operating hours per day.
Minimum crew requirements as well as the corresponding duty time and rest requirement for general self-propelled barges and self-propelled tank barges are declared in BINSCHUO (2008). In this respect, the crew for the presented representative vessel classes and a 16 hour day trip is formed by one skipper, one helmsman and one sailor. Their hourly wages are multiplied by 14 regular hours, complemented by shift surcharges for four additional hours and overtime surcharges for two hours as well as a night-work surcharge for two hours[188]. The total of this calculation is multiplied by labour cost surcharges of 23% of the actual wages paid, according to LOHNDIREKT (2012). In analogy with the road and rail freight costing scheme, a number of 20 days per annum to cover absences from work by drivers due to illness, mandatory driver training etc. is considered within the labour cost evaluation. Another 30 days are calculated for the number of paid holidays for IWW vessel crews[189]. When the labour cost table in BSV (2012) is interpreted to be set up for 228 paid working days per annum, the resulting 200 days at work lead to a total German IWW vessel crew labour cost rate of 69.69 € per hour. The related calculation is depicted in Table G-40.

---

[188] Wages for skippers are dependent on their working experience and are set to five and more years for simplification.
[189] Note that the minimum number of holidays for 40 hours of working time per week according to the German *Federal Holiday Act* is 24. In contrast, even 42 days of vacation per year are common for 40 weekly working hours of IWW vessel crew members according to e.g. IHK (2005).

Resembling the road freight cost evaluation, a share of foreign vessel operators is considered within the model. This concept is implemented not only for international but also for national IWW transport relations, since only about 69% of all national IWW transports are performed by German freight vessel operators[190]. The share of distribution of foreign IWW transport operators on international transport relations is depicted in Table G-7. This procedure is based upon the following assumptions:

- Both national and international IWW transport relations are either served by German or other European transport operators. Whereas national transport relations are served by freight vessels from multiple nations, international transport relations are exclusively served by German operators or vessels under the flag of the international origin or the destination of a modelled IWW transport.

Subsequently, the determined hourly crew cost rate is updated by the share of foreign freight vessel operators and corresponding IWW crew wage level for the focal OD pairing. The latter component is derived from wages paid within the EU-27, Switzerland and Russia for employees in the segment of *water transport* (CPA-50) in firms of 10 and more employees according to EUROSTAT (2014r) and EUROSTAT (2014s), as depicted in Table G-41. Each international price index is evaluated individually in relation to the reference average hourly crew wage of 69.69 € for German crews for each considered OD pairing. Additionally, the hourly base cost rate of 69.69 € for German crew wages paid on national transport relations is recalculated to a national weighted average of 69.51 € per hour in order to consider the share of crew wages paid for cabotage activities. The results are included in Table G-37[191].

IWW vessel energy consumption rates

The energy consumption per distance travelled and the corresponding cost factors for self-propelled barges are determined in line with general indicators – such as type and size of barge, load factor and travel speed – and certain specific components. The main specific energy consumption factors are the profile of the used waterway and the direction of travel. These factors take

[190] Cf. Table G-36.

[191] Note that the impact of cabotage activities for German IWW navigation on crew labour costs in total is almost insignificant due to almost compensated higher and lower rates.

into account that a waterway's current is different for channels and natural rivers that, in turn, differ from an ascent and a descent. These aspects are considered for fleet measurements of diesel fuel consumption rates of average loaded vessel as taken by KRANKE (2011, p. 205)[192].

For the presented model, an average fuel consumption rate of the *European vessel* and the *Large motor vessel* is calculated with 0.0061 litres per tonne-kilometre for natural rivers. Thus, a fuel consumption of 11.08 and 11.31 litres per thousand tonne-kilometres for a load capacity of 1,826 and 1,864 tonnes results for complete loads of the selected vessel classes[193]. For an empty journey, e.g. an empty return trip, an average fuel consumption rate of 4.02 litres per kilometre is considered[194].
For a subsequent implementation within the model, a commodity class specific load factor allows to determine specific energy consumption rates for different vessel- and load-configurations.

To determine an average national price for barge diesel fuel, the study BMVI (2014a, p. 185) recommends the integration of prices for light heating oil of large volume consumers in Germany. For 2012, DESTATIS (2014r, p. 42) indicates an average market price of 0.7533 € per litre. The international price level within the EU-27 is determined equivalently to Diesel prices for the road freight costing in accordance with the oil bulletin, published by the EC (2014). Excerpts for 2012 can be found in Table G-42.

It is assumed that foreign IWW vessels serving German origins and/or destinations exclusively fill up tanks in the country with the lowest energy costs. The case of refuelling when traversing a third country with comparatively lower or higher fuel prices is not considered. For example, a Dutch barge serving a German national *Rhine* OD relation – hence, exercising cabotage – will be capable of taking advantage of the lower fuel costs of only 93% of the German price level in its port of registry.

---

[192] See Table G-39 for corresponding results.
[193] Both selected values are close to 11.4 litres per 1,000 tonne-kilometres as referenced in KRANKE (2011, 203 f.) that serves e.g. as an input to the German *TREMOD Transport Emission Model*.
[194] Therefore, an average of the own weight of 467 tonnes for the *European vessel* and 860 tonnes for the *Large motor vessel* serves as a calculation basis.

The corresponding distance travelled results from a *shortest path* routing request from WSV (2015), complemented by data from BINNENREEDEREI (2014).

IWW vessel speed

In accordance with the selected basic energy cost calculation scheme, an average freight vessel speed of 12 km per hour is assumed for the model – measured as the riverside related speed as a simplified aggregate, independent of a waterway's location, river training, course or water current – as derived from the existing legislative framework for the German inland navigation in WSV (2010). An additional surcharge for lock waiting times and other travel time relevant factors is not considered due to an absence of adequate references.

IWW infrastructure charges

For Germany, two components for the use of infrastructure apply, namely pierage costs for each loaded and/or unloaded tonne as well as a mileage dependent charging per tonne kilometre. Both constituents are evaluated commodity specifically.

Compared to rail and especially compared to road freight infrastructure charges, the IWW charging system is very specific. Each port charges individually and for the majority of the national waterway system specific mileage dependent tolls apply, whereas certain sections – such as the international *Danube* and *Rhine* – are not charged at all. Nevertheless, a generic cost structure for a German IWW infrastructure usage is indicated in WSDW (2012, p. 14). Therein, average infrastructure charges for different groups of commodities in transport are presented. These groups are related to outdated NSTR goods classification standard. For an analysis of only a limited number of IWW OD relations instead, more specified charges are given in the German IWW online charging information system, the WSV (2014a).
For the presented study the more comprehensive – and at the same time less specified – source of information on IWW network charges from WSDW

(2012) is consulted. It is complemented by a correspondence table from EUROSTAT (2014a)[195], as depicted in Table G-43.

The second cost component for a use of the IWW infrastructure covers pierage costs and port dues. These costs only apply for intermodal transports in accordance with the modelled network node configuration [196]. An approach to model pierage costs and port dues is presented in BMVI (2014b, S. 191), where an average port charge per commodity class is built up from port operator interviews. For containers, a value of 0.44 € per tonne is adopted from an indication for grouped goods (ibid.). Although collected in 2010, the suggested cost rate is directly incorporated into the presented study due to an absence of further background information on the composition of this value. For an average load per TEU of 10.54 (cf. Table G-32), pierage costs and port dues are calculated to 4.64 € per TEU.

Since large sections of international IWW transports are assumed to be performed along the rivers *Rhine*, *Danube*, *Elbe* and *Oder* that are exempted from infrastructure fees, for international transports no IWW infrastructure fees are considered at all within the presented model.
Within the model, load-specific infrastructure costs are evaluated individually for each vessel configuration according to Table G-37. For containerised loads an average cost rate is modelled instead according to Table G-44 (cf. next section).

---

[195] In practice, difficulties arise from the fact that the given NSTR nomenclature for IWW infrastructure charges is not unambiguously transferable to the two-digit CPA classification scheme. As a simplification, the lowest reference for a NSTR classified commodity type is converted into the present CPA systematic. For instance, what is classified as *tractors; agricultural machinery and equipment, whether or not assembled; parts thereof* (NSTR-920) either corresponds to *machinery and equipment* (CPA-28) or *motor vehicles, trailers and semi-trailers* (CPA-29). Accordingly, an NSTR-920 entry is assigned to CPA-28 and vice-versa. Since different NSTR groups are related to one two-digit CPA code, a second simplification applies. Thereafter, the highest number of assigned cost factors is allocated. For instance, for CPA-01, all 6 infrastructure cost factors are applicable for its subgroups. However, a dominant share of 40% of all CPA-01-subgroups are related to NSTR groups that are assigned to price group one in WSDW (2012). Consequently, price group one is processed as transports of CPA-01 on the German IWW network.

[196] In this respect, only containerised commodity transports are performed multimodally. Transports of other non-containerised commodities are taken to be originated in or destined at the port within the analysed region. These ports are assumed to be private ports or public ports that are directly related to the origin or destination of a commodity flow. For these ports no additional infrastructure usage fees apply. Thus, the IWW infrastructure use is comparable to loading and unloading facilities that are required for modes road and rail.

IWW vessel energy consumption rates, infrastructure charges and transhipment costs for containerised loads

A load specification for non-containerised transport is set up per CPA-two-digit commodity class. Since the number of observations for containerised transports by freight vessels is, to a large extend, unspecified – hence, assigned to unspecified commodity classes – an overall average is processed instead. The same applies for an infrastructure charging of containerised loads.

Analogously to the rail freight transport cost calculation, additional transhipment efforts are considered for intermodal IWW transports. Therefore, container transhipments for transport links with, at least, an origin or a destination other than a seaport, are charged with16.67 € per TEU handling and a time surcharge of 0.04 hours per TEU transhipment in line with the findings of section 14.3.3.

The relevant number of TEU per IWW vessel is determined as an average according to Table G-44. Therein, likewise results for containerised load surcharges are given, complemented by Table G-45[197].

Empty freight vessel shares

In analogy to the presented rail freight transport cost calculation scheme, the potential for return loads on a given transport relation is additionally considered for charging outward going loads. Suitable loads are identified for each observed OD pairing within the model accordingly. Although vessels transporting *coal and lignite* (CPA-05) are potentially qualified to return containerised commodities, it is generally not a case observed in practice. Turnaround costs including time surcharges for a necessary thorough cleansing of the vessel's hull will not be cost effective compared to an immediate empty return in the given transport market[198]. Therefrom, the following assumption is derived:

---

[197] As a reminder, an average load factor in terms of TEU is only built due to a low number of observations for commodity specific containerised IWW transports that would otherwise lead to unlikely load configurations.

[198] A similar consideration of IWW return loads can be found in BMVI (2014a, p. 195)

- Return loads for inland vessels are only feasible for either identical commodities or similar transport units, whereas the latter is only observed for containerised IWW transports.

As a result, for each outward directed transport of x_{ij} the demand for an applicable return load of x_{ji} is evaluated according to DESTATIS (2013a)[199].

Load configuration assumptions

Compared to rail freight, IWW transports are reported by operators on German waterways in more detail. This allows the German *Federal Bureau of Statistics* to collect information on the annual number of trips for inland navigation. This information is helpful to determine an average commodity specific load factor for freight vessels[200]. Since the collected number of trips does not allow for a separation of multiple loads and/or unloads along a journey, the following assumption applies:

- The number of collective shipments on IWW within the annual statistical report for the number of trips is insignificantly small and therefore negligible.

According to this, average load configurations and related fuel consumption levels per freight vessel class are determined. The results for non-containerised commodity transports are depicted in Table G-37 and for containerised transports in Table G-44.

14.4.2. Overall IWW freight cost calculation scheme

Total link costs for IWW freight transports are calculated according to an empty return surcharge and the assumptions of the previous section, in line with the general format of equation (23), as:

[199] Within the model's computation, this setting is processed by a single vessel type per commodity class, complemented by one type for containerised transports. In total, 30 types are individually evaluated.

[200] This dataset is available upon request from DESTATIS (2014n). This dataset is based on the NST classification standard that can be transformed to the CPA classification scheme according to Table G-9, Table G-10 and Table G-11.

$$C_{s,ij}^{IWW}(x) = \left\lceil \frac{x}{NLV_s} \right\rceil * \left(P_{s,ij}^{IWW} + H_{s,ij}^{IWW} * T_{ij}^{IWW} + S_{w,ij}^{IWW}\right) \tag{36}$$

with:

$$S_{s,ij}^{IWW} = NBT_{s,ij} * \left(c_s^{trs} + t_s^{trs} * \left(c_s^{\Sigma hfc} + c_s^{cpc}\right)\right) \tag{37}$$

and:

$$\begin{aligned} P_{w,ij}^{IWW} = {} & \left(1 + SER_{s,ij}\right) * \; c_s^{wc} * d_{ij}^{IWW,6} + \left(d_{ij}^{IWW} - d_{ij}^{IWW,6}\right) \\ & * \left(c_s^{fcl} + SER_{s,ij} * c^{fcu}\right) * min(FPI^v) * c^{afp} \end{aligned} \tag{38}$$

and:

$$H_{s,ij}^{IWW} = c_s^{\Sigma hfc} + c_s^{cpc} + c_{ij}^{tcw} * \left(\left(1 - SFS_{ij}\right) + SFS_{ij} * LCI_{ij}^{IWW}\right) \tag{39}$$

and:

$$T_{ij}^{IWW} = \left(1 + SER_{s,ij}\right) * \left(\frac{d_{ij}^{IWW}}{v^{IWW}}\right) \tag{40}$$

where:

$C_{s,ij}^{IWW}(x)$	total IWW freight transport costs for a volume x in tonnes of the modelled commodity classes p, assigned to a vessel of configuration s from origin i to destination j, in: [€]
NLV_s	net load per IWW vessel of configuration $s \neq 30$ according to Table G-37 and for vessels of configuration $s = 30$ according to Table G-44, in $[t]$
$P_{s,ij}^{IWW}$	pecuniary costs for an IWW vessel of configuration s from origin i to destination j, in: [€]
$H_{s,ij}^{IWW}$	costs per time unit for an IWW vessel of configuration s from origin i to destination j, in: $[€/h]$
T_{ij}^{IWW}	total travel time from origin i to destination j, in: $[h]$

$S_{w,ij}^{IWW}$ transhipment costs for an IWW vessel of configuration $s = 30$ from origin i to destination j, in: [€]

$NBT_{s,ij}$ number of transhipments for an IWW vessel of configuration $s = 30$ from origin i to destination j, with:

- 0, for $i \wedge j \neq ID25, ID26, ID2000, \dots, ID1058$
- 1, for $i = ID25, ID26, ID2000, \dots, ID1058 \wedge j \neq ID25, ID26, ID2000, \dots, ID1058 \vee i \neq ID25, ID26, ID2000, \dots, ID1058 \wedge j = ID25, ID26, ID2000, \dots, ID1058$
- 2, for $i = ID25, ID26, ID2000, \dots, ID1058 \wedge j = ID25, ID26, ID2000, \dots, ID1058$

c_s^{trs} transhipment costs for an IWW vessel of configuration $s = 30$ of 101.69 €, according to Table G-45

t_s^{trs} transhipment time for an IWW vessel of configuration $s = 30$ of 2.44 hours, according to Table G-45

$c_s^{\Sigma hfc}$ sum of hourly fixed costs without wages for an IWW vessel of configuration s as indicated in Table D-22 in $[€/h]$

$SER_{s,ij}$ share of empty returns for transports by IWW vessel of configuration s from origin i to destination j

c_s^{wc} IWW network infrastructure charges per vessel of configuration s according to Table G-43, in $[€/km]$

$d_{ij}^{IWW,6}$ *shortest path* IWW freight distance in Germany ($v = 6$) from origin i to destination j as determined from routing procedure, in: $[km]$

d_{ij}^{IWW} overall *shortest path* IWW freight distance from origin i to destination j as determined from routing procedure, in: $[km]$

c_s^{fcl}	kilometre specific fuel consumption for a loaded IWW vessel of configuration s according to Table G-37 and Table G-44, in $[l/km]$
c^{fcu}	kilometre specific fuel consumption for a loaded IWW vessel of 4.02 l/km
FPI^v	fuel price index according to Table G-42 if a foreign country v is involved when travelling from origin i to destination j for $v = i$ or $v = j$ and 1, otherwise
c^{afp}	average fuel price for Germany ($v = 6$) of 0.7533 $€/l$
c_{ij}^{tcw}	hourly base cost rate for crew labour costs of 69.69 $€/h$
SFS_{ij}	share of international shipping companies that is involved when travelling from origin i to destination j according to Table G-7[201]
LCI_{ij}^{IWW}	labour cost index for crew wages paid, with 0.997 for $i \wedge j \leq ID2000$ – hence, an origin i and destination j within Germany – and otherwise, as indicated in Table G-41
c_s^{cpc}	average container provision costs for an IWW vessel of configuration $s = 30$ of 6.10 of $€/h$, according to Table G-45
v^{IWW}	average IWW vessel velocity of 12.00 km/h

Subsequently, the presented IWW freight cost calculation is processed for each modelled *direct* as well as for each *combined* IWW freight network link as follows:

[201] Note that for $i \wedge j \leq 2000$ the processed value for SFS_{ij} is 100%, since the evaluated average labour cost of national and international operators on national transport relations is considered within the related aggregated LCI_{ij}^{IWW} to obtain a recalculated national weighted labour cost average of 69.51 €.

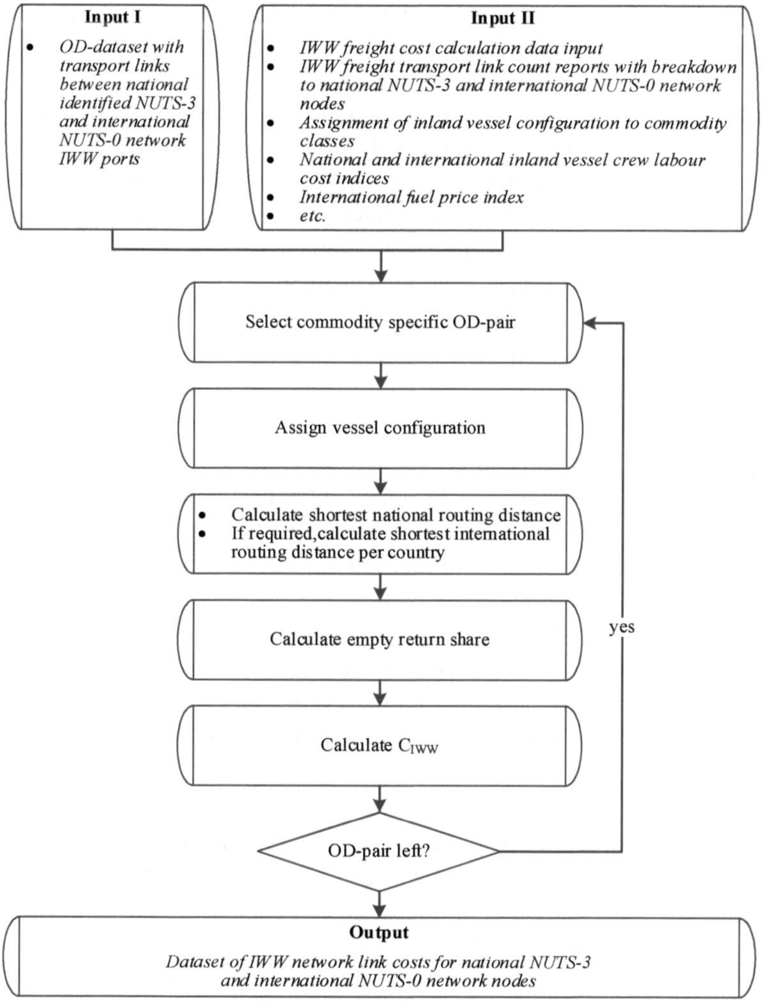

Fig. D-14 Simplified processing scheme to determine IWW freight transport network link costs

Exemplary application

An exemplary setting for an application of the presented IWW freight costing scheme comprises a *containerised transport* from a German origin in

Leverkusen (NUTS-ID: 05316) to an international destination, the *Port of Antwerp* (NUTS-ID: 26).

$SER_{s,ij}$ = 0.29, since for transports from i = 05316 (related to NUTS-2 *DEB3: Rheinhessen-Pfalz*) to j = 26 (related to NUTS-1 BE: *Belgium*) a total annual containerised transport volume of about 196.520 tonnes is reported; the equivalent return trip volume is rounded to 405.460 tonnes – hence, 0% of the return trips are assumed to be empty

d_{ij}^{IWW} = 363.03 km

v^{IWW} = 12 km/h

T_{ij}^{IWW} = 30.25 h

$c_s^{\Sigma hfc}$ = 104.50 €/h, since for container transports a self-propelled barge with a configuration of s = 30 is selected

c_s^{cpc} = 6.10 €/h, since for container transports a self-propelled barge with a configuration of s = 30 is selected

c_{ij}^{tcw} = 69.69 €/h

SFS_{ij} = 0.82, since for transports from i = 05316 to j = 26 about 82% of all transports are performed by Belgian crews

LCI_{ij}^{IWW} = 1.59,since since for transports from i = 05316 to j = 26 to the share of foreign crews a different wage is paid

$H_{s,ij}^{IWW}$ = 214.10 €/h

c_s^{wc} = 21.11 €/km, since for container transports a self-propelled barge with a configuration of s = 30 is selected

$d_{ij}^{IWW,6}$ = 149 km of infrastructure cost relevant navigation in Germany ($v = 6$)

c_s^{fcl} = 7.02 litres/km, since for container transports a self-propelled barge with a configuration of s = 30 is selected

c^{fcu} = 4.02 litres/km

$min(FPI^v)$ = 0.59, since for a transport from i = 05316 to j = 26 different fuel prices for Belgium are considered

$c^{fp,v}$ = 0.75 €/litre

$P_{s,ij}^{IWW}$ = 3514.24 €

$NBT_{s,ij}$	= 1, since for a transport from i = 05316 to j = 26 one transhipment is included into the costing scheme
c_s^{trs}	= 101.69 €, since for container transports a self-propelled barge with a configuration of s = 30 is selected
t_s^{trs}	= 2.44 h, since for container transports a self-propelled barge with a configuration of s = 30 is selected
$S_{w,ij}^{IWW}$	= 371.55 €
NLV_s	= 633.35 t, since for container transports a self-propelled barge with a configuration of s = 30 is selected
$C_{s,ij}^{IWW}(1)$	= 10,362.86 € for one exemplary containerised tonne

For an increasing transport volume within the exemplary setting, the course of the cost function is depicted in the following figure:

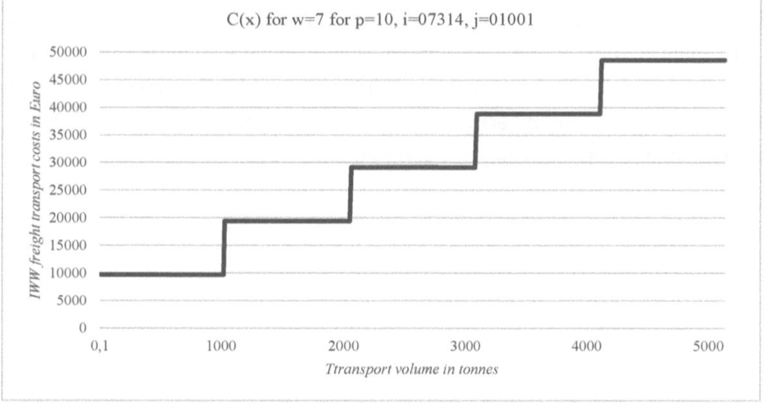

Fig. D-15 Exemplary IWW freight transport cost curve for *containerised transports* from *Leverkusen* within Germany to the *Port of Antwerp*

14.4.3. Validation

Given the diverse nature of the IWW network in Germany and Europe – e.g. comparing the river *Rhine*, the river *Elbe* or the German *Mittelland Canal* – it is difficult to state average factors as an input for a transport cost analysis, such as for an average vessel type, travel speed, fuel consumption or infrastructure usage costs. A major reason for this circumstance is that most of these specifications cannot be identified in a straightforward fashion. Their relevance for the final outcome, however, is partly higher when compared to similar inputs for the road and rail freight cost evaluation. In this context, the selected procedure has to be regarded as a second-best solution. When compared to the rail freight and especially the road freight costing scheme, the IWW transport cost calculation is considerable as the 'weakest chain link' within the presented context.

The high level of uncertainty when applying a generic costing on individual IWW transport relations in Germany is presumably a reason for the few number of reference calculations given in the corresponding literature, since the only identified study on this subject is the one given in WSDO (2007). Therein, several selected national and international IWW transport relations that are suited for a comparison with the results of the presented costing are evaluated, as depicted within the following table:

Origin	*Destination*	*Type of transport:*	Empty return share reported	**Transport costs in EUR per tkm reported**	Empty return share calcalculated	**Transport costs in EUR per tkm calculated**	**Dev.**
Hamburg	*Decin* *(Ústi nad Labem, CZE)*	(CPA-01)	0.20	0.014	0.99	0.031	121%
Hamburg	*Salzgitter*	(CPA-05)	1.00	0.023	0.00	0.016	-31%
Port of Rotterdam	*Duisburg*	(CPA-05)	1.00	0.013	0.00	0.011	-16%
Port of Rotterdam	*Großkrotzenburg* *(Main-Kinzig-Kreis)*	(CPA-05)	1.0	0.015	1.00	0.027	78%
Linz *(AUT)*	*Nürnberg*	(CPA-24)	1.0	0.027	1.00	0.037	37%
Port of Antwerp	*Ludwigshafen*	(CPA-20)	0.11	0.009	1.00	0.048	469%
Port of Rotterdam	*Duisburg*	*container*	0.09	0.015	0.12	0.035	133%
Port of Rotterdam	*Stuttgart*	*container*	0.24	0.017	1.00	0.056	236%
Hamburg	*Berlin*	*container*	1.00	0.025	1.00	0.058	133%
Hamburg	*Decin* *(Ústi nad Labem, CZE)*	*container*	1.00	0.019	1.00	0.042	118%

Table D-23 Comparison of calculated IWW freight costs with reference freight fares [incl. data from WSDO (2007, p. 268 ff.)]

Apparently, the level of deviation for most comparisons is high. One of the given examples covers a transport of *chemicals and chemical products* (CPA-20) from the *Port of Antwerp* to *Ludwigshafen.* The result of the presented cost evaluation is significantly higher than its counterpart within the reference study. A reason for the high magnitude of deviation is the variant return load factor, since for the year 2012 no return loads are observed in DESTATIS (2013a). Most likely, other cost factors for the overall result have changed over time as well. For this specific example, total costs are dominated by transport time dependent factors that stem from provision costs of the vessel. This component – related to costs for maintenance, operating material, insurance, overhead, administration and capital costs – covers about 75% of the entire transport costs for the evaluated setting. Because these costs correspond to the residual value for a replacement of an IWW vessel that, in turn, significantly increased within the last 10 years, vessel provision costs within the presented study are notably higher than that of the older reference. In other words, from a perspective in the year 2012, a steep price increase for the replacement of a vessel within the last few years requires an elevation of the annual fix cost component when compared to earlier calculations, such as in WSDO (2007). For transport relations with a high relevance of transport time related costs components, this has a large impact on the overall transport cost calculation.

Other important factors that lead to the presented deviations are different evaluated vessel load configurations. For instance, a transport of containers from *Hamburg* to the *Czech Republic* is assumed to be performed with a similar representative vessel type as applicable for a transport of *products of agriculture, hunting and related services* (CPA-01). Since for both the load factor differs significantly, the resulting costs per tonne-kilometre vary almost analogously. Since the assumed average load of 748 tonnes – or equivalent, 71 TEU for 10.54 tonnes per TEU – is derived as a weighted average for all reported containerised IWW transports in 2012, a different load factor for the presented framework is not considered[202].
Unfortunately, a practicable validation of the presented costing scheme with transport costs or prices paid in practice could not be realised. The relevance of the annual fixed cost calculation leads to a considerable scope of variant results, e.g. when comparing older and newer vessels with a divergent remaining operating life time or vessels with advanced technical equipment in contrast to a basic hull configuration with variant purchasing prices. Furthermore, the IWW segment of the transport market is less competitive compared to its modal alternatives. In consequence, individual transport costs are often not accessible to the public. Moreover, transport costs may differ significantly from transport prices paid by representatives of the transport demand. These interactions impose additional burdens on identifying reliable average

[202] Since the number of IWW trips is only statistically reported for the total of containerised and non-containerised transport, whereas the transported volume is indicated separately for both elements, an identical load to trip ratio is assumed for containerised and non-containerised transport. This circumstance and the potential impact of collective shipments on the reported number of trips may still cause a different average load factor in practice. This effect is assumed to be negligible due to an absence of any further indications.

values within a generic IWW costing scheme. For instance, assuming an average travel speed for vessels generates a significantly higher uncertainty in the context of hourly cost rates than within road and rail freight cost evaluations. As a result, a set of 30 variant IWW vessel configurations with e.g. individual load and fuel consumption rates is processed to further lower the level of uncertainty. In turn, the overall result is comparable to references of the related literature to a limited extent, only. Nevertheless, for a national transport model of the year 2012, the elaborated IWW costing is assumed to be at least comparable to its modelled modal alternatives.

15. Combined modal split and network assignment module

The outcome of the previous sections 13 and 14 is on the one hand:

- a dataset describing interlinked production and consumption sites, including their spatial distribution as well as their trade flow volume in terms of shipments: the PC data. Therein, a transport demand as a number of distinct annual shipments from origins of production to destinations of consumption is indicated.

As well as on the other hand:

- an intermodal freight transport network to connect production and consumption sites. This network consists of direct transport links that are potential transport paths at the same time as well as link combinations that form alternative intermodal transport paths. Each of which relates to individual transport costs.

The physical transport – performed by a transport service supply – is not captured within the initial PC dataset. This information, in turn, refers to a description of the actual chosen transport paths between origins and destinations. These transport paths are related to a selection among uni- or multimodal transport alternatives from origins of production to destinations of consumption, possibly via intermediate transfers, all together indicated in an origin/destination (OD) matrix. Representing this *real-world* setting is the goal of the subsequent section. The following depictions give an overview on the input PC and output OD data of the subsequent mode choice and assignment modelling sequence:

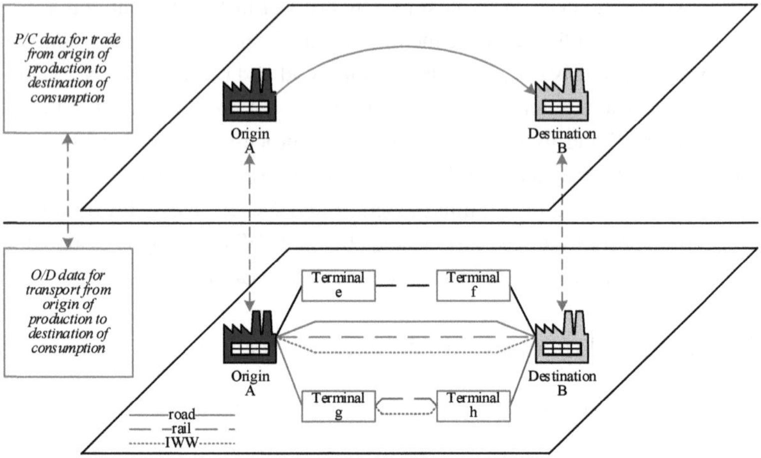

Fig. D-16 Basic comparison of PC trade and multimodal OD transport data

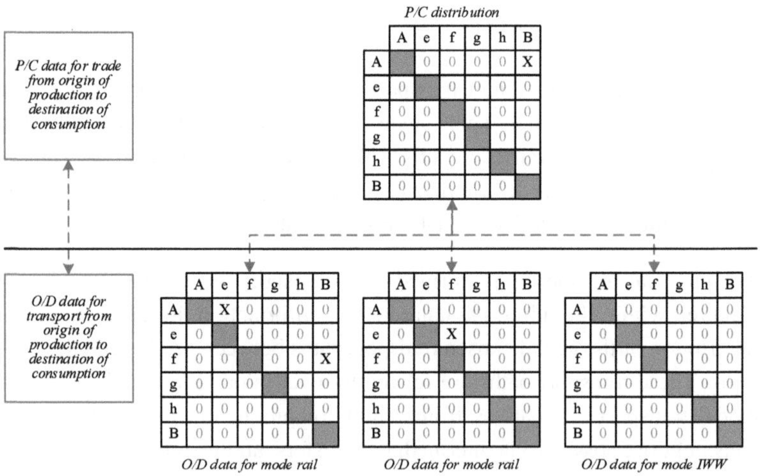

Fig. D-17 Basic transformation from PC trade to multimodal OD transport data

The outcome of a transformation from PC trade to OD transport data is equivalent to a *combined modal split and network assignment* procedure, referring to the last steps of the four-stage concept. An OD dataset contains the overall goal of the presented freight transport model: the information on commodity specific flow volumes along distinct transport paths that relate to a mode of transport.

15.1. Transport flow estimation

A transport flow configuration is the result of a transformation from the PC distribution to a multimodal OD network assignment. To obtain this set of information, an initial simplification is implemented:

- Within the presented context, only direct unimodal and intermodal paths are considered, instead of also explicit intramodal assignment configurations.

The major difficulty of a transformation from PC to OD data is the structured allocation of the transport demand among alternative transport paths and modes. A commodity flow survey – as e.g. implemented within the *ADA concept for Norway and Sweden* (cf. section 6) – could be helpful at this point. No such data is at hand for Germany – however, there are certain unexplored sources of information accessible, namely mode specific counts of transport volumes for at least sequences of paths from origins of production to destinations of consumption. These are published by the German *Federal Bureau of Statistics*, as presented in section 4.4.3.
Although they do not represent an entire commodity flow observation, they are of great value for a transformation of the given PC relations into OD data[203]. To bridge the remaining lack of information on mode and route choices, the PC and mode specific information need to be combined systematically.

[203] In case the complete information on transport flow volumes for every leg and mode for each path from an origin of production to a destination of consumption is at hand, a basic algebraic matching between the PC and OD data would be feasible – given the *correctness* of both inputs.

15.1.1. Input data combination principles

Starting from PC and fragmentary OD input data, an assignment to different transport paths for various modal alternatives is feasible in line with the following assumptions:

- Transport market actors – be they shippers, receivers or transport service providers – seek to minimise the overall costs of their transport activities.
- Within the presented model, individual transport costs for market actors are independent from each other. Hence, the choice of a distinct path and mode has no impact on the path and mode choice of another actor, apart from fixed components incorporated within the transport costing scheme[204].

Although not all individual cost components for the diversity of transport activities are at hand, certain major factors can be identified. Having the minimum cost principle in mind, two different *combined modal split and network assignment* strategies are conceivable:

- In case of a given PC dataset and unknown mode specific OD data, a *top down* approach, based upon a cost minimizing mode choice and network assignment strategy, will lead to a commodity (path) flow allocation which partly allows one to identify the missing mode specific OD information.
- For another case, where there are complete mode specific OD data at hand and the PC-relations are unknown, the missing information is for some parts derivable from a *bottom-up* interpretation of the network flow allocation that, in turn, results from a transport flow conservation principle.

15.1.2. Examples for combination strategy

Both assignment and mode choice procedures are applicable for an exemplary one-product transport scenario of two sites of production, two sites of

[204] See e.g. discussion of return loads or the impact of traffic congestions and time delays due to similar path choices of various actors that are incorporated to a large extent within the costing structure in section 14. An additional interrelationship is not modelled since a gain of information with respect to the additional computational efforts is assumed to be relatively low.

consumption and two intermodal transfer nodes. This exemplary setting of nodes is connected via seven directed links that allow for freight transports by two modes, one direct road freight and one combined transport option, with e.g. a rail freight main haul.

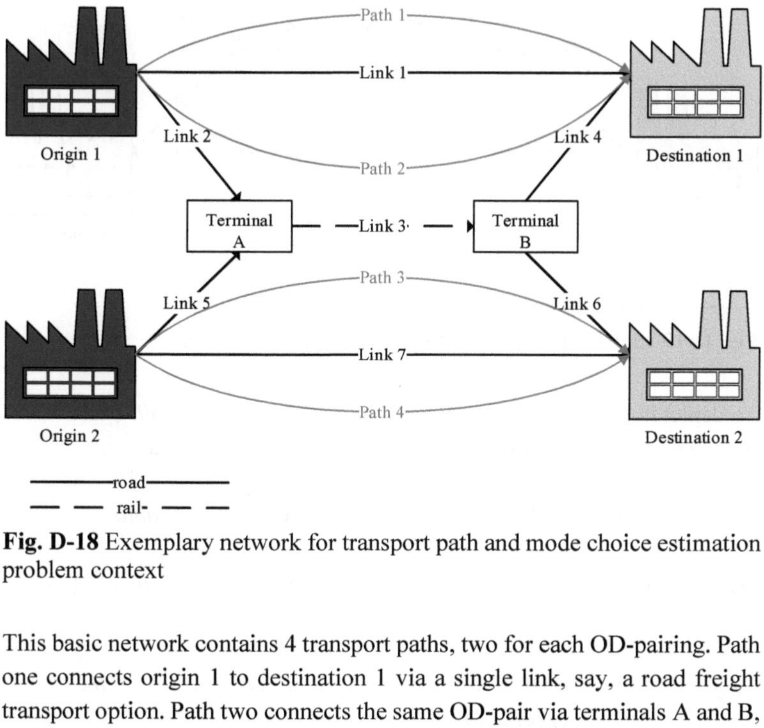

Fig. D-18 Exemplary network for transport path and mode choice estimation problem context

This basic network contains 4 transport paths, two for each OD-pairing. Path one connects origin 1 to destination 1 via a single link, say, a road freight transport option. Path two connects the same OD-pair via terminals A and B, along links 2, 3 and 4. Here, links 2 and 4 represent a pre- and post-haul by road, link 3 an exemplary rail freight main-haul. An identical configuration is set up for origin 2 and destination 2 with the specification that link 3 is part of both path sets[205]. The related exemplary transport activity cost factors for each link are indicated in squared brackets within the following setting:

[205] The set of paths from origins to destinations via network nodes along a set of links is predefined by an OD-pattern. That is why, for the given example, transports from origin 1 to destination 2 as well as transports from origin 2 to destination 1 are not evaluated.

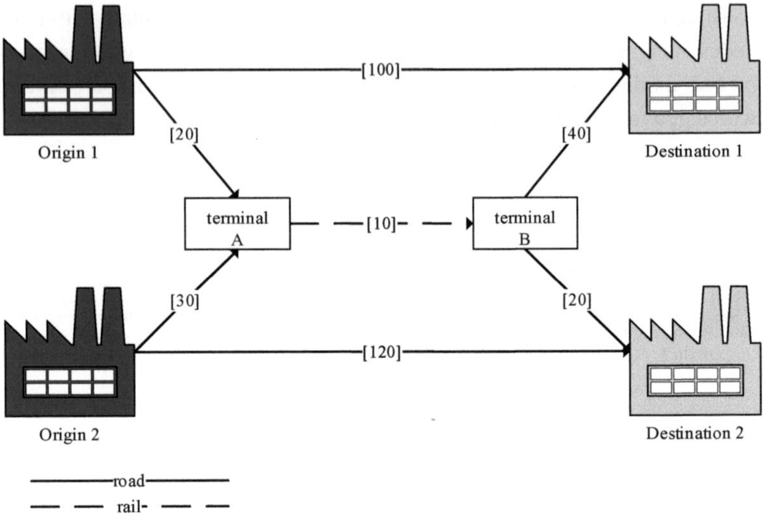

Fig. D-19 Link costs for exemplary network configuration

To depict both aforementioned different *combined modal split and network assignment* principles, the following notations are introduced:

$\hat{g}_{od}$ observed OD transport demand as *PC trade flow data*

$\hat{v}_a$ observed flow volume on a network link a as *mode specific counts of transport volumes per link*

g_{od} estimated OD transport demand as *one specification of the output OD transport data*

v_a estimated flow volume on link a as *another specification of the output OD transport data*

$c_a(\cdot)$ link cost function for transport service of v_a

Along with this exemplary network, five different scenarios are discussed to depict variant input data combination principles in more detail.

Scenario 1 - $\hat{g}_{od}$ at hand and $\hat{v}_a$ unknown:

For a first assignment procedure, a transport demand $\hat{g}_{od}$ of 100 tonnes from origin 1 to destination 1 and 200 tonnes from origin 2 to destination 2 is assumed. The lowest overall transport cost assignment procedure then results in a choice of path 2 along links 2, 3 and 4 for OD-pairing 1 as well as a choice of path 3 along links 5, 3 and 6 for OD-pairing 2. Subsequently, a transport volume of 100 tonnes is assigned to links 2 and 4. A volume of 200 tonnes is assigned to links 5 and 6. And a total of 300 tonnes is assigned to link 3. This scenario is resumed as follows[206]:

Input:

$\hat{g}_{od}$		$\hat{v}_a$	
od-pair	$\hat{g}$	link *a*	$\hat{v}$
11	100	1	n. v.
22	200	2	n. v.
		3	n. v.
		4	n. v.
		5	n. v.
		6	n. v.
		7	n. v.

Path choice:

f_r	
path *r*	f
1	0
2	100
3	200
4	0

Output:

g_{od}		v_a	
od-pair	g	link *a*	v
11	100	1	0
22	200	2	100
		3	300
		4	100
		5	200
		6	200
		7	0

Table D-24 An exemplary modal split and network assignment for scenario 1

The selected output within this table reflects a path choice that meets the transport demand in line with a *top down* strategy, based upon a cost minimizing mode choice and network assignment.

Scenario 2 - $\hat{g}_{od}$ unknown and $\hat{v}_a$ at hand:

A second transport network assignment example results from at hand link flow counts for modes road and rail. This input is constituted by flow counts of 100 tonnes in transport on link 1 as well as reports for 200 tonnes on links 3, 5 and 6. From this starting point an assignment that results from a flow conservation principle leads to an identical result as for scenario 1 where path 2 is identified to meet a transport demand within OD-pair 1. And the flow

206 For entries of the *PC trade flow data* as well as the *mode specific counts of transport volumes per link* within Table D-24 to Table D-28 either the result of an observation is indicated or no observation took place – hence, no value (n. v.) is available.

volume on path 3 is chosen to fulfil a transport demand of destination 2 from origin 2. The following table depicts these results:

Input:

$\hat{g}_{od}$		$\hat{v}_a$	
od-pair	$\hat{g}$	link *a*	$\hat{v}$
11	n. v.	1	0
22	n. v.	2	100
		3	300
		4	100
		5	200
		6	200
		7	0

Path choice:

f_r	
path *r*	f
1	0
2	100
3	200
4	0

Output:

g_{od}		v_a	
od-pair	*g*	*link a*	*v*
11	100	1	0
22	200	2	100
		3	300
		4	100
		5	200
		6	200
		7	0

Table D-25 An exemplary modal split and network assignment for scenario 2

The aim of this second exemplary scenario is to present an alternative network assignment and mode choice procedure that, although dealing with a different input data, may complement the setting of scenario 1 and vice versa. On the one hand, the dataset of an existing trade flow from an origin of production to a destination of consumption does not contain indications on the mode specific realisation along a given network structure. On the other hand, information on mode specific transport relations do not per se contain the information on an overall origin of production or a destination of consumption, e.g. for a transport network including intermodal transfers.
However, given the assumption that transport market actors seek to minimise their transport costs, both approaches complement each other ideally to outcome unknown mode and path choice specifications within a target OD transport dataset.

Scenario 3 - $\hat{g}_{od}$ (incomplete) at hand and $\hat{v}_a$ (incomplete) at hand, part 1:

Apart from an ability to deal with partly incomplete information on either PC trade data ($\hat{g}_{od}$) or mode specific OD transport data ($\hat{v}_a$), the merged application of both assignment procedures additionally offers the opportunity to cope with conflicting inputs.
An example of a joint assignment strategy under incomplete and inconsistent information is depicted for a given $\hat{g}_{11} = 100$ tonnes as well as a transport demand of 200 tonnes from origin 2 to destination 2. It is complemented by

fragmentary link flow counts for a combined main-haul transports of $\hat{v}_3 = 500$ tonnes. This setup is depicted as follows:

Input:

$\hat{g}_{od}$ *od*-pair	$\hat{g}$	$\hat{v}_a$ link *a*	$\hat{v}$
11	100	1	n. v.
22	200	2	n. v.
		3	500
		4	n. v.
		5	n. v.
		6	n. v.
		7	n. v.

Path choice:

f_r path *r*	f
1	0
2	166.67
3	266.67
4	0

Output:

g_{od} *od*-pair	g	v_a *link a*	v
11	166.7	1	0
22	266.7	2	166.7
		3	433.3
		4	166.7
		5	266.7
		6	266.7
		7	0

Table D-26 An exemplary modal split and network assignment for scenario 3

Here, either the link counts per mode and/or the PC trade data are incorrect. Such exemplary input configuration is of high relevance for the presented modelling framework. Therein, a data conflict between $\hat{g}_{od}$ and $\hat{v}_a$ is most likely the result of incorrect PC trade data, as they are 'only' a derivative of various statistical reports.

The presented data conflict is handled in a way that the estimated OD transport demand g_{od} and the estimated link flow volume v_a have the smallest offset to their corresponding inputs of $\hat{g}_{od}$ and $\hat{v}_a$. As a result, the output OD transport dataset of the given example considers a transport of about 166 instead of 100 tonnes from origin 1 to destination 1 as well as 266 instead of 200 tonnes from origin 2 to destination 2. The flow volume on link 3 is adapted from 500 tonnes reported, to around 433 tonnes estimated. Thus, scenario 3 refers to a basic data fitting strategy as the result of conflicting input data.

Scenario 4 - $\hat{g}_{od}$ (incomplete) at hand and $\hat{v}_a$ (incomplete) at hand, part 2:

Another conflict which affects the assumed cost minimising principle for the given setup is likely to occur. For *real-world* transport scenarios it may be the result of e.g. favouring a direct road transport instead of lower costs for a combined transport path. This circumstance is interpretable as the result of additional choice-related factors, such as a specific value of time for transports in addition to the inventory-based value of time evaluation, insufficient information on options for modal alternative, the approved road freight practice etc.

Within the given exemplary network, this data conflict will appear for instance for a transport demand of 200 tonnes from origin 2 to destination 2 and no observation for transports along link 3, as presented in the following table:

Input:

$\hat{g}_{od}$		$\hat{v}_a$	
od-pair	$\hat{g}$	link *a*	$\hat{v}$
11	0	1	n. v.
22	200	2	n. v.
		3	0
		4	n. v.
		5	n. v.
		6	n. v.
		7	n. v.

Path choice:

f_r	
path *r*	f
1	0
2	0
3	0
4	200

Output:

g_{od}		v_a	
od-pair	*g*	link *a*	*v*
11	0	1	0.0
22	200	2	0.0
		3	0.0
		4	0.0
		5	0.0
		6	0.0
		7	200.0

Table D-27 An exemplary modal split and network assignment for scenario 4

The conflict of input data that do not comply with the assumed cost minimising principle is solved with a distinct preference for a data fitting. Although transport via path 3 would lead to lower transport costs for OD-pair 2, the observation suggests that this path is not chosen in practice. As a result, 200 tonnes will be assigned to link 7, the direct road freight link. Then the estimated OD transport demand $g_{22} = 200$ will be equal to $\hat{g}_{22}$ and the estimated flow volume $v_3 = 0$ will be equal to $\hat{v}_3$.

Scenario 5 - $\hat{g}_{od}$ (incomplete) at hand and $\hat{v}_a$ (incomplete) at hand, part 3:

A third data conflict that requires a distinct solution strategy results from the fact that a basic data fitting will be not sufficient to model the German transport system adequately. A fitting, as presented within Scenario 3 and 4, aims at estimating the smallest offset of the target multimodal OD transport data and the inputs of $\hat{g}_{od}$ and $\hat{v}_a$. This differs from a scenario for the exemplary network with a given transport demand $\hat{g}_{11} = 100$ tonnes and $\hat{g}_{22} = 100$ tonnes as well as link flow counts $\hat{v}_3 = 100$ tonnes. Therefore, multiple outcomes that need to be evaluated are possible. Two of them are selected for the following depiction:

Input:

$\hat{g}_{od}$

od-pair	$\hat{g}$
11	100
22	200

$\hat{v}_a$

link a	$\hat{v}$
1	n. v.
2	n. v.
3	100
4	n. v.
5	n. v.
6	n. v.
7	n. v.

Path choice A:

f_r

path r	f
1	0
2	100
3	0
4	200

Output A:

g_{od}

od-pair	g
11	100
22	200

v_a

link a	v
1	0
2	100
3	100
4	100
5	0
6	0
7	200

Path choice B:

f_r

path r	f
1	100
2	0
3	100
4	100

Output B:

g_{od}

od-pair	g
11	100
22	200

v_a

link a	v
1	100
2	0
3	100
4	0
5	100
6	100
7	100

Table D-28 An exemplary modal split and network assignment for scenario 5

Case A of this scenario can be described as follows:

- An intermodal transport path is selected to meet $\hat{g}_{11}$ at minimum transport costs. For OD-pair 2 a direct road freight transport is chosen. As a result, the estimated flow volume $v_3 = 100$. Together with $g_{11} = 100$ and $g_{22} = 200$ the given input and estimated output data are ideally fitted to each other.

However, the best-feasible representation of transport market actors aiming at minimising their transport costs would result in a different network path and mode assignment of case B:

- This alternative solution involves a flow volume of 100 tonnes on path 1 as well as 100 tonnes for both path 3 and path 4. This network assignment leads to slightly higher costs for a transport between origin

1 and destination 1 and likewise significantly lower costs in order to meet $\hat{g}_{22}$ and as a result, to lower overall transport costs.

In other words, it is considered to be more likely that a *real-world* decision towards transport paths with higher transport costs is made when the divergence to alternatives of lower transport costs is comparatively small[207].

To obtain similar solutions for the presented national network model, the structure of a minimum-cost commodity flow network assignment and mode choice process is crucial within the context of a data fitting.

15.2. Formalisation of the transport flow estimation

The goal of this section is to present a structure for estimating the unknown multimodal commodity flow OD data, together with originally missing road freight OD information, that considers a given PC pattern as well as flow counts for two out of three modelled inland modes, the transports by train and IWW in Germany. In total, this process is based upon the following inputs:

- PC data on trade relations between origins of production and destinations of consumption,
- rail link flow counts,
    - containerised and non-containerised,
- IWW link flow counts,
    - Containerised and non-containerised,

to result into the subsequent outputs:

- estimated multimodal commodity OD flow data for transports via distinct transport paths and modes,
    - containerised and non-containerised,
- estimated truck link flows,
    - Containerised and non-containerised.

[207] Note that within the presented framework such a specific case is not only relevant for combined transports but also for an allocation of direct rail and direct IWW transports to a given transport demand. Here, the specific establishment as an origin of production or a destination of consumption within a NUTS-3 region for link flow observations cannot be identified (cf. section 14).

In order to obtain the desired outcome, a complementing transport network needs to be modelled. As a result, not only the unknown trucking OD matrix is identified, but also commodity classes that are presumably transported in containers along intermodal transport chains. Likewise initial origins and final destinations as well as potential locations of intermediate transfers for all modelled transports are estimated. The key for such a multimodal freight network estimation is an OD matrix model.

One of the early efforts in freight OD matrix estimation modelling[208] was carried out by LIST AND TURNQUIST (1994). They introduced a concept to estimate multi-class truck OD matrix from fragmentary observations and additional sources of information in a large-scale linear optimisation problem which was applied in a case study for New York City. CRAINIC ET AL. (2001) addressed the problem in a multimodal, multi-commodity context but they did not present a practical implementation. AL-BATTAINEH AND KAYSI (2007) introduced a genetic algorithm[209] to solve this type of transport flow configuration problem. Their framework incorporates input-output calculation data and a truck survey as inputs to estimate a truck movement OD matrix for a case study in the province of Ontario, Canada. The model is built on the problem formulation of LIST ET AL. (2002) and solved by a gradient-based optimisation technique.
PATTANAMEKAR ET AL. (2014) formulated a model for a multi-modal OD estimation for an extract of the Korean transport network and applied a GA as well. Instead of using a random population as initialisation, as e.g. in the study of AL-BATTAINEH AND KAYSI (2007), they applied a conditional gradient descent method to generate a seed for a genetic algorithm based solution[210]. Despite the well explored mathematical framework for linear as well as nonlinear network assignment conditions, neither a multimodal application that explicitly incorporates multiple commodity classes and their individual transport cost components, nor a large scale concept for a complete set of detailed transport network elements can be found in the literature. That

[208] Earlier work on OD matrix estimation/ adjustment modelling predominantly dedicated to passenger transport, e.g. in SPIESS (1990), CASCETTA AND NGUYEN (1988) and VAN ZUYLEN AND WILLUMSEN (1980) has been carried out.
[209] See also section 9.1 for an exemplary description of this search heuristic.
[210] A stepwise presentation of the gradient descent method, applied within the problem context, is given e.g. in NORIEGA AND FLORIAN (2009).

is why a context specific model formulation and solution search is required to estimate a transport flow configuration for the given context.

Modell formulation

The overall objective of the presented multimodal OD matrix estimation problem is to minimize the error between initial input values – here, a given PC structure as well as rail and IWW link flow counts[211] – in order to estimate output values for a resulting OD dataset at best. An additional sub-target is to select the transport paths for related PC and link flow data according to a *minimum transport cost path selection principle*. Thus, for PC and link flow inputs that have a corresponding counterpart for a data fitting, a path flow alternative that minimises the overall transport costs within the network is preferable.
In this context, the observed OD-counts indicate entire or sequences of commodity flows for a multimodal network that is denoted as $G = (N, A)$ with a set of directed links $a \in A$ and a set of related nodes $n \in N$ that are either origins $o \in O$, destinations $d \in D$ or transfer nodes $t \in T$. Therefore, a volume $\hat{g}_{od}^{p}$ of commodity $p \in P$ that is to be transported from an origin $o \in O$ to a destination $d \in D$ forms a cell of the given PC matrix $\hat{g}_{OD}^{p}$. Link flow counts with a specification of commodity $p \in P$ per mode $m = 1\ (road), 2\ (rail), 3\ (IWW), 4\ (road_{container}), 5\ (rail_{container}), 6\ (IWW_{container})$ are denoted as $\hat{v}_{a}^{pm}$. Additionally, for all links a mode and commodity specific cost function $c_{a}^{pm}(\cdot)$ is assigned – a function that incorporates line-haul and potential transfer costs.
This setting is consolidated with an overall preference on data-fitting – hence, a *combined modal split and network assignment* that is, first of all, as close as possible to the observed and determined input data. In a subordinate instance the cost-minimisation principle for selection among routing alternatives applies. Accordingly, a notion of the context-specific multimodal multi-commodity OD matrix estimation has the format of a bi-level linear optimisation problem as follows[212]:

---

[211] To take adequate account of the fact that the PC calculation itself is rather an estimation compared to flow counts, weighting elements are introduced in the modell for possible future adaptions.
[212] In case of a network assignment with link costs that are not independent of the capacity usage, a non-linear program is required, as discussed e.g. within CRAINIC ET AL. (2001), NORIEGA AND FLORIAN (2009) and PATTANAMEKAR ET AL. (2014). This refers to a consideration of e.g. traffic jams on a specific link, rendering the respective link more or less *attractive*. It can be justified that a

$$Min\ Z(v) = \sum_{p\in P}\sum_{a\in A} v_a^p * c_a^p(v) \quad (41)$$

where v_a^p is the result of:

$$Min\ F(g,v) = \lambda \sum_{p\in P}\sum_{a\in A}(\hat{v}_a^p - v_a^p)^2 + (1-\lambda)\sum_{p\in P}\sum_{o\in O}\sum_{d\in D}\left(\hat{g}_{od}^p - g_{od}^p\right)^2 \quad (42)$$

subject to:

$$v_a^p = \sum_{p\in P}\sum_{a\in A}\sum_{r\in R}\delta_{ar}^p * f_r^p \quad (43)$$

$$\sum_{r\in R} f_r^p = g_{od}^p \quad (44)$$

$$\delta_{ar}^p = \begin{cases} 1, & if\ a \in r \\ 0, & otherwise \end{cases} \quad (45)$$

with:

$$\hat{g}_{od}^p \geq 0,\ \hat{v}_a^p \geq 0, f_r^p \geq 0 \quad (46)$$

for all:

$$a \in A, \quad o \in O, \quad d \in D, \quad p \in P, \quad r \in R.$$

Therein λ is a weighting parameter and R represents a set of paths from origin o to destination d. A flow on path r is denoted f_r and $\delta_{a,r}^p$ is a Kronecker delta to consider the fact that a link flow either contributes to a path flow and vice versa or not. The notations used are given in the following overview.

---

freight transport model includes link performance evaluations. However, not only the extensive computational effort of finding a solution within the resulting non-linear optimisation problem for the presented context precludes this approach but also the absence of reliable indications on individual link performance evaluation parameters that could justify the additional efforts.

Network:

N - set of network nodes, with $N = \{O \cup T \cup D\}$ for a set of transfer nodes $t \in T$, destinations $d \in D$ and origins $o \in O$

A - set of links between network nodes $n \in N$ for transport modes $m =$ $1\ (road), 2\ (rail), 3\ (IWW), 4\ (road_{container}), 5\ (rail_{container}),$ $6\ (IWW_{container})$ with $A = A_1 \cup A_2 \cup A_3 \cup A_4 \cup A_5 \cup A_6$

P - set of products, with $p = 1\ (CPA01),$ $2\ (CPA02), \dots, 31\ (CPA32)$

R - set of paths from origin o to destination d via network nodes n along links a

Input variables:

$\hat{g}_{od}^{p}$ - OD transport demand per commodity class p

$\hat{v}_{a}^{p}$ - link flow counts for mode m per commodity class p

Output variables:

g_{od}^{p} - estimated OD transport for mode m per commodity class p

v_{a}^{p} - estimated link flow for mode m per commodity class p

Decision variables

f_{r}^{p} - commodity flow of product p on path r from origin o to destination d

Additional variables

$c_a^p(\cdot)$ - cost function related to transports of v_a^p, including potential transfer activity costs

λ - weighting parameter with values within the range of $\{0,1\}$

δ_{ar}^p - binary variable with 1 if link a is part of path r and 0, otherwise

Problem solving

The proposed formal model needs to be executed for about 43 million modelled firm-to-firm trade relations of the overall German supply chain network. These, in turn, are connected via at least one of three direct transport modes as well as a very large set of potential combinations for intermodal transports (cf. section 14).
To handle this complex problem, a decomposition is proposed to handle instances for each commodity class p seperately. Nevertheless, the computational burden remains very high for this particular setting. That is why the following simplification applies:

- The goal of the presented problem solving algorithm is not to find the most accurate solution according to the objective function in equation (41), but at least a feasible one.

This attempt assures that the overall objective of the given multimodal OD matrix estimation problem – to estimate output values with a minimum deviation to input values at hand – is reasonably approached. It furthermore leads to an OD data estimation that, in case of no data fitting options, assigns the transport path alternative of the lowest comparative costs to a given transport demand or observation. However, the overall transport network cost optimum, as e.g. discussed exemplarily in Scenario 5 of section 15.1.2, will presumably not be met. As a result, a transport by e.g. mode rail is assigned to a distinct firm-to-firm pair although it relates to another one[213].

[213] This refers to the consideration that it is more likely to observe a real-world decision towards transport paths with higher transport costs when the divergence to alternatives of lower transport costs is comparatively small according to section 15.1.2.

A strategy to estimate the magnitude of this effect and likewise achieve a more accurate result for the objective function in equation (41) is the application of an iterative data processing and a stepwise solution evaluation, as e.g. a genetic algorithm presented in AL-BATTAINEH AND KAYSI (2007) and PATTANAMEKAR ET AL. (2014). This, however, would result in an unprecedented large computational burden for the presented context.
In line with the suggested decomposition of the problem into distinct instances, the solution algorithm is devided into the following two steps:

- First, for a set of commodities p, all related $\hat{v}_a^p$ are determined according to a data fitting with given observations v_a^p and $\hat{g}_{od}^p$ by g_{od}^p, respectively, in line with equation (42) and the corresponding constraints.
- Subsequently, the remaining given transport demand $\hat{g}_{od}^p$ as well as the given observations v_a^p for the selected set p are assigned to a path r at minimum transport costs, in line with equation (41) and the related constraints.

A simplified structure of the proposed search for a feasible solution to the OD data estimation problem is given in Fig. D-20.

The set of feasible paths, including each related modal alternative, that relates to a processing in line with equation (45) as well as the cost functions for $c_a^{pm}(v)$ are separately determined, as presented in section 14.

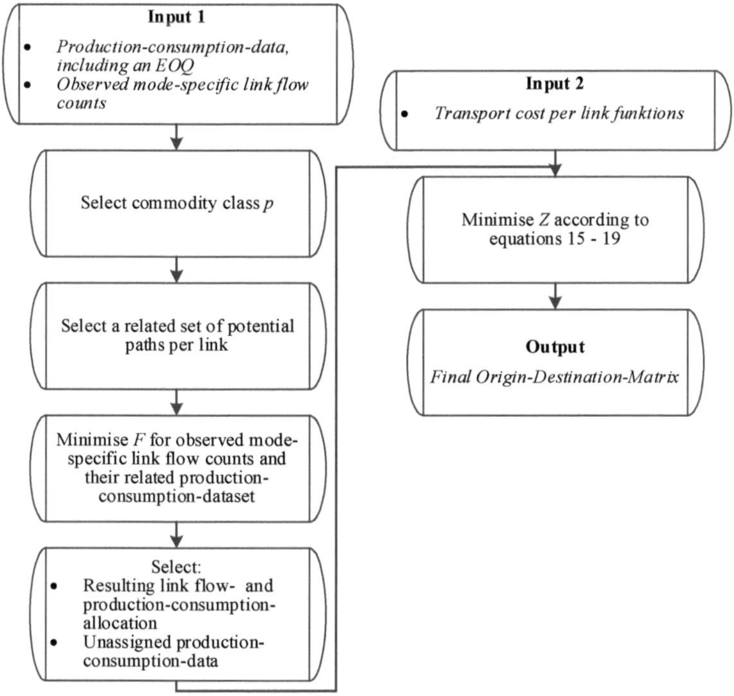

Fig. D-20 Simplified computational two-stage structure for a problem solving of the given OD data estimation problem

15.3. Execution within the overall model's context

With link specific transport costs per mode and commodity class at hand, complemented by a link-to-path assignment dataset for each modelled OD trade relation, the required inputs for the *combined modal split and network assignment* module are complete. Subsequently, the module is realised for each modelled OD trade relation in consideration of up to 3 direct transport alternatives as well as potentially 8 different intermodal transport paths. An overview on these transport path alternatives is given as follows:

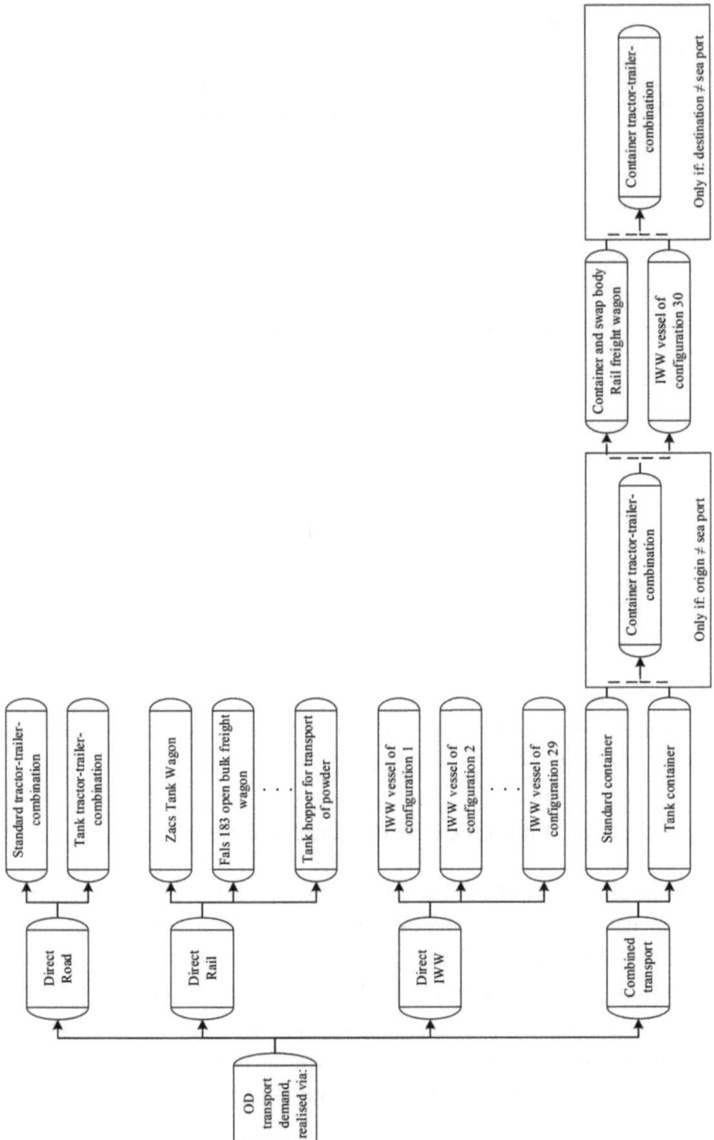

Fig. D-21 Mode and path choice alternatives for an exemplary transport setting

A cost rate as the result of the modal and path choice specifications is then assigned to each commodity specific OD pair. This includes a set of inputs that are:

- a set of modelled firms with a location and economic activity based on labour statistics,
- a level of production by weight and volume for producing firms that relates to the predefined economic activity based on *production statistics*,
- a set of additional sources for imports based on *import* and *production statistics*,
- a level of consumption for each modelled firm, derived from national input output distributions in line with the related economic activity and
- a set of additional sinks for exports that results from national input output accountings and export statistics.

This initial configuration is the outcome of the *freight generation* module. It results in an allocation of producing firms and synthetic sources of production and – at first only loosely interlinked – consuming firms and synthetic sinks. Further inputs are:

- a commodity specific criteria set for an evaluation of supply chain partnerships and
- an average shipment size for each individual trade relation.

This subsequent result of input data transformations relates to a *freight distribution*, specified as trade flows. These trade flows relations between origins and destinations are complemented by:

- a set of alternative transport paths in the format of link combinations for multimodal modes that are feasible of each trade relation,
- costs for each transport path and relevant loading and/or unloading as well as transhipment activities and
- an incomplete set of mode specific link flow counts from statistics.

The latter inputs are required to derive the overall *modal split and network assignment* distribution, together with results of the preceding *freight distribution* module.

As a result, a dataset of individual firm-to-firm transport flows is determined. It contains a two-digit CPA commodity class specification for a three-digit NST spatial resolution. These transport flows are furthermore specified by the mode of transport as well as a corresponding OD network path.

15.4. Calibration

A conclusion on the quality of the outcome of the *combined modal split and network assignment* module – hence, the overall model's outcome – is still pending. Results of the *freight generation* module, measured in a total annual transport volume in terms of tonnes, are already discussed in section 13.5. For the subsequent *freight distribution* and *combined modal split and network assignment* module an isolated validation is not feasible. Both the respective trade flow as well as the final freight flow assignment are correlated. Moreover:

- a validation of the *freight distribution* and the *combined modal split and network assignment* results will not be feasible at all.
- There is no such dataset available that allows one to adequately reference the modelled entirety of German freight transport activities.
- Instead, a calibration is implemented for the last two modelling stages.

References for a calibration of the *combined modal split and network assignment* module's results will be the total reported transport distance in Germany for 2012, measured in tonne-kilometres as given in Table G-8. For an additional validation there will be no further reference data available that is both sufficiently encompassing and likewise detailed to mirror the model's overall outcome.

The total transport volume per mode and commodity class in terms of tonne-kilometres, as given in Table G-8, will be used to reference the overall result and likewise help to adjust the trade flow assignment input, as discussed within sections 13.1 to 13.6. In this regard, the input for the distance-related supply chain synthesis evaluation criteria is questioned – a variable that results from an interpretation of the average straight-line[214] distance between potentially interlinked firms (cf. section 13.2). Unfortunately, no adequate

[214] Note that the average distance travelled by mode in terms of tkm is not congruent to the average distance from suppliers and consumers, but correlated.

counterpart for the size-related supply chain synthesis evaluation criteria is identified. As a result:

- The calibration of the overall model's result will only be performed for one out of two elementary input variables for the chosen firm-to-firm trade flow distribution modelling section.

A pre-testing on the distance-related input parameter magnitude reveals that the initial setup leads to higher overall transport distances, compared to references in Table G-8. Instead, a calibrated *global average trade flow distance adjustment parameter* mostly ranges below 50% of the initial value[215]. Therefore, three test runs of the model are performed, starting with a distance-related supplier performance measure that is lowered by 50% and further. An overview on the corresponding test results is given in the following table:

[215] As a reminder, the initial input for a distance-related comparative supplier evaluation is not a *fixed value* but a reference unit to separate either more local or rather scattered producer-consumer pairings.

CPA	NST	Transport volume in mill. of tkm for NST-01 to NST-13 *total reported*	Determined transport volume in millions of tkm for CPA-01 to CPA-32 grouped by target value for initial sourcing-distance parameter a) 50% *total*	*dev.*	b) 40% *total*	*dev.*	c) 35% *total*	*dev.*	*min{|a|,|b|,|c|}* *dev.*
1 2 3	1	39,492	46,915 7,420 120	38%	41,504 6,318 111	21%	38,684 5,796 106	13%	13%
5 6	2	15,149	73,362 0	384%	71,340 0	371%	70,712 0	367%	367%
7 8	3	59,231	19,997 85,309	78%	19,493 74,349	58%	19,502 67,817	47%	47%
10 11 12	4	68,222	35,684 9,959 378	-33%	22 8,585 369	-87%	27,223 7,888 366	-48%	-33%
13 14 15	5	3,865	614 246 67	-76%	526 233 60	-79%	477 222 57	-80%	-76%
16 17 18	6	42,729	13,093 18,618 341	-25%	11,338 16,608 278	-34%	10,107 15,189 242	-40%	-25%
19	7	28,948	50,235	74%	45,331	57%	43,714	51%	51%
20 21 22	8	55,585	42,882 2,745 4,739	-9%	37,492 2,735 4,012	-20%	242 2,728 3,551	-88%	-9%
23	9	39,487	42,698	8%	36,331	-8%	32,223	-18%	8%
24 25	10	54,161	34,914 7,185	-22%	32,050 6,105	-30%	30,601 5,396	-34%	-22%
26 27 28	11	16,697	642 2,113 6,282	-46%	531 1,866 5,676	-52%	493 1,706 5,182	-56%	-46%
29 30	12	27,287	13,328 2,759	-41%	12,125 2,538	-46%	11,674 2,364	-49%	-41%
31 32	13	7,256	1,396 386	-75%	1,190 350	-79%	1,071 306	-81%	-75%
TOTAL		458,107	524,426	14%	439,468	-4%	405,641	-11%	-4%
Weighted absolute deviation				48%		55%		57%	40%
Weighted absolute deviation for limited scope a)				37%		44%		46%	29%
Weighted absolute deviation for limited scope b)				30%		41%		46%	26%

a) excluding CPA-05 and CPA-06 that relate to NST-02
b) excluding CPA-05 and CPA-06 that relate to NST-02 as well as CPA-07 and CPA-08 that relate to NST-03

Table D-29 Comparison of total transport volume reported to calculated counterparts for Germany 2012 in millions of tonne-kilometres [incl. data from Table G-8]

Deploying the feedback loop

Each target value adaption for the distance-related sourcing criteria represents a complete iteration within the model from the second stage of the freight distribution module – referring to the feedback loop as indicated in Fig. C-1. The result of these iterations is a global recalibrated target distance of about 40% of the overall weighted average transport distance between all eligible firm-to-firm pairs. This outcome best reflects the reference total transport volume indications for Germany, as given in Table G-8[216]. It may be argued accordingly that:

- the present state of the disaggregate freight transport model for Germany is best configured with a *global average trade flow distance adjustment parameter* of 40% of the overall average straight-line distance within a pre-selected trade flow network for each modelled commodity class,
- the model's performance is further increased with commodity class specific adjustment parameters for the trade flow distance – hence, each set of CPA classified commodity relating to a NST classification heading shall be *individually adjusted* for further elaborations of the model[217].

An *individual trade flow distance adjustment parameters* leads to a weighted absolute deviation converging to zero. The gain from this computational effort will be further increased with additional data inputs and corresponding model advancements as e.g. suggested in section 16.1.

[216] Note that a validation of the first stage of the *freight distribution* module refers to an estimation of the total annual freight transport demand in terms of tonnes (cf. Table G-20). Thereafter, the total transport demand for *metal ores* (CPA 07) and *other mining and quarrying products* (CPA 08) is not adequately represented within the model. These commodities, however, refer to outstandingly short average transport distances (compare related entries in Table A-3 and Table G-8 and see also discussion in 13.5). That gives reason to the selected level of calibration of the model's total transport volume in terms of tonne-kilometres (see impact on selective total in Table D-29). A further precession within the iteration processing requires additional input data, as suggested within the following section.

[217] Cf. the level of deviation from a *global average trade flow distance adjustment parameter* with a configuration of *individual adjustments* per NST group within the last column of Table D-29.

E Résumé

As a result of the model's implementation and calibration, several interpretations of the overall outcome as well as suggestions for future developments are conceivable.

16. Interpretation of the model

A calibration of the distance- as well as the size-related modelling parameters is crucial. But assembling the required referencing units is difficult to perform. Furthermore, a weighting between these two input parameters will be required. However, until now, neither an adequate nation-wide and likewise detailed analysis of supply chain configurations in practice nor, at least, firm size-related supply chain decision criteria are at hand[218]. That is why the presented model's overall result can be calibrated only partially.

Nevertheless, the chosen framework allows to limit the level of uncertainty. The first out of three modelling stages, the *freight generation* procedure, is based on statistically evaluated input data. Inherent data gaps are, to a large extent, treated with a deterministic data retrieval processing. The impact of estimates on the subsequent supply chain configuration and the overall model's result are likewise limited. The concept of including additional statistically evaluated inputs – here, link flow counts for two out of three modal alternatives – within the final output data generation enables a further enhancement of the respective data output quality.

As a result, the model generates a detailed nation-wide transport freight flow configuration that, on the one hand, results from an appropriately validated *firm generation* and *supply chain synthesis* and, on the other hand, is the outcome of a *trade flow generation* as well as a *combined modal split and network assignment* procedure that adequately reflects a set of specified network link flow observations[219].

---

[218] Cf. discussions ensuing section 13.1.

[219] See discussion on statistical data collection in section 2 and the resulting reasoning for selecting the level of detail per specification by mode, region as well as commodity class as well as the referenced model structure in section 10.

Nevertheless, it is indispensable to further refine the model to prove useful for operational, strategic and tactical freight transport organisation measures.

16.1. Strategies for future developments

First of all, the model can be recalibrated for individual commodity classes according to the findings of section 13.5 and 15.4. A potential starting point could be an introduction of *commodity class specific correction factors*, resulting from additional analysis and/or observations, such as for:

- a breakdown of the total annual production and import volume of *coal and lignite* (CPA-05) to a freight transport demand for the modelled set of modes
- and an estimation of the total annual transport demand for *other mining and quarrying products* (CPA-08) as an annual outcome of all *mining, quarrying and earthmoving* activities that relate not only to firms of CPA-08 but likewise to firm of e.g. *buildings and building construction works* (CPA-41) and *constructions and construction works for civil engineering* (CPA-42)[220].

Such fundamental future development will be significantly more effective with a complementing improvement of the data quality for a disaggregate German freight transport model that, in turn, can be achieved in at least two different ways:

- either collecting additional link flow counts for mode road or
- compiling further details on supply chain configuration decisions.

The first option leads to an increase of the counterbalance within the process of minimising the offset between estimated and observed data as part of the proposed *combined modal split and network assignment* module. As a result, potential caveats within the preceding *freight distribution* module can be identified and iteratively resolved. A potential data source would be:

[220] An implementation of *commodity class specific correction factors* within the model can be subsumed for instance under measure 11, as indicated in Table E-1.

- the use of unpublished German toll statistics data with a sufficient level of detail that result from a road freight infrastructure user charges for distinct network segments of particular road freight lorries, covering e.g. the exemplary evaluated tractor trailer combinations along motorways[221] and/or
- an implementation of commercial freight exchange data that – although not adequate as a transparent high quality input data allowing for future reproductions of the model – may serve as additional calibration or even validation references[222].

The second indicative trajectory leads to a different calibration approach, since a large set of input factors for the highly individual supply chain configuration decisions will generally be required. Bearing in mind that a representative coverage is the overall target number of observations that will not be within reach in a limited time frame with reasonable efforts, a model comparable to the presented *combined modal split and network assignment* framework is suggested. With a number of references for trade flow observations, relevant inadequacies might be identified within the overall framework of the presented disaggregate German freight transport model.
A first step towards an evaluation of sectoral specific trade flow patterns is performed e.g. by:

- SCHUH AND STICH (2013) as well as SCHIEGG (2005), who aimed at developing a typology and formal description of sourcing and distribution networks within the German industry, based on interviews.

This approach needs to be further refined to lead to sufficiently comprehensive and likewise detailed results for a certain time span. It may be argued that gathering such dataset on individual trade flow patterns is an extremely burdensome task. That is why another option to compile further details on supply chain configuration decisions must be conceived. This contains:

---

[221] For further details see e.g. BAG (2015) and BAG (2013b) as well as discussion on national statistical publications in section 12.1 and 12.2. One way to make these confidential data accessible to scientific evaluations, would be given by the well-defined legislative framework of the German *Research Data Centres of the Federal Statistical Office*, as presented in ZWICK (2007) and FDZ (2016).
[222] For operators of commercial freight exchanges, incentives to gain access to their transport service request log files could be provided through an exchange with the presented model's *freight distribution* outcome – a dataset that could serve to prioritise particular freight exchange demands and offers within an overall national context.

- a shift of the present effort on surveying road freight forwarders towards a survey of the transport demand side. They will be questioned on their supply chain network configuration, focussing on the commodity specific number of suppliers, size relations, distances etc. The result of this survey may not only be valuable for the context of freight transport modelling but also for other research fields, such as e.g. economics.

Both approaches, in turn, call for another field of research that enables a profound weighting of estimated and observed inputs[223] that allows for the derivation of distinct measures for identified caveats. In other words, when estimated and observed inputs fall apart, the question arises of which adjusting parameter is to use and in which manner. Some preliminary steps towards such concept had already been taken, for instance in:

- NORIEGA AND FLORIAN (2009, p. 7 ff.), who present an attempt to estimate the role of a weighting between estimates and observations within an OD data adjustment mode – a field of research that is extremely promising in the context of enhancing the calibration quality of large scale freight transport models.

Further potential refinements of the model

If distribution centres and consolidation network nodes have to be modelled explicitly on a large scale, the model may continue certain concepts of the elementary methods from *operations research*, as discussed e.g. in FRIEDRICH (2010). These methods could be implemented within specific transport cost schemes that are e.g. conditional to average utilisation rates and transport schedules for different types of distribution centres and/or rail freight yards and/or intermodal freight flow consolidation nodes. Corresponding preparatory work is published in:

- THALLER ET AL. (2013), wherein e.g. a typology for different consolidation and distribution is presented.

[223] This formally refers to a seizing of λ in equation (42).

Furthermore, for an implementation of e.g. *secondary raw materials; municipal wastes and other wastes* (NST-14), *goods moved in the course of household and office removals* (NST-17), *mail, parcels* (NST-15) and the large proportions of *grouped goods* (NST-18) as well as *unidentifiable goods* (NST-19) or the German bottle deposit system, the principle of core and complementary peripheral model of FRIEDRICH (2010) could be helpful as well[224].

Besides, a joint transport model with explicit interactions along particular transport links for freight and passenger traffic is conceivable as a valuable further development to e.g. evaluate the priority of different national infrastructure investments[225].

As a further objective, an MRIO dataset would not only allow for a representation of geographic heterogeneity within Germany, it could also be helpful for the second supply chain configuration step of the model, e.g. a validation of the multiregional firm-to-firm pairings[226]. Although a large amount of resources might be required to assemble regional specific data on economic activities throughout the country, the balancing effect of an increasing freight transport modelling quality is strong.

Ideally, an even more complex dataset is at hand, capturing each section of the various shipments from an origin of production to a destination of consumption for each commodity class. This involves detailed specifications of each firm along a commodity flow configuration, including information on the respective role within a supply chain as well as the supplier-customer ratio. One step further, this dataset also includes shipment characteristics, such as value and weight as well as specific time sensitivity e.g. due to just in time and just in sequence production strategies or additional risk assessment factors etc. – a set of information that could be, at least in part, provided by a commodity flow survey[227]. Such a survey, covering sufficiently the activities of the German economy, requires a high number of professional resources. In return, it will enable a significantly more profound understanding

[224] Cf. section 7.1.
[225] An exemplary framework for a corresponding passenger traffic model is given in REICHE AND ZADEK (2015).
[226] Cf. sections 6.2.2 and 6.2.3.
[227] Cf. sections 6.4.2, 9.2 and 9.3.

of freight transport activities – e.g. based on a comprehensively calibrated and validated version of the presented modelling approach.

Apart from context specific refinements of the presented model, certain structural elements of its framework can be transferred to other national contexts. Although the model is specified for Germany and the corresponding data at hand, especially the IPF method enabling to reconstruct relevant anonymised statistical data as well as the proposed concept for the *combined modal split and network assignment* module hold considerable potential with a view to their universal applicability. This will require distinct research efforts that, in turn, might offer new perspectives to apply and, where appropriate, refine the presented model.

Overview on further developments of the model

A prioritisation of potential future enhancements of the presented model in the context of the overall usefulness strives to identify the most relevant measures to approach unresolved issues[228]:

[228] Note that both the effort for a realisation and the corresponding impact of each suggested measure are only estimated in comparison to each other as well as the already established modelling components.

		Estimated effort (low, med., high)	Estimated impact (low, med., high)
1)	introducing an MRIO database for Germany and including it in the presented model	medium	high
2)	implementation of detailed German toll statistics data	medium	high
3)	developing an adequate typology of German sourcing and distribution networks	high	high
4)	introducing economic producer-consumer trade flow data for Germany and including it in the presented model	high	high
5)	estimating the role of a weighting between estimates and observations within the PC/OD data processing	low	medium
6)	implementation of commercial freight exchange data	medium	medium
7)	expanding the present road freight forwarder survey for Germany and including it in the presented model	high	medium
8)	identifying a more refined costing scheme for different trade flow consolidations via road freight consolidation centres as well as for rail freight wagon load configurations and intermodal transport paths	medium	low
9)	integration of yet excluded commodity classes	medium	low
10)	implementing economic producer-consumer trade flow data for a regional scale	low	low
11)	implementing economic producer-consumer trade flow data for selected economic branches	low	low

Table E-1 Comparison of exemplary measures for a further development of the presented model

16.2. Usefulness of the overall results

Compared to the status quo of mainly survey based aggregate data on freight traffic activities, the most noteworthy use of the presented disaggregate German freight transport model's outcome is provided to structuring freight traffic policy measures – an effect that is manifold.

16.2.1. Freight generation measures

Within the model, a distinct relation is represented between economic activities – as the result of the *freight generation* module[229] – and the resulting freight transport activities – as an output of the *combined modal split and network assignment* module [230]. This feature of a disaggregate freight transport modelling framework enables a systematic freight transport governance at its source, namely the configuration of the freight transport demand.

The *decoupling* of both streams, an economic growth and an equally aligned increase of freight traffic, is a much discussed topic for researchers as well as for practitioners[231]. Since the presented modelling framework does not start from aggregate freight traffic data inputs and aims to backwards derive an initial transport demand – still a common practice[232] – but instead pursues a contrary concept, a contribution to the progress in this scientific and practical field is reasonable to expect. Especially, when taking into consideration the detailed specification of economic sectors and their corresponding output and input volumes in terms of tonnes.
In this regard, effects of a possible growth or decline of a certain economic sector – e.g. the segment of *motor vehicles, trailers and semi-trailers* (CPA-29) or *electricity, gas, steam and air conditioning* (CPA-35) – may accordingly be evaluated with a focus on its impact on other economic branches and, in consequence, the freight traffic allocation within the modelled system on a ceteris paribus basis. One way to exemplarily depict the data base for such analysis if given as follows[233]:

[229] Cf. section 13.
[230] Cf. section 15.
[231] Cf. section 1.2.
[232] See e.g. discussion of the German *Federal Transport Infrastructure Plan* in section 7.2.
[233] Note that the chosen format displays the modelled trade flow interactions via weighted directed arcs from the perspective of sinks that source from various other national economic sectors or international sources. The latter component is therefore indicated with CPA headings 101 to 129 as substitutes.

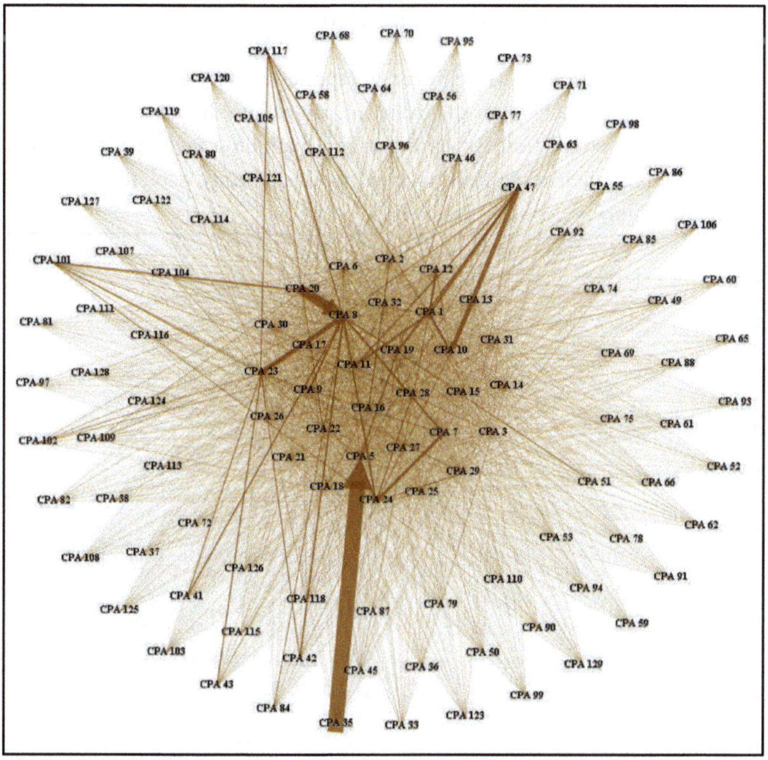

Fig. E-1 Depiction of weighted and directed total trade flow interactions between modelled economic branches in terms of tonnes per year as a result of the freight generation module

Since the modelled transport system for Germany additionally comprises distinct linkages to international trade partners, an assessment of impacts on freight transport flows due to international trade restrictions is also conceivable.

In summary, this is only a brief outlook on what might be feasible in the context of freight transport policy measures as the result of an additional insight into a disaggregate setup of the German *freight generation* process.

16.2.2. Freight distribution measures

Another important gain is achieved for policy measures that address the *freight distribution*. This results from an insight into the allocation of interlinked sites of production and consumption, including their spatial distribution as well as their trade flow volume, as exemplarily depicted within the following figure[234]:

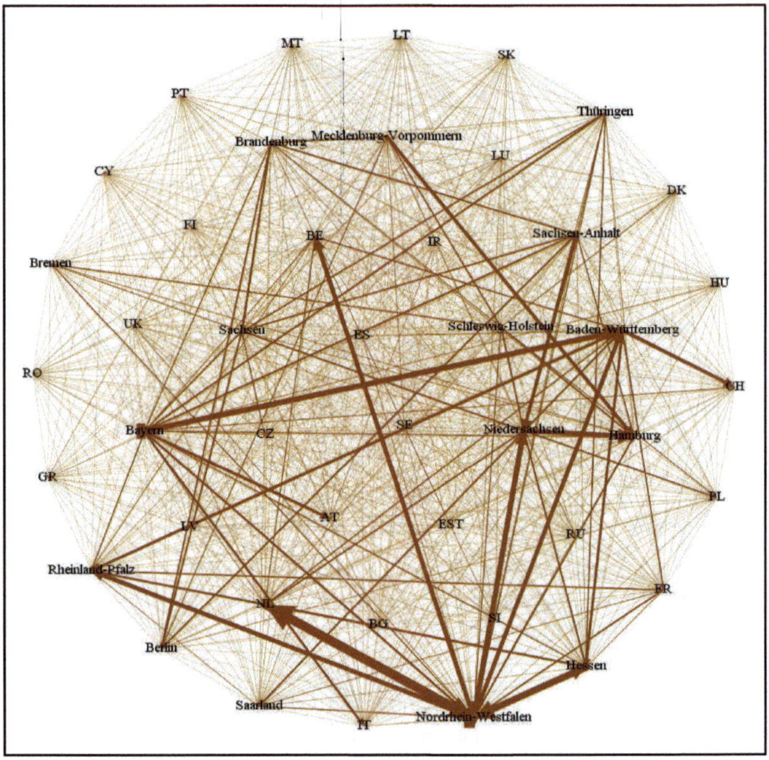

Fig. E-2 Depiction of weighted and directed total trade flow interactions between regional aggregates in terms of tonnes per year as a result of the freight distribution module

[234] Note that the chosen format aggregates all commodity class specifications and displays the modelled trade flow interactions for a NUTS-1 level of detail via weighted directed arcs from the perspective of spatially distributed sinks. In line with EUROSTAT (2011) national NUTS-1 states are indicated with their original German label.

This insight may introduce the possibility to assess general economic and societal trends, for instance, the expansion of global or local sourcing strategies as well as particular aspects of social transformations, such as e.g. a demographic development with different regional manifestations.
A more profound understanding likewise leads, for instance, to a comparison of regional proportions for local burdens from freight transport activities. These burdens reflect all societal and environmental costs relating to e.g. air pollution, noise pollution, accidents and congestions. With an emphasis on transit regions, the principles of a burden sharing can be discussed. Transit regions experience burdens of freight transport activities that are not directly related to an economic and societal well-being compared to that of production and/or consumption regions. For such analysis, the data at hand on incoming and outgoing trade flows for aggregate region pairings are complemented by a list of transit regions[235]. Therefrom, the share of transit trade flows on the overall total leads to the following result for German NUTS-1 states:

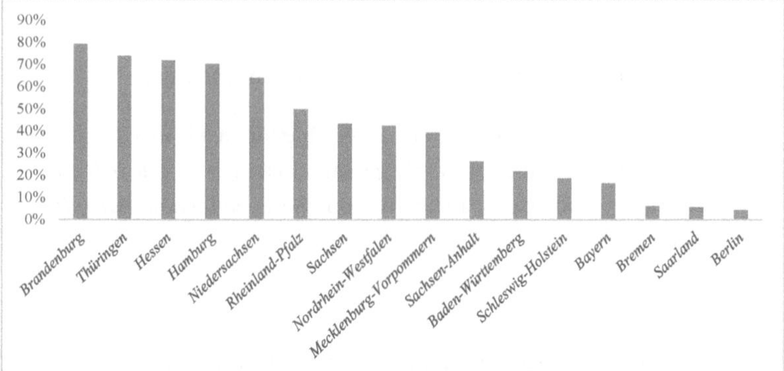

Fig. E-3 Share of transit on overall freight traffic volume for German NUTS-1 regions in tonnes for the year 2012

A conceivable analysis in this context is exemplarily interpretable as follows:

[235] This listing reflects an estimation of likely path choices on an aggregated level. For further refinements, a consultation of the *combined modal split and network assignment* module's results is proposed.

- the German NUTS-1 state of *Bayern* is among the top regions when comparing the total incoming and/or outgoing trade flows and the contrary is true for the German NUTS-1 state of *Brandenburg* (see Fig. E-2),
- the state of *Brandenburg* has one of the largest share of transit trade flows, the opposite is true for the state of *Bayern*.

As a result, a more differentiated freight infrastructure policy is implementable, especially when the spatial resolution is increased to a two-digit or higher level of detail (c.f. section 4.4.1). Therefore, a basic aggregation of e.g. data on rail and IWW freight activities from DESTATIS (2013b) DESTATIS (2013a) with results from the German road freight survey data, as e.g. given in KBA (2012b), is no longer possible, as a consequence of a finite level of detail within the latter[236].
If the goal is to focus one step further, including an additional commodity and/or modal specification, a consultation of the *combined modal split and network assignment* results is suggested – hence, the overall model's outcome.

16.2.3. Modal split and network assignment measures

The overall result of the model will be applicable for an advanced systematic freight transport governance of *freight generation* and *freight distribution* related elements in a context of the corresponding *combined modal split and network assignment* process.

An example

The scope of these measures is best illustrated with an exemplary backwards oriented analysis for a particular basic setting for e.g. transports of *machinery and equipment* (CPA-28) from the German city of *Magdeburg* (NUTS-ID: 15003) to the *port of Hamburg* (NUTS-ID 01), wherefore the presented model leads to the following findings:

---

[236] C.f. section 4.4.3. An example for a further regional focus is the analysis for potential distributions of the total freight flow volume along the German *Rhine valley* and its corresponding alternative transport paths.

- the total transport volume for the year 2012 is estimated to nearly 24,000 tonnes, thereof:
  - 0.4% are transported by mode IWW via the *inland port of Magdeburg* and the *port of Hamburg*,
  - 0% are transported via one of the four modelled intermodal transport path combinations for the nearest intermodal IWW transfer terminals (*Magdeburg Hafen*, *Haldensleben*, *Hamburg-Billwerder* and *Bützfleth*),
  - 0% are transported by mode rail via the *freight yard of Magdeburg* and the *freight yard of Hamburg*,
  - 0% are transported via one of the four modelled intermodal transport path combinations for the nearest intermodal rail transfer terminals (*Magdeburg Hafen*, *Haldensleben*, *Hamburg-Billwerder* and *Bützfleth*),
  - 99.6% are transported by mode road,
- the origin of this transport flow is a defined set of 40 firm-to-firm pairings trading products of the economic segment of CPA-28 with distinct firms of production within the city of *Magdeburg* and international EU and non-EU state-aggregates for the related consumption and
  - for 0% of these firm-to-firm pairs a transport by road freight alternatives is favoured in the context of economic shipment size considerations for comparatively small annual trade flow totals[237].

Exemplary tactical freight policy measures

Against this background, effective freight transport policy measures could be installed on a local basis, where global trade flows would otherwise not be made transparent – at least not according to a harmonised systematic for Ger-

[237] See Table D-4 for value per tonne of products from CPA-28 and section 13.7 for details on economic shipment size considerations within the presented model. See also section 14 for a configuration of the related transport cost alternatives. Notwithstanding, there is a reported annual transport volume of 0.4% (105 tonnes) on this exemplary transport link – an observation that gives reason to the relaxation of the cost-minimising principle for an implementation of the *combined modal split and network assignment* module as discussed in section 15.1.1. Note that this particular volume is close to the total load of a single vessel configured for a transport of CPA-28.

many. Accordingly, measures enhancing the attractiveness of one of the exemplary rail freight yards, inland ports or intermodal transfer terminals will not only be the result of an isolated planning foundation but a broader context.

For instance, a lookup within the presented model's database for firms neighbouring the *inland port of Magdeburg* that send products of CPA-28 to international destinations via the *port of Hamburg* with e.g. outstandingly large annual transport volumes could be helpful to avoid a waste of recourses, at least, within the planning process of freight policy measures. A planning that relies on an increasing transport demand for CPA-28 by IWW has to be related to an identification of a number of eligible firms other than the one deploying this particular link at present. According to the model's result, such estimate is identifiable with manageable expenses. Furthermore, a significant increase of the corresponding firms' output level with an international destination via the *port of Hamburg* or the location of new firms will support the exemplary planning processes, wherefore the required magnitude can be estimated according to the presented model.

Especially an identification of complementary trade flows of different commodity specifications that are eligible for intermodal transports on a particular transport link, such as the exemplary one, could be identified with reasonable efforts[238].

Exemplary strategic freight policy measures

Besides rather *operational* and *tactical* interventions, a freight transport policy could comprise business settlement measures – i.e. the spatial location of firms. For the given example, adequate governmental incentive strategies could result in locations of firms that produce commodities of CPA-28 for mainly international consumers in an immediate reach of the *inland port of Magdeburg* instead of e.g. more distant *green field* sites with little attractiveness for road freight alternatives – if the goal is to shift road freight volumes

[238] Note that the presented transport cost analysis will also be helpful for a detailed assessment of infrastructure related freight transport policy measures that address the increase of e.g. the travel speed or travel time along a distinct transport link or entire transport path. Another analysis asses the economic impact of freight infrastructure user charges. Alternatively, this type of analysis contributes to an assessment of an economic impact of e.g. trade barriers within continental Europe that affect the travel time of freight traffic – hence, the overall transport costs. These exemplary cases are then further analysable e.g. in terms of a monetary deadweight loss. Another result of the transport cost analysis is the possibility to compare international freight transport fares that compete with each other, for instance in the context of a fiscal policy.

of CPA-28 to modal alternatives. The effectiveness of such measures will increase with the size of the planning foundation in terms of considered trade flows, based on multiple firm pairings. Especially the commodity class specification of directed trade flows will be helpful to effectuate this class of policy measures.

Even though the presented model's outcome is an estimation that – especially for higher resolutions on particular firms and their individual shipments – comes along with a level of uncertainty, it is a viable starting point for more efficient or at least, more effective, freight transport policy measures compared to ones commonly used at present[239].

Exemplary internal freight transport market measures

Apart from different governmental measures, the transport market is likewise enabled to enhance its organisation internally[240]. An immediate gain is achievable on an *operational* level, e.g. as the result of an increased transparency on potential return load volumes. They are determinable for various national OD pairings and commodity class specifications according to the presented model, i.e. the overall outcome on mode and route specific transports that, in turn, result from a spatial distribution of the overall identified transport demand. Consequently, the usage on this particular *operational* level will help to decrease the large share of empty vehicle, train and vessel return loads in Germany – thus, will help to increase the efficiency of freight transport organisations. This communication is conceivable as a result of the given *modal split and network assignment* module's output, complemented by the detailed transport cost analysis[241].

On a *tactical and strategic* level, a gain in knowledge about the volume and frequency of freight shipments on a particular transport link is valuable for

[239] Cf. discussion of the present German *Federal Transport Infrastructure Plan* (BVWP 2030) in section 6.
[240] Note that internal freight transport market measures may comprise, similar to freight policies, a corresponding *operational*, *tactical* and *strategic* level. For example, a firm strategically relocates as the result of political incentives to reduce the overall freight traffic along a distinct transport link.
[241] A further gain within this field of application is achievable in the context of an interconnected freight transport demand organisation – e.g. resulting from the aspired *digitisation of the economy*, as the German government declares one of its central long-term industry policy goals e.g. in BMWI (2016) that, in turn, will not be achieved without significant improvements on '(...) *the logistics side* (...)'.

transport service providers of competing modes. For the given exemplary transport link from the city of *Magdeburg* (NUTS-ID: 15003) to the *port of Hamburg* (NUTS-ID 01) a relevant volume of *basic metal* commodities (CPA-24) is directly transported by mode rail. Nevertheless, a volume almost 5 times the rail freight volume is transported by mode road. For this particular link, an enhanced – or at least more transparent – rail freight transport service supply offered to related firms within the origin region might potentially lead to an increased rail freight transport demand for CPA-24.
As another result of this study, more transparency about road freight alternatives is provided to the transport demand side. The presented mode and commodity class specific costing scheme for individual transport links and paths may likewise result in an increasing awareness on the intermodal competitiveness. Transferred to the exemplary transport link, large volumes of *wood and of products of wood and cork* (CPA-16) are transported by mode road despite the fact that for certain corresponding firm-to-firm pairs direct rail freight transports might be more cost effective. Although the presented freight transport costing is only an estimation, it may indicate the potential of transport alternatives that otherwise were not considered among actors of the freight transport market demand and/or supply.

A generalisation

The final outcome of the presented model will be most of all helpful to make further progress concerning the chicken-and-egg problem of further improving a freight transport infrastructure for alternatives to road freight transports. This relates to the question of what should be a starting point, an adequate number of alternative transport modes or a sufficient transport demand for the respective OD pair. A question that can now be answered more qualified: An efficient deployment of the freight transport infrastructure system takes place, where there is already a specific transport demand or will be in future[242]. The outcome of the disaggregate freight transport model for Germany will help to identify the corresponding transport demand that is spread among various types of transport means at present. It will furthermore be useful for

[242] The specification is required since neither all commodity classes nor all undifferentiated aggregate transport network relations are equally subject to modal split considerations.

a more specified freight transport *modal split and network assignment* forecast – compared to the present practice for the German *Federal Transport Infrastructure Plan* as presented in BMVI (2016a)[243].

In summary, more efficient – at least more effective – modal shift policy measures as well as advanced freight transport infrastructure flow organisation measures cannot be implemented without a detailed understanding of:

- *where* freight traffic occurs – a question that is answered by the *network assignment* component of the presented model – as well as
- *how* it takes place – that is answered by the *modal split* component – with both elements including a detailed spatial and commodity class specification.

From this perspective, the usefulness of the overall results is not only given for freight transport policy decision makers, but also for the *transport supply* side, i.e. freight forwarders, as well as the *transport demand*, i.e. shippers and receivers and ultimately, the final consumer. Regardless if the future development favours a *decoupling* of an economic prosperity and a freight traffic increase or an increasing request for *more sustainable transports solutions* – a gain of information on the freight transport demand and freight flow organisation is essential.

16.2.4. Overview on the model's overall usefulness

An overview on the variety of potential benefits from the model's overall result is given as follows:

[243] See also related discussion in section 7.2.

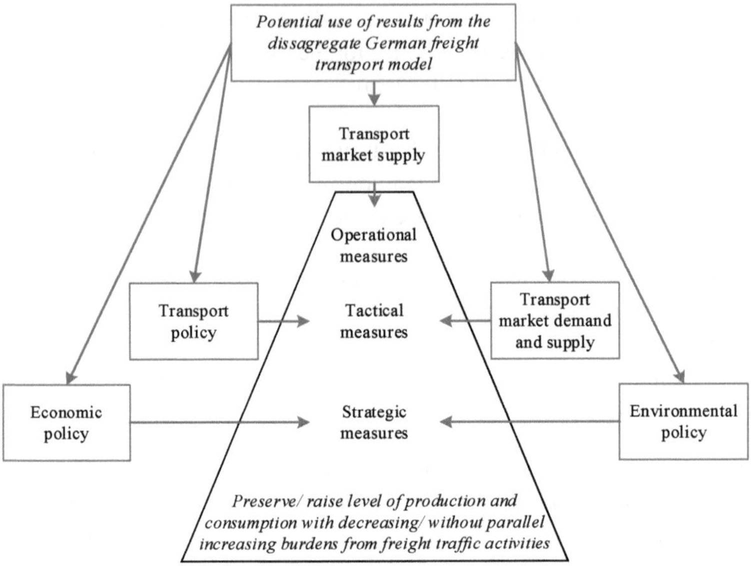

Fig. E-4 Depiction of potential benefits from the model's overall informational gain to different types of decision makers with reference to the subject

Therein, the subsequent three supported levels for freight traffic organisational decisions are broken down:

- *Operational measures* that aim at increasing the freight traffic organisation's efficiency,
    - mainly referring to the model's findings from the mode, path and commodity specific freight costing scheme,
- *Tactical measures* seeking to most of all influence the choice of modes and related transport paths for given trade flow pairings,
    - mainly induced as the result of the model's *combined modal split and network assignment* module and
    - partly emanated from the model's mode, path and commodity specific freight costing scheme,

- *Strategic measures* that strive to influence the spatial freight distribution of origins and destinations of trade flows by location policy measures in order to preserve and/or raise economic development and social welfare with respect to negative social and environmental impacts or by fostering more sustainable trade flow solutions by a selective location of research and pilot projects to enable efficient infrastructure deployment measures,
  - mainly as the result of the model's *freight generation* and *freight distribution* modules and
  - partly as the result of the model's *combined modal split and network assignment* module.

Commonly for all three supported levels of decision making – regardless of an internal or external induced freight transport market transformation – a performance evaluation of related measures is feasible for regional, national and up to transnational specifications that are compatible with each other. Furthermore, such encompassing evaluation of related measures in contrast to the affected freight traffic organisation is complemented by consistent references to an economic development and vice-versa. As a result, the preliminary work for a more target-oriented coordination of the German freight traffic is significantly advanced with regard to a better *ecological compatibility* as well as *cost efficiency.*

16.3. Summary and Conclusion

If it is an aim of the German government to maintain or increase the quality of the German freight traffic organisation, efficient infrastructure investments are essential. One step towards this goal is taken by implementing the present German *Federal Transport Infrastructure Plan*, as presented in BMVI (2016a). The plan earmarks an investment in total of about EUR 269.6 billion up to year 2030 for an upgrading of the existing federal railway infrastructure, federal trunk roads and federal waterways, used for both passenger and freight traffic. Unfortunately, the relevant freight transport related investment is to a large extent derived from an aggregate freight traffic model of the year 2010. This model, in turn, derives from highly consolidated traffic flow volumes. As a result, the underlying economic structure of trade interactions – the origin of the freight traffic demand – is hardly captured within this model. Thus, a large part of explanatory elements of the national freight

traffic configuration is lost. When compared to other national freight transport models, wherein different disaggregate modelling elements are considered, the German instance can likewise no longer be considered to be state of the art. Reflecting this initial situation, the demand for a more advanced freight transport model is given – a distinct need that is also expressed within a module of the presented freight model, the transport cost analysis of BMVI (2014a, p. 63 ff.).
Besides the context of reasonable infrastructure investments, considering the negative impacts on the environment and the society, it is of common interest that freight traffic activities should be organised as efficiently as possible. To enable a high standard for the freight demand and supply organisation – either by internal freight transport market measures and/or external governmental measures – a state of the art freight transport model is indispensable.

Therefore, a disaggregate freight traffic model was developed that uses a three-stage structure to make the *real-world* freight traffic genesis significantly more transparent – an essential prerequisite for effective measures to achieve a better freight traffic organisation. The three-stage structure relates to the 'classic' four steps of scientific traffic modelling, combining the last two into one step. This particular framework allows for a target-oriented implementation of input data and corresponding data processing strategies from various practical and scientific backgrounds.
First, a set of more than 2.1 million firms is generated as part of the *freight generation* that is to a large extent based on a deterministic input data configuration. These firms are spatially located and exhibit a level of production and consumption, respectively. Although the latter feature is based on an average distribution, a valuable dataset is obtained to describe and analyse the freight traffic genesis using only public information – a principle that is pursued along all further processing stages. As a second guiding principle, the level of detail within the model is raised as far as possible at multiple instances with regard to a related trade-off for handling confidential information by deterministic methods.
Second, the *freight distribution module* identifies likely trade flow interactions within the firm dataset. Since each individual modelled firm has a distinct level of consumption and selected firms likewise have a specified level of production, a freight flow distribution is feasible along with selected supply chain modelling determinants. The resulting flow configuration covers

around 43 million firm-to-firm pairs that have a commodity class specific relation. In addition, specific economic lot size decisions are modelled aiming at reflecting the *real-world* breakdown of an annual trade flow volume to particular shipments. With a complex context on the one hand and only limited information on the other hand, the result of the second modelling stage includes a significant level of uncertainty regarding data validity.
This is where the third and last stage continuous to deal with the *modal split and transport network assignment*. Hence, this stage answers the questions *where* and *how* freight traffic occurs as the result of an economic trade flow pattern within the German transport infrastructure system – with the latter referring to the question of *why* freight traffic takes place at all. Therefore, a set of 400 thousand alternative transport network paths is evaluated concerning the expected transport costs. Among these alternatives are the generic direct transport modes as well as forms of intermodal transport. Instead of reducing the module to a cost-based network allocation, a different concept is chosen in line with the aspiration of the presented project of getting the maximum benefit from the limited data at hand. Accordingly, the third module is elaborated in order to fit a transport cost minimisation as well as to fit specific *real-world* traffic flow observations. Thereby, inevitable errors within the supply chain modelling of the second stage are reduced and the model's overall outcome is likewise further refined by an iteration between these two modelling stages.

In total, a very large dataset on commodity class specific shipments for each modelled firm-to-firm pairs with a further specification on the mode and the related transport path is obtained – the specific *why, where* and *how* of the German freight traffic organisation. On this basis, a specific freight traffic evaluation with a very high spatial resolution and likewise commodity class specification is feasible, such as between two individual firms. Such detailed freight transport evaluation, in turn, comes along with an unidentifiable level of uncertainty for the respective data due to a missing equivalent for a calibration and validation. Alternatively, a broader context covering e.g. transports of distinct commodity specifications for sets of firms within two focal regions of origin and destination can be analysed with a comparatively lower level of uncertainty. For a further level of aggregation – such as the total annual traffic flow volume per mode and commodity class – the overall result

becomes significantly less uncertain as it can be referenced to statistical reports and calibrated iteratively. This opportunity likewise enables substantial further refinements of the model's output data quality in future.
A variety of measures is therefore suggested in order to make this outcome stepwise a *ready-to-use* solution for various applications and levels of detail within the complex field of freight traffic analysis. For instance, a standardisation of economic related and traffic related statistical report adds significant value to the modelling practice. Furthermore, in case of additional information on elementary trade flow interactions as the outcome of individual supply chain configurations in practice – that may result from a variety of different additional input data – a further step towards a validated disaggregate freight transport model for Germany can be made.

Besides, at its present stage, the final dataset is significantly more extensive than results obtained from aggregate freight traffic models for Germany. The presented model performs adequately to represent large segments of the German freight transport market. Moreover, compared to the current state of advanced international nation-wide multimodal freight traffic models, the presented approach not only serves as an innovative concept, but yields a reasonable outcome for present and further practical use.

Accordingly, alternative freight traffic policy measures can be evaluated thoroughly. With the model at hand, freight traffic organisation measures can be derived affecting the process from freight generation to mode specific network distribution effectively. What is more, these measures can be specified for various distinct regions as well as particular economic sectors as it would not be feasible without a disaggregate freight traffic model. As a result, the German freight traffic may be organised more efficiently by decision makers with reference to the subject.

Since this result is of a general interest, a future German freight traffic organisation plan should comprise extensive further investments in research on disaggregate freight transport modelling.

F Bibliography

AGORA (2014): Intermodal Terminals in Europe. Frankfurt am Main. Available online at http://www.intermodal-terminals.eu/database/, checked on 11/21/2014.

Aissaoui, Najla; Haouari, Mohamed; Hassini, Elkafi (2007): Supplier selection and order lot sizing modeling: A review. In *Operations Research and Outsourcing,* 34 (12), pp. 3516–3540.

Al-Battaineh, Omar; Kaysi, Isam A. (2007): Genetically-optimized origin-destination estimation (GOODE) model: application to regional commodity movements in Ontario. In *Canadian Journal of Civil Engineering,* 34 (2), pp. 228–238.

Alises, Ana; Vassallo, José Manuel (2015): Comparison of road freight transport trends in Europe. Coupling and decoupling factors from an Input–Output structural decomposition analysis. In *Transportation Research Part A: Policy and Practice,* 82, pp. 141–157.

Ambrosini, C.; Routhier, J. L. (2004): Objectives, Methods and Results of Surveys Carried out in the Field of Urban Freight Transport. In *Transport reviews,* 24 (1), pp. 57–77.

Arnold, Dieter (2008): Handbuch Logistik. 3., neu bearbeitete Aufl. Berlin: Springer (VDI-Buch).

BA Statistik (2012): Sozialversicherungspflichtig Beschäftigte nach Wirtschaftsabteilungen der WZ 2008 und Größenklassen. Deutschland. Kreise und kreisfreie Städte. Arbeitsmarkt in Zahlen -Beschäftigungsstatistik. Nürnberg.

BA Statistik (2013a): Betriebe nach Wirtschaftsabteilungen der WZ 2008 und Größenklassen; 2012. Deutschland. Kreise und kreisfreie Städte. Arbeitsmarkt in Zahlen -Beschäftigungsstatistik. Nürnberg.

BA Statistik (2013b): Statistik der sozialversicherungspflichtigen und geringfügigen Beschäftigung 2012. Version 7.4. With assistance of Thomas Frank, Agnes Dundler, 8/9/2013. Available online at http://statistik.arbeitsagentur.de/Navigation/Statistik/Statistik-nach-Themen/Beschaeftigung/Beschaeftigung-Nav.html, checked on 11/18/2013.

Bacharach, Michael (1970): Biproportional matrices & input-output change. Cambridge [Eng.]: University Press (University of Cambridge. Dept. of Applied Economics, Monographs, 16).

BAFA (2014): Drittlandskohlepreis 2012, 9/24/2014. Available online at http://www.bafa.de/bafa/de/energie/steinkohle/drittlandskohlepreis/, checked on 9/25/2014.

BAG (2012): Marktbeobachtungen Güterverkehr - EU-Osterweiterung. Mögliche Auswikrungen der Kabotagefreigabe für Bulgarien und Rumänien zum 1. Januar 2012 auf den deutschen Güterverkehrsmarkt. Köln, checked on 7/30/2015.

BAG (2013a): Marktbeobachtungen Güterverkehr. Bericht Herbst 2013. Köln. Available online at http://www.bag.bund.de/DE/Navigation/Verkehrsaufgaben/Marktbeobachtung/Herbst_und_Jahresberichte/_functions/herbst_und_jahresberichte_table.html?nn=13060, checked on 10/3/2014.

BAG (2013b): Mautstatistik Jahrestabellen 2012/2011. Köln, 2013. Available online at https://www.bag.bund.de/SharedDocs/Downloads/DE/Statistik/Lkw-Maut/Jahrestab_12_11.html?nn=13100, checked on 2016.

BAG (2015): Toll Statistics. Methodical Explanations. Köln. Available online at https://www.bag.bund.de/DE/Navigation/Verkehrsaufgaben/Statistik/Mautstatistik/Methodische_Erlaeuterungen.html?nn=13100, checked on 8/21/2016.

Baumgarten, Helmut; Darkow, Inga-Lena; Zadek, Hartmut (2004): Supply Chain Steuerung und Services. Logistik-Dienstleister managen globale Netzwerke - Best Practice. Berlin: Springer.

Baumol, W. J.; Vinod, H. D. (1970): An Inventory Theoretic Model of Freight Transport Demand. In *Management Science,* 16 (7), pp. 413–421.

Ben-Akiva, Moshe; Lerman, Steven R. (1985): Discrete choice analysis. Theory and application to travel demand. Cambridge (Massachusetts): MIT (MIT Press Series in Transportation Studies, 9).

Ben-Akiva, Moshe; Palma, André; McFadden, Daniel; Abou-Zeid, Maya; Chiappori, Pierre-André; Lapparent, Matthieu et al. (2012): Process and context in choice models. In *Marketing Letters,* 23 (2), pp. 439–456.

Ben-Akiva, Moshe E.; Jong, Gerard de (2013): The aggregate-Disaggregate (ADA) Freight Model System, pp. 69–90. In Moshe E. Ben-Akiva, Hilde Meersman, Eddi van de Voorde: Freight transport modelling (2013). Bingley, UK: Emerald.

Ben-Akiva, Moshe E.; Meersman, Hilde; van de Voorde, Eddi (2013): Freight transport modelling. 1st ed. Bingley, UK: Emerald.

Bendul, Julia Christine (2011): Performance-oriented Integration of Combined Transport into Supply Chain Concepts. Dissertation. Universität St.Gallen, 10/9/2015. Available online at http://www1.unisg.ch/www/edis.nsf/vEDIS-ByTitleEN/D8A9E7CB33DE5ABDC12578FD004E82D0?OpenDocument&lang=en, checked on 11/3/2015.

Bergrath, Jan (2010): Fahrerlöhne in Deutschland - Jeder für Sich. In *Fernfahrer; Motor Presse Stuttgart,* 2, pp. 16–19.

Beuthe, M.; Vandaele, E.; Witlox, F. (2004): Total logistics cost and quality attributes of freight transportation. selected papers of the 10th World Conference on Transport Research, Istanbul.

Bing Maps (2016): Bing Maps. Redmond WA, 5/20/2016. Available online at https://www.bing.com/mapspreview, checked on 5/20/2016.

Binnenreederei (2014): Das Netz europäischer Wasserstraßen, 2014. Available online at http://www.binnenreederei.de/docs/2_Das_Netz_europaischer_Wasserstrasen.pdf, checked on 12/15/2014.

BinSchUO (2008): Verordnung über die Schiffssicherheit in der Binnenschifffahrt. Available online at http://www.gesetze-im-internet.de/binschuo2008anh_xi/index.html#BJNR245071008BJNE002100000, checked on 11/19/2014.

Blauwens, Gust; Baere, Peter de; van de Voorde, E. (2008): Transport economics. 3rd. ed. Antwerpen: De Boeck.

BMEL (2013): Statistisches Jahrbuch über Ernährung, Landwirtschaft und Forsten 2014. Tabelle1000200; Landwirtschaft und Ernährung. 1. Aufl. Münster, Westf, 2013. Available online at http://www.bmel-statistik.de/de/service/archiv/statistisches-jahrbuch/, checked on 9/9/2014.

BMVBS (2010): Aktionsplan Güterverkehr und Logistik. Berlin. Available online at http://www.bmvbs.de/SharedDocs/DE/Anlage/VerkehrUndMobilitaet/aktionsplan-gueterverkehr-und-logistik-anlage.pdf?__blob=publicationFile, checked on 10/16/2013.

BMVI (2014a): Beratergruppe Verkehr + Umwelt GmbH; TNS Infratest GmbH - Entwicklung eines Modells zur Berechnung von modalen Verlagerungen im Güterverkehr für die Ableitung konsistenter Bewertungsansätze für die Bundesverkehrswegeplanung. Vorläufiger Endbericht, 2014. Available online at http://www.bmvi.de/SharedDocs/DE/Artikel/UI/bundesverkehrswegeplan-2015-methodische-weiterentwicklung-und-forschungsvorhaben.html?nn=121406, checked on 9/19/2014.

BMVI (2014b): Bundesverkehrswegeplan 2015. Verkehrsprognose 2030, 2014. Available online at http://www.bmvbw.bund.de/SharedDocs/DE/Artikel/UI/verkehrsprognose-2030.html?nn=121406, checked on 4/15/2014.

BMVI (2015): Logistikstandort Deutschland. Verkehr und Mobilität; Verkehrspolitik; Güterverkehr und Logistik. Berlin, 5/5/2015. Available online at http://www.bmvi.de/SharedDocs/DE/Artikel/G/logistikstandort-deutschland.html?linkToOverview=js, checked on 10/31/2016.

BMVI (2016a): Bundesverkehrswegeplan 2030. Gesamtplanentwurf. Berlin. Available online at http://www.bmvi.de/DE/VerkehrUndMobilitaet/Verkehrspolitik/Verkehrsinfrastruktur/Bundesverkehrswegeplan2030/InhalteHerunterladen/inhalte_node.html.

BMVI (2016b): Transeuropäische Verkehrsnetze (TEN-V) im Überblick. Hauptziele der TEN-Verordnung; Hauptziele (Förderschwerpunkte) der CEF; Position der Bundesregierung. Berlin, 6/2/2016. Available online at http://www.bmvi.de/SharedDocs/DE/Artikel/G/transeuropaeische-verkehrsnetze-im-ueberblick.html?linkToOverview=js, checked on 10/31/2016.

BMWI (2013): Der Bergbau in der Bundesrepublik Deutschland 2012 - Bergwirtschaft und Statistik - Dokumentation. 64. Jahrgang. Berlin. Available online at http://www.bmwi.de/DE/Mediathek/publikationen,did=611948.html, checked on 3/25/2014.

BMWI (2016): Industrie 4.0 - The Digitisation of the Economy. Berlin. Available online at http://bmwi.de/EN/Topics/Economy/Industrial-policy/industrie-4-0,did=708234.html, checked on 9/12/2016.

BNetzA (2013): Tätigkeitsbericht Eisenbahnen 2012. Bonn, checked on 10/14/2014.

Bochynek, C.; Menge, J.; Schneider, S.; Venus, M. (2009): Erstellung und Verwendung einer synthetischen Wirtschaftsstruktur zur dissagregierten Modellierung der Wirtschaftsverkehrsnachfrage, pp. 23–37. In Uwe Clausen: Wirtschaftsverkehr 2009 (2009). Dortmund: Verl. Praxiswissen.

Bossel, Hartmut (2004): Systeme, Dynamik, Simulation. Modellbildung, Analyse und Simulation komplexer Systeme. Norderstedt: Books on Demand.

Bowersox, Donald J. (2013): Supply chain logistics management. 4th ed. New York: McGraw-Hill.

Bröcker, Johannes (1998): Operational spatial computable general equilibrium modeling. In *The Annals of Regional Science,* 32 (3), pp. 367–387.

BSV (2012): Lohn- und Gehaltstabellen für die deutsche Binnenschifffahrt. Güterschiffahrt; Continuefahrt; gültig ab 1. Januar 2012. Duisburg.

Bühler, Georg (2006): Verkehrsmittelwahl im Güterverkehr. Eine Analyse ordnungs- und preispolitischer Maßnahmen. In *Verkehrsmittelwahl im Güterverkehr*.

Bundesbank (2014): Interest rates and yields. Yields on debt securities outstanding issued by residents / Total / Monthly average (BBK01.WU0017). Available online at http://www.bundesbank.de/Navigation/EN/Statistics/Money_and_capital_markets/Interest_rates_and_yields/Tables/table_zeitreihenliste.html?id=71408, checked on 10/9/2014.

Bundesstatistikgesetz (1/22/1987): Gesetz über die Statistik für Bundeszwecke, revised BGBl. I S. 462, ber. 565.

Buss Capital (2011): Kompendium. Wissenswertes zu Containern, Leasind and Fonds. Hamburg. Available online at www.buss-capital.de, checked on 10/10/2014.

Buss Capital (2013): Buss Capital: Zwei neue Container-Angebote von Buss – darunter das erste währungsgesicherte Euro-Direktinvestment. Hamburg, 9/23/2013. Available online at http://www.buss-capital.de/Pressemitteilungen.77+M5ce418e3d87.0.html, checked on 10/27/2014.

Button, Kenneth (2010): Transport economics. 3rd edition. Cheltenham: Elgar.

Cascetta, Ennio (2009): Transportation systems analysis. Models and applications. 2nd ed. New York: Springer (Springer optimization and its applications, v. 29).

Cascetta, Ennio; Marzano, Vittorio; Papola, Andrea; Vitillo, Roberta (2013): A Multimodal Elastic Trade Coefficients MRIO Model for Freight Demand in Europe, pp. 45–68. In Moshe E. Ben-Akiva, Hilde

Meersman, Eddi van de Voorde: Freight transport modelling (2013). Bingley, UK: Emerald.

Cascetta, Ennio; Nguyen, Sang (1988): A unified framework for estimating or updating origin/destination matrices from traffic counts. In *Transportation Research Part B: Methodological,* 22 (6), pp. 437–455.

Chan, F. T.; Chan, H. K.; Ip, R W L; Lau, H C W (2007): A decision support system for supplier selection in the airline industry. In *Proceedings of the Institution of Mechanical Engineers, Part B: Journal of Engineering Manufacture,* 221 (4), pp. 741–758.

Charnes, A.; Cooper, W. W.; Rhodes, E. (1978): Measuring the efficiency of decision making units. In *European Journal of Operational Research,* 2 (6), pp. 429–444.

Che, Z. H.; Wang, H. S. (2008): Supplier selection and supply quantity allocation of common and non-common parts with multiple criteria under multiple products. In *Computers & Industrial Engineering,* 55 (1), pp. 110–133.

Clausen, Uwe; Friedrich, Hanno; Thaller, Carina; Geiger, Christiane (2015): Commercial Transport. Proceedings of the 2nd Interdiciplinary Conference on Production Logistics and Traffic 2015. Cham: Springer International Publishing AG (Lecture Notes in Logistics).

Combes, François (2009): The choice of shipment size in freight transport. Ph.D. Thesis. Université Paris-Est.

Combes, François (2012a): An empirical evaluation of EOQ model for choice of shipment size in freight transport. In *Transportation Research Record: Journal of the Transportation Research Board* (2269), pp. 92–98.

Combes, François (2012b): An empirical evaluation of EOQ model for choice of shipment size in freight transport. In *Transportation Research Record: Journal of the Transportation Research Board,* 2269, pp. 92–98.

Combes, François; Leurent, Fabien (2007): Advances in freight transport demand modelling: an assessment with research perspectives. Paper presented at the European Tranpsport Conference, Leeuwenhorst, The Netherlands.

Combes, François; Tavasszy, Lóránt A. (2016): European Journal of Transport and Infrastructure Research. Inventory theory, mode choice and network structure in freight transport, 16 (1), pp. 38–52.

Crainic, Teodor Gabriel; Dufour, Gina; Florian, Michael; Larin, Diane; Leve, Zvi (2001): Demand matrix adjustment for multimodal freight networks. In *Transportation Research Record: Journal of the Transportation Research Board,* 1771 (1), pp. 140–147.

DB Cargo (2012): DIUM. Uniform distance table for international freight traffic: list of railways stations - list of the railways places of acceptance/delivery. Edition of 01 Juli 2012: Mainz, 7/1/2012. Available online at http://www.rail.dbschenker.de/rail-deutschland-de/start/e_rail/gueterbahnhoefe.html, checked on 5/2/2013.

DB Cargo (2015a): DIUM. Entfernungssuche, 2015. Available online at http://dium.dbschenker.com/dium/profisuche.do?initContext=1&style=stinnes, checked on 2/5/2015.

DB Cargo (2015b): DIUM. Güterbahnhofs-und Ladestellensuche, 2015. Available online at http://dium.dbschenker.com/dium/bahnhofsuche.do?initContext=1&style=stinnes, checked on 2/5/2015.

DB Netz (2012): Bahnstrompreisregelung ab 01.01.2012. Available online at https://www.dbenergie.de/file/2542424/data/bahnstrom_preisblatt.pdf, checked on 10/13/2014.

DB Netz (2013): Netznachrichten Juni 2013. Gleisanschlüsse: Der Bund hilft mit Zuschüssen. Frankfurt am Main. Available online at www.dbnetz.de/file/3969274/data/netznachrichten_juni_2013.pdf, checked on 11/21/2014.

DB Netz (2014): 835 m long freight trains between Padborg (DK) and Hamburg Maschen planned. Frankfurt/ Main, 12/15/2014. Available online at http://fahrweg.dbnetze.com/fahrweg-en/technic/innovations/longer_freight_trains.html, checked on 8/1/2015.

DB Schenker (2012a): Güterwagenkatalog. Behälterwagen. Duisburg, 12/20/2012. Available online at http://www.gueterwagenkatalog.rail.dbschenker.de/gwk-de/start/gattung_u/3027640/cs_11.html?start=0, checked on 10/16/2014.

DB Schenker (2012b): Güterwagenkatalog. Drehgestellflachwagen mit sechs Radsätzen. Duisburg, 12/20/2012. Available online at http://www.gueterwagenkatalog.rail.dbschenker.de/gwk-de/start/gattung_s/3156592/samms.html, checked on 10/14/2014.

DB Schenker (2012c): Güterwagenkatalog. Drehgestellflachwagen mit vier Radsätzen. Duisburg, 12/20/2012. Available online at http://www.gueterwagenkatalog.rail.dbschenker.de/gwk-de/start/gattung_r/3155916/cs_08.html, checked on 10/14/2014.

DB Schenker (2012d): Güterwagenkatalog. Gedeckte, großräumige Schiebewandwagen. Duisburg, 12/20/2012. Available online at http://www.gueterwagenkatalog.rail.dbschenker.de/gwk-de/start/gattung_h/3179302/hbis.html, checked on 10/14/2014.

DB Schenker (2012e): Güterwagenkatalog. Kraftfahrzeugtransportwagen. Duisburg, 12/20/2012. Available online at http://www.gueterwagenkatalog.rail.dbschenker.de/gwk-de/start/kfz-transport/3027642/Laekks.html?start=0, checked on 10/14/2014.

DB Schenker (2012f): Güterwagenkatalog. Offene Schüttgutwagen. Duisburg, 12/20/2012. Available online at http://www.gueterwagenkatalog.rail.dbschenker.de/gwk-de/start/gattung_f/3151730/Falns_121.html, checked on 10/14/2014.

DB Schenker (2012g): Güterwagenkatalog. Offene Schüttgutwagen. Duisburg, 12/20/2012. Available online at http://www.gueterwagenkatalog.rail.dbschenker.de/gwk-de/start/gattung_f/3151732/faals_151.html, checked on 10/14/2014.

DB Schenker (2012h): Güterwagenkatalog. Offene Wagen mit vier Radsätzen. Duisburg, 12/20/2012. Available online at http://www.gueterwagenkatalog.rail.dbschenker.de/gwk-de/start/gattung_e/3033522/4radsaetze.html, checked on 10/14/2014.

DB Schenker (2012i): Preise und Konditionen der DB Schenker Rail Deutschland AG. Allgemeine Bestimmungen für Gütertransportleistungen mit Allgemeiner Preisliste. Mainz, 1/1/2012. Available online at http://www.db-intermodal.com/file/intermodal-de/2299058/j-azO09jvxom0YMzeJJz9XUylrc/2349160/data/dbschenker-rail_preise_konditionen_2012.pdf, checked on 9/7/2015.

Deming, W. Edwards; Stephan, Frederick F. (1940): On a Least Squares Adjustment of a Sampled Frequency Table When the Expected Marginal Totals are Known. In *Ann. Math. Statist.*, 11 (4), pp. 427–444.

Destatis (2008a): Güterverzeichnis für Produktionsstatistiken. Ausg. 2009. Wiesbaden, 2008. Available online at https://www.destatis.de/DE/Methoden/Klassifikationen/GueterWirtschaftklassifikationen/Content75/KlassifikationGP09.html.

Destatis (2008b): Klassifikation der Wirtschaftszweige 2008 (WZ 2008), December 2008. Available online at https://www.destatis.de/DE/Methoden/Klassifikationen/GueterWirtschaftklassifikationen/Content75/KlassifikationWZ08.html, checked on 1/31/2014.

Destatis (2010): Input-Outputrechnung im Überblick. Wiesbaden, checked on 3/25/2014.

Destatis (2012): Warenverzeichnis für die Außenhandelsstatistik. Ausgabe 2012. Wiesbaden, 2012. Available online at https://www.destatis.de/DE/Methoden/Klassifikationen/Aussenhandel/warenverzeichnis_downloads.html, checked on 9/13/2012.

Destatis (2013a): E3-Schifffahrt, Güterbeförderung nach regionaler Gliederung. Sonderauswertung für die Güterverkehrsstatistik der Binnenschifffahrt. Wiesbaden.

Destatis (2013b): Eisenbahngüterverkehr, Güterbeförderung nach regionaler Gliederung. Sonderauswertung für die Güterverkehrsstatistik des Eisenbahngüterverkehrs. Datenbestellung vom 21.10.2013. Wiesbaden.

Destatis (2013c): Erzeugung in Aquakulturbetrieben - Fachserie 3 Reihe 4.6 - 2012, 7/8/2013. Available online at https://www.destatis.de/DE/Publikationen/Thematisch/LandForstwirtschaft/Fischerei/Aquakulturbetriebe2030460127004.pdf?__blob=publicationFile, checked on 9/25/2014.

Destatis (2013d): Gegenüberstellung der Meldenummern des GP 2009 mit den Warennummern des Warenverzeichnisses für die Außenhandelsstatistik, Ausgabe 2013 (WA 2013). Wiesbaden, checked on 3/25/2014.

Destatis (2013e): Luftverkehr auf ausgewählten Flugplätzen - Fachserie 8 Reihe 6.1 - 2012, checked on 9/29/2014.

Destatis (2013f): Produzierendes Gewerbe - Fachserie 4 Reihe 3.1. Produktion des Verarbeitenden Gewerbes sowie des Bergbaus und der Gewinnung von Steinen und Erden. Wiesbaden (2012), 2013. Available online at https://www.destatis.de/DE/Publikationen/Thematisch/IndustrieVerarbeitendesGewerbe/Konjunkturdaten/ProduktionVj.html, checked on 3/22/2014.

Destatis (2013g): Seeschifffahrt - Fachserie 8 Reihe 5 - 2012. Wiesbaden, 8/4/2014. Available online at https://www.destatis.de/DE/Publikationen/Thematisch/TransportVerkehr/Schifffahrt/SeeschifffahrtJ2080500127004.html, checked on 12/15/2014.

Destatis (2013h): Wirtschaftsbereiche - Fischerei - Fischerei. Preise für ausgewählte Fischarten nach Vermarktungswegen 2012, 9/25/2014. Available online at https://www.destatis.de/DE/ZahlenFakten/Wirtschaftsbereiche/LandForstwirtschaftFischerei/Fischerei/Tabellen/Aqua-Preise.html, checked on 9/25/2014.

Destatis (2014a): Eisenbahnverkehr Betriebsdaten des Schienenverkehrs (mit Daten zur Schienennetzfrequentierung) - Fachserie 8 Reihe 2.1 - 2012. Wiesbaden, checked on 10/13/2014.

Destatis (2014b): Genesis-Online Datenbank. Beförderte Güter (Binnenschifffahrt): Deutschland, Jahre, Hauptverkehrsbeziehungen, Flagge des Schiffes, Güterverzeichnis (Abteilungen). 2012. Wiesbaden, 12/3/2014. Available online at https://www-genesis.destatis.de/genesis/online, checked on 12/9/2014.

Destatis (2014c): Genesis-Online Datenbank. Beförderte Güter (Eisenbahngüterverkehr): Deutschland, Jahre, Hauptverkehrsbeziehungen. 2012. Wiesbaden, 12/3/2014, checked on 12/9/2014.

Destatis (2014d): Genesis-Online Datenbank. Beförderte Gütermenge und Beförderungsleistung (Straßengüterverkehr): Deutschland, Jahre, Verkehrswege. 2012. Wiesbaden, 12/2/2014. Available online at https://www-genesis.destatis.de/genesis/online, checked on 12/9/2014.

Destatis (2014e): Genesis-Online Datenbank. Einsteiger, Aussteiger, Frachteinladungen, Frachtausladungen (OFOD): Deutschland, Jahre, Einsteiger, Aussteiger, Frachteinladungen, Frachtausladungen (OFOD): Deutschland, Jahre, Berichtsflughafen. requested for year 2010 and for 2012. Wiesbaden, 9/6/2014. Available online at https://www-genesis.destatis.de/genesis/online, checked on 9/6/2014.

Destatis (2014f): Genesis-Online Datenbank. Empfang und Versand von Gütern (Seegüterumschlag deutscher Häfen): Deutschland, Jahre, Güterabteilungen und -gruppen. 2012. Wiesbaden, 12/12/2014. Available online at https://www-genesis.destatis.de/genesis/online, checked on 12/12/2014.

Destatis (2014g): Genesis-Online Datenbank. Empfang und Versand von Gütern bzw. Ladeeinheiten (Seeverkehr): Deutschland, Monate, Ausgewählte Häfen, Güterabteilungen und -gruppen. 2012. Wiesbaden, 9/6/2014. Available online at https://www-genesis.destatis.de/genesis/online, checked on 9/6/2014.

Destatis (2014h): Genesis-Online Datenbank. Empfang von Gütern (Luftverkehr): Deutschland, Monate (bis 2010), Herkunftsland, Zielflughafen. Wiesbaden, 12/18/2014. Available online at https://www-genesis.destatis.de/genesis/online/data, checked on 12/18/2014.

Destatis (2014i): Genesis-Online Datenbank. Produktionswert und Unternehmen der Vierteljährlichen Produktionserhebung: Deutschland, Jahre, Güterverzeichnis (2-/4-Steller). requested for 2012 for 2-digit and 4-digit classification. Wiesbaden, 3/17/2014. Available online at https://www-genesis.destatis.de/genesis/online, checked on 3/17/2014.

Destatis (2014j): Genesis-Online Datenbank. Produzierendes Gewerbe - Aus- und Einfuhr (Außenhandel). Wiesbaden, 2014. Available online at https://www-genesis.destatis.de/genesis/online/link/tabellen/51000*, checked on 3/31/2014.

Destatis (2014k): Genesis-Online Datenbank. Umgeschlagene Güter (Seeverkehr): Deutschland, Jahre, Versandküstengebiet, Empfangsküstengebiete, Hauptverkehrsbeziehungen, Gütergruppen. requested for EU-27 countries for German Baltic Bea and North Sea receivals and sendings for 2012. Wiesbaden, 12/20/2014. Available online at https://www-genesis.destatis.de/genesis/online, checked on 12/20/2014.

Destatis (2014l): Genesis-Online Datenbank. Versand von Gütern (Luftverkehr): Deutschland, Monate (bis 2010), Herkunftsflughafen, Zielland. Wiesbaden, 12/18/2014. Available online at https://www-genesis.destatis.de/genesis/online/data, checked on 12/18/2014.

Destatis (2014m): Güterverkehr. Zahlen & Fakten, Wirtschaftsbereiche, Transport & Verkehr, Güterverkehr, Verkehr, 9/22/2014. Available online at https://www.destatis.de/DE/ZahlenFakten/Wirtschaftsbereiche/TransportVerkehr/Gueterverkehr/Tabellen/GueterbefoerderungLR.html, checked on 9/22/2014.

Destatis (2014n): Güterverkehrsstatistik der Binnenschifffahrt. Sonderauswertung zur Anzahl der Fahrten nach Güterarten NST-2007 2012. Datenbestellung vom 28.07.2015. Wiesbaden.

Destatis (2014o): Güterverkehrsstatistik der Binnenschifffahrt - Fachserie 8 Reihe 4 - 2012. Wiesbaden, checked on 10/21/2014.

Destatis (2014p): Index der Erzeugerpreise gewerblicher Produkte (Inlandsabsatz) nach dem Güterverzeichnis für Produktionsstatistiken, Ausgabe 2009 (GP 2009). Wiesbaden, 4/19/2014, checked on 10/9/2014.

Destatis (2014q): Kombinierter Verkehr - Fachserie 8 Reihe 1.3 - 2012. Wiesbaden, checked on 10/26/2014.

Destatis (2014r): Preise für leichtes Heizöl, schweres Heizöl, Motorenbenzin und Dieselkraftstoff. Wiesbaden, checked on 10/1/2014.

Destatis (2014s): Reallohnindex und Nominallohnindex - 2. Vierteljahr 2014. Available online at https://www.destatis.de/DE/Publikationen/Thematisch/VerdiensteArbeitskosten/ReallohnNetto/ReallohnindexPDF_5623209.pdf?__blob=publicationFile, checked on 10/1/2014.

Destatis (2014t): Volkswirtshaftliche Gesamtrechnungen - Input-Output-Rechnung 2012. Fachserie 18 Reihe 2 - 2012 (Revision 2014). Wiesbaden, checked on 3/25/2016.

Destatis (2014u): Wirtschaftsbereiche - Wald und Holz - Wald und Holz - Statistisches Bundesamt (Destatis). Gesamteinschlag nach Holzgruppen. 2012, 9/25/2014. Available online at https://www.destatis.de/DE/ZahlenFakten/Wirtschaftsbereiche/LandForstwirtschaftFischerei/WaldundHolz/Tabellen/GesamteinschlagHolzartengruppen.html, checked on 9/25/2014.

Deutsch, Andreas (2013): Verlagerungseffekte im containerbasierten Hinterlandverkehr. Analyse, Bewertung, Strategieentwicklung. Bamberg: Univ. of Bamberg Press (Logistik und Supply Chain Management, 8).

Ding, Hongwei; Benyoucef, Lyès; Xie, Xiaolan (2006): A simulation-based multi-objective genetic algorithm approach for networked enterprises optimization. In *Special Section on Innovative Production Machines and Systems (I*PROMS),* 19 (6), pp. 609–623.

Dios Ortúzar, Juan; Willumsen, Luis G. (2011): Modelling transport. 4th ed. Oxford: Wiley-Blackwell.

Dommer, Martin (2013): Binnenschifffahrt: Energiewende auf dem Wasser. Köln, 7/12/2013. Available online at http://www.rundschau-online.de/lokales/binnenschifffahrt-energiewende-auf-dem-wasser,15185494,23697278.html, checked on 10/21/2014.

dybas (2008): Die Bahnseiten. Güterwagenladung; Gefahrgut; Kohlenwasserstoffgas, 8/30/2014. Available online at http://www.dybas.de/dybas/gw_allg/ladung/gefahr/gefahr_1965.html, checked on 10/14/2014.

EC (2003a): Commission Recommendation of 6 May 2003 concerning the definition of micro, small and medium-sized enterprises, 5/6/2003.

Available online at http://eur-lex.europa.eu/legal-content/EN/TXT/?qid=1477907114578&uri=CELEX:32003H0361, checked on 7/22/2014.

EC (2003b): RECORDIT; Real Cost Reduction of Door-to-door Intermodal Transport; Final Report. Supported by the Commission of the European Communities – DG TREN Key Action 2: Sustainable Mobility and Intermodality Accompanying Measure. With assistance of Ian Black, Roger Seaton, Andrea Ricci, Riccardo Enei, 2/20/2003. Available online at http://www.transport-research.info/project/real-cost-reduction-door-door-intermodal-transport, checked on 9/19/2014.

EC (2014): Oil bulletin. Time series 2005 onwards for 2012, 10/2/2014. Available online at http://ec.europa.eu/energy/observatory/oil/bulletin_en.htm, checked on 10/4/2014.

Eurostat (2008): Statistical classification of economic activities in the European Community. NACE Rev. 2. Methodologies and Working papers. Luxemburg (Methodologies and working papers), 2008. Available online at http://ec.europa.eu/eurostat/web/products-manuals-and-guidelines/-/KS-RA-07-015, checked on 2/2/2013.

Eurostat (2009): RAMON - Correspondence Tables List. Correspondence table between the Statistical Classification of Products by Activity in the European Economic Community, 2008 (CPA-2008) and the Standard goods classification for transport statistics, 2007 (NST-2007), 1/12/2009. Available online at http://ec.europa.eu/eurostat/ramon/relations/index.cfm?TargetUrl=LST_REL&StrLanguageCode=EN&IntCurrentPage=4, checked on 8/13/2015.

Eurostat (2010): Introduction to Prodcom, 3/19/2010. Available online at http://epp.eurostat.ec.europa.eu/portal/page/portal/prodcom/introduction, checked on 6/11/2013.

Eurostat (2011): Regions in the European Union. Nomenclature of territorial units for statistics : NUTS 2010/EU-27. 2011 ed. Luxembourg: EUR-OP (Theme : General and Regional Statistics), 2011. Available online at http://ec.europa.eu/eurostat/documents/3859598/5916917/KS-RA-11-011-EN.PDF, checked on 5/20/2015.

Eurostat (2012): Statistics Explained. Glossary - Statistical classification of products by activity (CPA), 2012. Available online at http://ec.europa.eu/eurostat/statistics-explained/index.php/Glossary:Statistical_classification_of_products_by_activity_(CPA), checked on 11/11/2013.

Eurostat (2013): Statistical classification of products by activity, 11/25/2013. Available online at http://epp.eurostat.ec.europa.eu/statistics_explained/index.php/Glossary:Statistical_classification_of_products_by_activity_%28CPA%29, checked on 1/31/2014.

Eurostat (2014a): Correspondence table NST-2007 - CPA 2008, 2014. Available online at http://ec.europa.eu/eurostat/ramon/relations/index.cfm?TargetUrl=LST_REL&StrLanguageCode=ES&IntCurrentPage=9, checked on 7/24/2014.

Eurostat (2014b): Eurostat Data Explorer. Annual cross-trade road freight transport by link, group of goods and type of transport (1 000 t), from 2008 onwards. for European Union (27 countries) and Germany in 2012, 9/5/2014. Available online at http://epp.eurostat.ec.europa.eu/portal/page/portal/statistics/search_database, checked on 12/10/2014.

Eurostat (2014c): Eurostat Data Explorer. Catches - Total all fishing areas in Germany 2012, 9/2/2014. Available online at http://appsso.eurostat.ec.europa.eu/nui/show.do?dataset=fish_ca_00&lang=en, checked on 9/25/2014.

Eurostat (2014d): Eurostat Data Explorer. Electricity prices for industrial consumers, from 2007 onwards - bi-annual data for 2012, 10/13/2014. Available online at http://epp.eurostat.ec.europa.eu/portal/page/portal/statistics/search_database, checked on 10/14/2014.

Eurostat (2014e): Eurostat Data Explorer. International annual railway transport from the reporting country to the unloading country. for Germany in 2012, 11/5/2014. Available online at http://epp.eurostat.ec.europa.eu/portal/page/portal/statistics/search_database, checked on 12/2/2014.

Eurostat (2014f): Eurostat Data Explorer. International annual road freight transport by country of loading and unloading with breakdown by reporting country (1 000 t, Mio Tkm) for 2012, 9/5/2014. Available online at http://epp.eurostat.ec.europa.eu/portal/page/portal/statistics/search_database, checked on 10/4/2014.

Eurostat (2014g): Eurostat Data Explorer. National annual road transport by group of goods and type of transport (1 000 t, Mio Tkm), from 2008 onwards. for Germany in 2012, 9/5/2014. Available online at http://epp.eurostat.ec.europa.eu/portal/page/portal/statistics/search_database, checked on 9/10/2014.

Eurostat (2014h): Eurostat Data Explorer. Railway transport - Goods transported, by group of goods - from 2008 onwards based on NST-2007 (1 000 t, million tkm). for Germany 2012, 11/5/2014. Available online at http://epp.eurostat.ec.europa.eu/portal/page/portal/statistics/search_database, checked on 12/10/2014.

Eurostat (2014i): Eurostat Data Explorer. Railway transport - Goods transported, by type of transport. for Germany in 2012, 10/30/2014, checked on 12/2/2014.

Eurostat (2014j): Eurostat Data Explorer. Road cabotage by reporting country and country in which cabotage takes place (1000t ; 1 000 tkm) - as from 1999 (Regulation (EC) 1172/98). for Germany in 2012, 9/5/2014. Available online at http://epp.eurostat.ec.europa.eu/portal/page/portal/statistics/search_database, checked on 9/10/2014.

Eurostat (2014k): Eurostat Data Explorer. Road cabotage transport by country in which cabotage takes place (1000t ; 1 000 tkm) - as from 1999. for Germany in 2012, 9/5/2014, checked on 12/2/2014.

Eurostat (2014l): Eurostat Data Explorer. Top 20 ports - Gross weight of goods handled in each port, by direction. 2012, 12/12/2014. Available online at http://epp.eurostat.ec.europa.eu/portal/page/portal/statistics/search_database, checked on 12/23/2014.

Eurostat (2014m): Eurostat Data Explorer. Transport by nationality of vessel (country/regional flows from 2007 onwards) for 2012, 10/20/2014. Available online at http://epp.eurostat.ec.europa.eu/portal/page/portal/statistics/search_database, checked on 10/20/2014.

Eurostat (2014n): Eurostat Data Explorer. Transport by nationality of vessel for 2012, 7/15/2014. Available online at http://epp.eurostat.ec.europa.eu/portal/page/portal/statistics/search_database, checked on 10/20/2014.

Eurostat (2014o): Eurostat Data Explorer. Transport by type of vessel for 2012. for Germany in 2012, 7/15/2014. Available online at http://epp.eurostat.ec.europa.eu/portal/page/portal/statistics/search_database, checked on 10/20/2014.

Eurostat (2014p): Eurostat Data Explorer: International annual road freight transport - goods loaded in reporting country, by group of goods and type of transport (1 000 t), from 2008 onwards. for Germany 2012, 9/5/2014. Available online at http://epp.eurostat.ec.europa.eu/portal/page/portal/statistics/search_database, checked on 12/9/2014.

Eurostat (2014q): Eurostat Data Explorer: International annual road freight transport - goods unloaded in reporting country, by group of goods and type of transport (1 000 t), from 2008 onwards. for Germany 2012, 9/5/2014. Available online at http://epp.eurostat.ec.europa.eu/portal/page/portal/statistics/search_database, checked on 12/9/2014.

Eurostat (2014r): Eurostat Data Explorer: Labour cost, wages and salaries, direct remuneration - NACE Rev. 2. for 2012, 9/24/2014. Available online at http://epp.eurostat.ec.europa.eu/portal/page/portal/statistics/search_database, checked on 10/4/2014.

Eurostat (2014s): Eurostat Data Explorer: Labour cost, wages and salaries, direct remuneration - NACE Rev. 2 for 2008, 10/2/2014. Available online at http://epp.eurostat.ec.europa.eu/portal/page/portal/statistics/search_database, checked on 10/4/2014.

Eurostat (2014t): PRODCOM. annual sold Germany 2012, 9/25/2014. Available online at http://epp.eurostat.ec.europa.eu/portal/page/portal/prodcom/data/database, checked on 9/25/2014.

Eurostat (2014u): Statistics Explained. Glossary - Statistical classification of economic activities in the European Community (NACE), 1/30/2014. Available online at http://epp.eurostat.ec.europa.eu/statistics_explained/index.php/Business_economy_by_sector_-_NACE_Rev._2, checked on 1/31/2014.

Eurostat (2015a): Administrative units / Statistical units. NUTS 2010 Europe; Scale: 1:60 Million, 2015. Available online at http://ec.europa.eu/eurostat/web/gisco/geodata/reference-data/administrative-units-statistical-units, checked on 6/18/2015.

Eurostat (2015b): Eurostat Data Explorer. Transport by type of good (from 2007 onwards with NST-2007). 2012, 13.11.14. Available online at http://epp.eurostat.ec.europa.eu/portal/page/portal/statistics/search_database, checked on 05.02.15.

Eurostat (2015c): NUTS - Nomenclature of territorial units for statistics. Regional typologies and local information corresponding to NUTS 3. Available online at http://ec.europa.eu/eurostat/web/nuts/overview, checked on 6/11/2015.

EWI (2014): Ecological Transport Information Tool for Worldwide Transports. Methodology and Data Update. Berne, Hannover, Heidelberg, checked on 9/1/2015.

FDZ (2016): Research data centres of the Federal Statistical Office and the statistical offices of the Länder. Düsseldorf, 12/5/2002. Available online at http://www.forschungsdatenzentrum.de/en/data_supply.asp, checked on 10/12/2016.

Feldbinder (2014): Feldbinder Spezialfahrzeugwerke GmbH (Feldbinder): Produkt Tankfahrzeuge. Tanksattelanhänger für Gefahrgüter. Winsen (Luhe). Available online at http://www.feldbinder.com/de/Tanksattelanhaenger-fuer-Gefahrgueter__92/, checked on 10/29/2014.

Fliegl (2014): Fliegl Trailer: Fahrzeugbörse. Triptis, 10/29/2014. Available online at http://www.fliegl-fahrzeugbau.de/index.cfm?cid=1544&cmp.usedvehicles.id=944, checked on 10/29/2014.

Fratar, T. (1954): Vehicular trip generation by successive approximations. In *Traffic Quaterly,* 8, pp. 53–65.

Friedrich, Hanno (2010): Simulation of logistics in food retailing for freight transportation analysis. accepted dissertation. Karlsruhe.

Fries, Nikolaus (2008): Modal split functions for a swiss national freight transport model. Paper presented at the European Transport Conference.

Furness, K. P. (1965): Time function iteration. In *Traffic Engineering Control,* 7, pp. 458–460.

GDV (2014): Transport-informations-Service. Containertypen. Berlin, 7/2/2014. Available online at http://www.tis-gdv.de/tis/containe/inhalt2.htm, checked on 10/27/2014.

Guneri, Ali Fuat; Yucel, Atakan; Ayyildiz, Gokhan (2009): An integrated fuzzy-lp approach for a supplier selection problem in supply chain management. In *Expert Systems with Applications,* 36 (5), pp. 9223–9228.

Hansen, Wiljar (2011): Does it matter if trade or transport data are used in SCGE modelling? Paper presentet at the European Transport Conference.

Hanssen, Thor-Erik Sandberg; Mathisen, Terje Andreas; Jørgensen, Finn (2012): Generalized Transport Costs in Intermodal Freight Transport. In *Procedia - Social and Behavioral Sciences,* 54, pp. 189–200.

Harris, Ford W. (1913): How many parts to make at once. reprinted in Operations Research 38(6): 947-950, 1990. In *Factory - The magazine of management,* 10 (2), pp. 135–152.

Hensher, David A.; Button, Kenneth John (2000): Handbook of Transport Modelling: Pergamon.

Ho, William; Xu, Xiaowei; Dey, Prasanta K. (2010): Multi-criteria decision making approaches for supplier evaluation and selection: A literature review. In *European Journal of Operational Research,* 202 (1), pp. 16–24.

Holguín-Veras, José; Thorson, Ellen; Zorrilla, Juan C. (2010): Commercial Vehicle Empty Trip Models With Variable Zero Order Empty Trip Probabilities. In *Networks and Spatial Economics,* 10 (2), pp. 241–259.

Holland, John H. (1992): Adaptation in natural and artificial systems. An introductory analysis with applications to biology, control, and artificial intelligence. Cambridge, Mass: MIT Press.

HPE (2014): Holzpreisindex. Available online at http://www.holzpreisindex.de/, checked on 9/25/2014.

Hunt, John Douglas; Abraham, J. E. (2005): Design and Implementation of PECAS. A Generalised System for Allocating Economic Production, Exchange and Consumption Quantities, pp. 253–274. In Martin Lee-Gosselin, Sean T. Doherty: Integrated land-use and transportation models (2005). Amsterdam, Boston: Elsevier.

hwh (2013): Schnittstelle Rampe. Lösungen zur Vermeidung von Wartezeiten. Schlussbericht. Karlsruhe. Available online at http://www.bmvi.de/SharedDocs/DE/Anlage/VerkehrUndMobilitaet/laderampe-schlussbericht-schnittstelle-rampe.pdf?__blob=publicationFile, checked on 6/1/2015.

IfM (2014): SME-definition of IfM Bonn. Bonn, 7/22/2014. Available online at http://en.ifm-bonn.org/definitions/sme-definition-of-ifm-bonn/.

IHK (2005): Ausbildungsvergütung in der Binnenschifffahrt. Oldenburg. Available online at http://www.ihk-oldenburg.de/aus-_und_weiterbildung/ausbildung/unternehmen/verguetung_urlaub_und_foerderungen/verguetung_und_urlaubsanspruch_von_auszubildenden/ausbildungsverguetung/binnenschifffahrt.php, checked on 10/16/2015.

INFRAS (2004): Handbuch Emissionsfaktoren des Straßenverkehrs 2.1. Dokumentation. With assistance of Mario Keller, Peter de Haan. Bern/Heidelberg/Graz/Essen. Available online at www.hbefa.net/d/documents/HBEFA21_Dokumentation.pdf, checked on 5/29/2015.

IRPUD (2005): IRPUD. NUTS-3 Regions. Dortmund, 3/18/2005. Available online at http://www.raumplanung.tu-dortmund.de/irpud/pro/ten/nutsys3_e.htm, checked on 6/18/2015.

Ivanova, Olga (2007): On the development of the new version of the RAEM model for the Netherlands (RAEM 3.0). Paper presented on the 47th Congress of the European Regional Science Association, Paris.

Janic, Milan (2007): Modelling the full costs of an intermodal and road freight transport network. In *Transportation Research Part D: Transport and Environment,* 12 (1), pp. 33–44.

Jong, G. de; Ben-Akiva, Moshe (2004): The Specification of Logistics in the Norwegian and Swedish National Freight Model Systems. Model scope, structure and implementation plan. Leiden/ Netherlands, 2004. Available online at http://www.ntp.dep.no/Transportanalyser/Transportanalyse+godstransport/_attachment/502967/binary/813901?_ts=1400fd902d8, checked on 2/20/2014.

Jong, Gerard; Vierth, Inge; Tavasszy, Lori; Ben-Akiva, Moshe (2013): Recent developments in national and international freight transport models within Europe. In *Transportation,* 40 (2), pp. 347–371.

Jong, Gerard de (2005): The Development of a Logistics Module in the Norwegian and Swedish National Freight Model Systems. Deliverable 4, Project 05078. Available online at http://www.ntp.dep.no/Transportanalyser/Transportanalyser+rapporter/_attachment/502963/binary/813900?_ts=1400fd2f028, checked on 10/28/2013.

Jong, Gerard de (2008): Value of freight travel-time savings, pp. 649–663. In David A. Hensher, Kenneth John Button: Handbook of transport modelling (2008). Amsterdam, London: Elsevier.

Jong, Gerard de; Gunn, Hugh; Walker, Warren (2004): National and international freight transport models. An overview and ideas for future development. In *Transport reviews,* 24 (1), pp. 103–124.

Jong, Gerard de; Vellay, C.; Houpée, M. (2001): Conference proceedings. European Transport Conference, 10-12 September 2001, Homerton College, Cambridge. Final version. London: PTRC for AET.

Kahraman, Cengiz; Cebeci, Ufuk; Ulukan, Ziya (2003): Multi-criteria supplier selection using fuzzy AHP. In *Logistics Information Management,* 16 (6), pp. 382–394.

Kaiser, Alexander; Zadek, Hartmut (2015): Verfahren zur Rekonstruktion einer LKW-Fahrtenkette auf Basis der Straßengüterverkehrsstatistik, pp. 70–75. In Hanne Wolf-Kluthausen: Jahrbuch Logistik 2015 (2015). Korschenbroich.

KBA (2012a): Methodische Erläuterungen zu Statistiken über den Verkehr deutscher Lastkraftfahrzeuge. Flensburg, 01.2012. Available online at http://www.kba.de/DE/Statistik/MethodischeErlaeuterungen/methodische_erlaeuterungen_node.html, checked on 7/15/2014.

KBA (2012b): Verkehr deutscher Lastkraftfahrzeuge - Verkehrsverflechtungen (VD 2). Flensburg, 10.2013. Available online at http://www.kba.de/DE/Statistik/Produktkatalog/produkte/Kraftverkehr/vd2_uebersicht.html, checked on 8/13/2015.

KBA (2013a): Verkehr deutscher Lastkraftfahrzeuge (VD1). Verkehrsaufkommen Jahr 2012. Flensburg, 2015. Available online at http://www.kba.de/DE/Presse/Presseportal/VD_Verkehr_deutscher_Lastkraftfahrzeuge/vd_verkehr_deutscher_lastkraftfahrzeuge_inhalt.html, checked on 7/28/2015.

KBA (2013b): Verkehr deutscher Lastkraftfahrzeuge (VD4). Güterbeförderung Jahr 2012. Flensburg, checked on 9/29/2014.

KBA (2014): Neuzulassungen von Lkw im Jahr 2012 nach ausgewählten Herstellern. Available online at http://www.kba.de/DE/Statistik/Fahrzeuge/Neuzulassungen/MarkenHersteller/2012/2012_n_herst_lkw_dusl_absolut_tabelle.html?nn=820998, checked on 10/9/2014.

Kille, Christian; Schwemmer, Martin (2013): Top 100 in European transport and logistics services. Market sizes, market segments and market leaders in the European logistics industry; a study by Fraunhofer Center for Applied Research on Supply Chain Services SCS. Ed. 2013/2014. Hamburg: DVV Media Group.

Kockelman, Kara M.; Jin, Ling; Zhao, Yong; Ruíz-Juri, Natalia (2005): Tracking land use, transport, and industrial production using random-utility-based multiregional input–output models: Applications for Texas trade. In *Journal of Transport Geography,* 13 (3), pp. 275–286.

Kranke, Andre (2011): CO2-Berechnung in der Logistik. Datenquellen, Formeln, Standards. München: Vogel (Verkehrs-Rundschau).

Krugman, Paul (1996): Gaston Eyskens lecture series. Geography and trade. 6. print. Leuven: Leuven Univ. Press.

Lafourcade, Miren; Thisse, Jaques-Francois (2011): New economic geography. the role of transport costs, pp. 67–96. In André de Palma, Robin

Lindsey, Emile Quinet, Roger Vickerman: A handbook of transport economics (2011). Cheltenham, Glos, U.K, Northampton, Mass: Edward Elgar.

Lee, Hau L.; Ng, Shu Ming (1997): Introduction to the special issue on global supply chain management. In *Production and Operations Management,* 6 (3), pp. 191–192.

Leontief, Wassily W. (1936): Quantitative Input and Output Relations in the Economic Systems of the United States. In *The Review of Economics and Statistics,* 18 (3), pp. 105–125.

Liao, Zhiying; Rittscher, Jens (2007): A multi-objective supplier selection model under stochastic demand conditions. In *International Journal of Production Economics,* 105 (1), pp. 150–159.

Liedke, G.; Schröder, S.; Zhang, L. (2013): Modelling the Emergence of Spatiotemporal Structures in Commodity Transport. In Moshe E. Ben-Akiva, Hilde Meersman, Eddi van de Voorde: Freight transport modelling (2013). Bingley, UK: Emerald.

Liedtke, Gernot (2006): An actor-based approach to commodity transport modelling. 1. Aufl. Baden-Baden: Nomos (Karlsruher Beiträge zur wirtschaftspolitischen Forschung, Bd. 20).

List, George F.; Konieczny, Laura A.; Durnford, Chanda L.; Papayanoulis, Vassilios (2002): Best-practice truck-flow estimation model for the New York City region. In *Transportation Research Record: Journal of the Transportation Research Board,* 1790 (1), pp. 97–103.

List, George F.; Turnquist, Mark A. (1994): Estimating truck travel patterns in urban areas. In *Transportation research record,* 26 (1430), pp. 1–9.

Lohndirekt (2012): Lohnnebenkosten für Arbeitgeber: Was kostet Sie Ihr Mitarbeiter wirklich? Flensburg, 5/9/2012. Available online at http://www.lohn-gehaltsabrechnung.com/lohnnebenkosten-fur-arbeitgeber-was-kostet-sie-ihr-mitarbeiter-wirklich/, checked on 10/9/2014.

LWKN (2014): Aktuelle Holzpreise im Privatwald Niedersachsen, Juli 2014, 7/15/2014. Available online at http://www.lwk-niedersachsen.de/index.cfm/portal/4/nav/0/article/15277.html, checked on 9/25/2014.

Manski, Charles F. (1977): The structure of random utility models. In *Theory and decision,* 8 (3), pp. 229–254.

McFadden, Daniel (1974): Conditional logit analysis of qualitative choice behavior. In *Frontiers in Econometrics, New York*, pp. 105–142.

McNelly, Michael G. (2000): The four-Step Model, pp. 35–41. In David A. Hensher, Kenneth John Button: Handbook of Transport Modelling (2000): Pergamon.

Meimbresse, B.; Sonntag, H. (2001): Modeling urban commercial traffic with the model Wiver. In *Etudes et Recherches – LET* (15), pp. 93–106.

Mest, Larissa (2011): Typologische Ordnung von Logistikstrategien und -strukturen für die Güterwirtschaftsverkehrsmodellierung. Dortmund: Verl. Praxiswissen (Logistik, Verkehr und Umwelt).

Miller, Ronald E.; Blair, Peter D. (2009): Input-output analysis. Foundations and extensions. 2nd ed. Cambridge [England], New York: Cambridge University Press.

Monzón, Andrés; Ballé Ndiaye, Alassane; Pfaffenbichler, Paul C.; Wegener, Michael (2010): Evaluation of TRANS-TOOLS Version 2. Report for the European Commission Joint Research Centre (JRC), Institute for Prospective Technological Studies (IPTS), Seville.

MRCE (2014): Data sheet. Bombardier - TRAXX F140. Amsterdam, 2014. Available online at http://www.mrce.eu/en/fleet/electric-locomotives/bombardier-traxx-f140.html, checked on 10/17/2014.

Müller, Stephan; Wolfermann, Axel; Huber, Stefan (2012): A Nation-wide Macroscopic Freight Traffic Model. In *Procedia - Social and Behavioral Sciences,* 54 (0), pp. 221–230.

Newton, S. (2008): Worldnet. Applying transport modelling techniques to long distance freight flows. In *European transport conference proceedings.*

Noriega, Yolanda; Florian, Michael Alexander (2009): Some enhancements of the Gradient method for OD Matrix adjustment, 04.2009. Available online at https://www.cirrelt.ca/DocumentsTravail/CIRRELT-2009-04.pdf, checked on 9/2/2014.

Notteboom, Theo (2013): Transport Markets, pp. 211–225. In Jean-Paul Rodrigue, Theo Notteboom, Jon Shaw: The SAGE handbook of transport studies (2013). London: SAGE.

OECD (2008): International Transport Forum. Charges for the use of Rail Infrastructure 2008, 2008. Available online at http://ec.europa.eu/transport/sites/transport/files/modes/rail/doc/2008_rail_charges.pdf, checked on 10/14/2014.

OECD (2010): Globalisation, Transport and the Environment. 1. Aufl. s.l.: OECD. Available online at http://site.ebrary.com/lib/alltitles/docDetail.action?docID=10373280.

OSMF (2014): OpenStreetMap. Open Source Routing Machine (OSRM), 05/2015. Available online at http://www.openstreetmap.org/, checked on 05/2015.

Pattanamekar, Parichart; Park, Dongjoo; Lee, Kang-Dae; Kim, Chansung (2014): Genetic Algorithm-Based Approach for Estimating Commodity OD Matrix. In *Wireless Personal Communications,* 79 (4), pp. 2499–2515.

Patterson, Zachary; Ewing, Gordon O.; Haider, Murtaza (2010): How different is carrier choice for third party logistics companies? In *Transportation Research Part E: Logistics and Transportation Review,* 46 (5), pp. 764–774.

Pfohl, Hans-Christian (2010): Logistiksysteme. Betriebswirtschaftliche Grundlagen. 8., neu bearb. und aktualisierte Aufl. Berlin: Springer.

PLANCO (2013): Gutachten zur Erhöhung der Wettbewerbsfähigkeit der Binnenhäfen. Endbericht. für das Bundesministerium für Verkehr, Bau und Stadtentwicklung. Essen, checked on 11/21/2014.

Port of Rotterdam Authority (2013): Port of Rotterdam; Global Hub & Industrial Cluster in Europe Expertgroup Euro-Asian Transport Linkages. With assistance of Sarah Olierook. Rotterdam. Available online at www.unece.org/fileadmin/DAM/trans/doc/2014/wp5/wp5-eatl/WP5_GE2_10th_session_Ms_Olierook_Port_of_Roterdam.pdf, checked on 12/23/2014.

Railco (2014): Railco a.s.: Zaes 65 m³ Tank Wagon. Prague, 10/14/2014. Available online at http://www.railco.eu/en/available-vehicles/zaes-65-m3-tank-wagon/, checked on 10/14/2014.

Reiche, Sascha; Zadek, Hartmut (2015): Schätzung des Verkehrsaufkommens für die Stadt Magdeburg auf Grundlage verkehrserzeugender Einrichtungen. Energie und Nachhaltigkeit. In Marco K. Koch: Stadtentwicklung und Mobilität (2015). Münster, Berlin: LIT.

Rich, J.; Holmblad, P. M.; Hansen, C. O. (2009): A weighted logit freight mode-choice model. In *Transportation Research Part E: Logistics and Transportation Review,* 45 (6), pp. 1006–1019.

Richardson, Anthony J.; Ampt, Elizabeth S.; Meyburg, Arnim H. (1995): Survey methods for transport planning. Melbourne, Vic., Australia: Eucalyptus Press.

Rodrigue, Jean-Paul; Comtois, Claude; Slack, Brian (2009): The geography of transport systems. 2nd ed. London: Routledge.

Rodrigue, Jean-Paul; Notteboom, Theo; Shaw, Jon (2013): The SAGE handbook of transport studies. London: SAGE.

Saaty, R. W. (1987): The analytic hierarchy process—what it is and how it is used. In *Mathematical Modelling,* 9 (Issue 3–5), pp. 161–176.

Saen, Reza Farzipoor (2007): Suppliers selection in the presence of both cardinal and ordinal data. In *European Journal of Operational Research,* 183 (2), pp. 741–747.

Samimi, Amir (2010): A behavioral mode choice microsimulation model for freight transportation in the U.S. Thesis (Ph.D.), University of Illinois at Chicago, 2010. 34-17358 UMI. Ann Arbor, Michigan: UMI Dissertation Publishing.

Samimi, Amir; Mohammadian, Abolfazl; Kawamura, Kazuya (2010): A behavioral freight movement microsimulation model: method and data. In *Transportation Letters: The International Journal of Transportation Research,* 2 (1), pp. 53–62.

Sarkar, Ashutosh; Mohapatra, Pratap K. J. (2006): Evaluation of supplier capability and performance: A method for supply base reduction. In *Journal of Purchasing and Supply Management,* 12 (3), pp. 148–163.

Schade, Wolfgang (2005): Strategic sustainability analysis. Concept and application for the assessment of European transport policy. 1. Aufl. Baden-Baden: Nomos (Karlsruher Beiträge zur wirtschaftspolitischen Forschung Karlsruhe Papers in Economic Policy Research, Bd. 17).

Schade, Wolfgang; Krail, Michael (2006): Modeling and calibration of large scale system dynamics models: the case of the ASTRA model. Paper presented at the 24th International Conference of the System Dynamics Society. Nijmegen, The Netherlands: Citeseer. Available online at http://citeseerx.ist.psu.edu/viewdoc/download?doi=10.1.1.408.7518&rep=rep1&type=pdf.

Schiegg, Philipp (2005): Typologie und Erklärungsansätze für Strukturen der Planung und Steuerung in Produktionsnetzwerken. Aachen: Shaker (Schriftenreihe Rationalisierung und Humanisierung, Bd. 75).

Schubert, Markus (2014): Verflechtungsprognose 2030. Schlussbericht. für das Bundesministerium für Verkehr und digitale Infrastruktur (Issue 96.0981/2011). Available online at http://www.bmvi.de/DE/VerkehrUndMobilitaet/Verkehrspolitik/Verkehrsinfrastruktur/Bundesverkehrswegeplan2030/Inhalte-Herunterladen/inhalte_node.html.

Schuh, Günther; Stich, Volker (2013): Logistikmanagement. Berlin, Heidelberg: Springer Berlin Heidelberg.

Simchi-Levi, David; Bramel, Julien; Chen, Xin (2005): The logic of logistics: theory, algorithms, and applications for logistics and supply chain management: Springer.

Spiess, Heinz (1990): A Gradient Approach for the O-DMatrix Adjustment Problem. In *Centre de Rechercher sur les Transports; Universite de Montréal; Montréal; Quebec; Canada* (693).

Städtestatistik (2012): Urban Audit Cities. Nürnberg, Stuttgart, 4/26/2012. Available online at http://www.staedtestatistik.de/, checked on 6/18/2015.

Stadtler, Hartmut; Kilger, Christoph (2008): Supply chain management and advanced planning. Concepts, models, software, and case studies. 4th ed. Berlin: Springer.

Statista (2013): Statista: Durchschnittliches Gewicht von Personenkraftwagen ausgewählter Hersteller im Jahr 2013 (in Kilogramm). Hamburg, 2015. Available online at http://de.statista.com/statistik/daten/studie/238004/umfrage/gewicht-von-pkw-nach-autoherstellern/, checked on 8/3/2015.

SZ (2012): Lokführer bekommen mehr Gehalt. Wegweisender Abschluss. München, 6/26/2012. Available online at http://www.sueddeutsche.de/wirtschaft/lokfuehrer-bekommen-mehr-gehalt-wegweisender-abschluss-1.1420620, checked on 10/13/2014.

Talluri, Srinivas; Sarkis, Joseph (2002): A model for performance monitoring of suppliers. In *International Journal of Production Research,* 40 (16), pp. 4257–4269.

Tavasszy, L. A. (2009): The Extended Generalized Cost Concept and its application in Freight Transport and General Equilibrium Modeling. Paper prepared for: Integration of Spatial Computable General Equilibrium and Transport Modelling - Bilateral Joint Seminar under agreement between NWO and JSPS, The University of Tokyo, SANJO-Hall.

Tavasszy, L. A.; Smeenk, B.; Ruijgrok, C. J. (1998): A DSS For Modelling Logistic Chains in Freight Transport Policy Analysis. In *International Transactions in Operational Research,* 5 (6), pp. 447–459.

Tavasszy, Lóránt A. (2008): Freight Modeling. An Overview of International Experiences. Washington, D.C.: Transportation Research Board (Transportation Research Board conference proceedings, 40).

Tavasszy, Lóránt A.; Bliemer, Michiel C.J. (2013): Transport Flow, Distribution and Allocation Models, Traffic Assignment and Forecasting, pp. 331–346. In Jean-Paul Rodrigue, Theo Notteboom, Jon Shaw: The SAGE handbook of transport studies (2013). London: SAGE.

Tavasszy, Lóránt A.; Ruijgrok, Kees; Davydenko, Igor (2012): Incorporating Logistics in Freight Transport Demand Models: State-of-the-Art and Research Opportunities. In *Transport reviews,* 32 (2), pp. 203–219.

Thaller, Carina; Klauenberg, Jens; Clausen, Uwe; Lenz, Barbara (2013): Charakterisierung logistischer Knoten mittels logistik-, verkehrs- und betriebsspezifischer empirischer Daten. In Uwe Clausen, Carina Thaller: Wirtschaftsverkehr 2013 (2013). Berlin: Springer Vieweg.

Train, Kenneth (2009): Discrete choice methods with simulation. 2nd ed. Cambridge, New York: Cambridge University Press.

UNECE (2012): Inland Transport Committee; Working Party on Intermodal Transport and Logistics. Intelligent Transport Systems (ITS): Opportunities and challenges for intermodal transport. Available online at http://www.unece.org/trans/wp24/wp24-themes/2012.html, checked on 9/19/2014.

Uniconsult (2009): Freie und Hansestadt Hamburg Behörde für Wirtschaft und Arbeit. Konzeptstudie zur Verkehrsverlagerung vom Lkw auf Binnenschiffe und zur Stärkung der Hinterlandverkehre. With assistance of Hartmut Beyer, Björn Pistol. Hamburg, 6/17/2009. Available online at http://www.hamburg.de/contentblob/1547328/efc7afe4759ed383c18cd99371186fef/data/kurzpraesentation-binnenschifffahrt.pdf, checked on 10/21/2014.

van Zuylen, Henk J.; Willumsen, Luis G. (1980): The most likely trip matrix estimated from traffic counts. In *Transportation Research Part B: Methodological,* 14 (3), pp. 281–293.

Verny, Jérôme (2007): The importance of decoupling between freight transport and economic growth. In *European Journal of Transport and Infrastructure Research,* 7 (2), pp. 113–128.

ViaMichelin (2013): Michelin route planner and maps, restaurants, traffic news and hotel booking. Boulogne Billancourt, 3/11/2015. Available online at http://www.viamichelin.com/, checked on 6/2/2015.

Vold, A.; Andersen, J.; Hovi, I. B.; Ivanova, O.; Jean-Hansen, V.; Lervåg, L.-E. et al. (2002): NEMO Nettverksmodell for godstransport innen Norge og mellom Norge og utlandet, Versjon 2. Oslo, Transportøkonomisk institutt. TØI-rapport 581/2002. Available online at http://worldcatlibraries.org/wcpa/oclc/70951694.

Vold, Arild; Jean-Hansen, Viggo (2007): Pingo. A model for prediction of regional and interregional freigth transport in Norway. Institute of Transport and Economics. In *TOI-rapport 899/2007*.

Wadhwa, Vijay; Ravindran, A. Ravi (2007): Vendor selection in outsourcing. In *Computers & operations research,* 34 (12), pp. 3725–3737.

Wagner, Harvey M.; Whitin, Thomson M. (1958): Dynamic version of the economic lot size model. In *Management Science,* 5 (1), pp. 89–96.

Wegener, Michael (2011): Transport in spatial models of economic development, pp. 46–65. In André de Palma, Robin Lindsey, Emile Quinet, Roger Vickerman: A handbook of transport economics (2011). Cheltenham, Glos, U.K, Northampton, Mass: Edward Elgar.

Weisbrod, Glen (2008): Models to predict the economic development impact of transportation projects: historical experience and new applications. In *Ann Reg Sci,* 42 (3), pp. 519–543.

Wermuth, M.; Wirth, R. (2005): Modelle und Strategien des Güterverkehrs, pp. 296–326. In Gerd Steierwald, H.-D Künne, Walter Vogt: Stadtverkehrsplanung (2005). Berlin: Springer.

Werner, Hartmut (2013): Supply Chain Management. Grundlagen, Strategien, Instrumente und Controlling. 5., überarb. und erw. Aufl. Wiesbaden: Gabler (Lehrbuch).

Wigan, Marcus Ramsay; Southworth, Frank (2006): What's wrong with freight models and what should we do about it? Washington, D.C. (Transport Research Board Annual Meeting). Available online at http://web.utk.edu/~tnmug08/TRB/freight.pdf, checked on 10/18/2013.

Wilson, A. G. (1970): Inter-regional Commodity Flows: Entropy Maximizing Approaches. In *Geographical Analysis,* 2 (3), pp. 255–282.

Wilson, Alan Geoffrey (1969): The use of entropy maximising models, in the theory of trip distribution, mode split and route split. In *Journal of transport economics and policy*, pp. 108–126.

Wilson, R. H. (1934): A Scientific Routine for Stock Control. In *Harvard Business Review,* 13, pp. 116–128.

Wittenbrink, Paul (2010): Transportkostenmanagement im Straßengüterverkehr. Grundlagen - Optimierungspotenziale - Green Logistic. 1. Aufl. Wiesbaden: Gabler.

WSDO (2007): Verkehrswirtschaftlicher und ökologischer Vergleich der Verkehrsträger Straße, Schiene und Wasserstraße - Schlussbericht. Auftraggeber Wasser- und Schifffahrtsdirektion des Bundes - vertreten durch die Wasser- und Schifffahrtsdirektion Ost. vorgelegt von PLANCO Consulting GmbH - in Zusammenarbeit mit Bundesanstalt für Gewässerkunde. Essen, checked on 10/13/2014.

WSDW (2012): Tarife für die Schifffahrtsabgaben auf den norddeutschen Bundeswasserstraßen im Binnenbereich. im Auftrag des Bundesministeriums für Verkehr, Bau und Stadtentwicklung. Fassung des 22. Nachtrags vom 05. Juni 2012, checked on 10/23/2014.

WSV (2010): Zulässige Höchstgeschwindigkeit für Fahrzeuge und Verbände gemäß Binnenschifffahrtsstraßenordnung, 10/4/2010. Available online at http://www.wsa-minden.wsv.de/schifffahrt/hoechstgeschwindigkeiten/index.html, checked on 10/9/2015.

WSV (2014a): Informationsberechnung der Schifffahrtsabgaben im Binnenbereich. Bonn, 10/20/2014. Available online at http://www.wsv.de/Schifffahrt/abgaben/infoberechnung.html, checked on 10/24/2014.

WSV (2014b): Verkehrsnetz der Bundeswasserstraßen. VerkNet-BWaStr. Metadaten. Bonn, 2014. Available online at https://www.wsv.de/service/karten_geoinformationen/verknet_bwastr/index.html, checked on 4/4/2014.

WSV (2015): Bundeswasserstraßen-Routenplaner (Beta). Berufsschifffahrt. Bonn, 1/30/2015. Available online at https://atlas.wsv.bund.de/routing/map/, checked on 5/4/2015.

Zadeh, Lotfi A. (1965): Fuzzy sets. In *Information and control,* 8 (3), pp. 338–353.

Zangemeister, Christof (1971): Nutzwertanalyse in der Systemtechnik: Wittemannsche Buchhandlung Munchen.

Zeitzen, Frank (2012): Lastauto-Omnibus-Katalog 2013. 2013rd ed. Stuttgart: EuroTransportMedia (42).

Zwick, Markus (2007): Forschungsdatenzentren – Nutzen und Kosten einer informationellen Infrastruktur für Wissenschaft, Politik und Datenproduzenten. Wiesbaden, 12.2006, checked on 10/12/2016.

G Appendix

	Volume in 1,000 tonnes	*Freq.*	*Volume in millions of tkm*	*Freq.*
Rail - train	366,141	9%	110,065	17%
IWW - barge	223,171	5%	58,488	9%
Sea - ship	295,103	7%	-	-
Air - plane	4,317	0%	1,420	0.2%
Crude oil - pipeline	87,898	2%	16,207	3%
Road - lorry	3,310,700	77%	447,000	71%
TOTAL	4,287,330		633,180	

Table G-1 Freight transport volume in tonnes and tonne-kilometres for Germany in 2012 [DESTATIS (2014m)]

German international transport volume (in thousands of tonnes) loaded/unloaded in:	*Sending*			*Receiving*		
	Total of forwarders in EU-27	Transport volume conducted by German forwarders	**Share of foreign forwarders**	Total of forwarders in EU-27	Transport volume conducted by German forwarders	**Share of foreign forwarders**
Belgium	19159	8855	**54%**	18867	7203	**62%**
Bulgaria	150	:	**(100%)**	614	:	**(100%)**
Czech Republic	10178	1192	**88%**	12560	1637	**87%**
Denmark	5938	3960	**33%**	5386	3113	**42%**
Estonia	158	:	**(100%)**	207	:	**(100%)**
Ireland	82	:	**(100%)**	98	:	**(100%)**
Greece	681	36	**95%**	750	56	**93%**
Spain	4554	608	**87%**	5954	492	**92%**
France	25171	12215	**51%**	21039	9651	**54%**
Italy	9925	3612	**64%**	10569	3257	**69%**
Cyprus	0	:	**(100%)**	24	:	**(100%)**
Latvia	256	:	**(100%)**	621	:	**(100%)**
Lithuania	923	:	**(100%)**	814	:	**(100%)**
Luxembourg	6158	2946	**52%**	3466	1442	**58%**
Hungary	2624	111	**96%**	2687	113	**96%**
Malta	:	:	**(100%)**	:	:	**(100%)**
Netherlands	38329	12763	**67%**	36520	10444	**71%**
Austria	18354	6904	**62%**	14609	4932	**66%**
Poland	24618	1268	**95%**	23323	693	**97%**
Portugal	680	140	**79%**	907	:	**(100%)**
Romania	1126	:	**(100%)**	1343	:	**(100%)**
Slovenia	1253	84	**93%**	1358	:	**(100%)**
Slovakia	2241	108	**95%**	2323	60	**97%**
Finland	92	:	**(100%)**	167	:	**(100%)**
Sweden	1506	839	**44%**	1367	650	**52%**
United Kingdom	3562	1159	**67%**	2631	590	**78%**
Total EU-27	177718	56800	**68%**	168204	44333	**74%**
Switzerland	9822	8396	**15%**	4117	3261	**21%**
Russia	755	:	**(100%)**	44	:	**(100%)**
others	982	407	**59%**	715	209	**71%**
TOTAL	189277	65603	**65%**	173080	47803	**72%**

: no value () estimate

Table G-2 Volume of sending and receiving of goods by road in tonnes for Germany in 2012 [EUROSTAT (2014f)]

German international transport volume (in millions of tkm) loaded/ unloaded in:	*Sending*			*Receiving*		
	Total of forwarders in EU-27	Transport volume conducted by German forwarders	**Share of foreign forwarders**	Total of forwarders in EU-27	Transport volume conducted by German forwarders	**Share of foreign forwarders**
Belgium	6562	2966	**55%**	7193	2612	**64%**
Bulgaria	281	:	**(100%)**	1342	:	**(100%)**
Czech Republic	5264	393	**93%**	5867	454	**92%**
Denmark	2693	1766	**34%**	2275	1232	**46%**
Estonia	300	:	**(100%)**	370	:	**(100%)**
Ireland	111	:	**(100%)**	134	:	**(100%)**
Greece	1404	65	**95%**	1303	75	**94%**
Spain	7886	1047	**87%**	10899	829	**92%**
France	12651	5353	**58%**	11228	4261	**62%**
Italy	8265	2859	**65%**	8769	2464	**72%**
Cyprus	0	:	**(100%)**	45	:	**(100%)**
Latvia	412	:	**(100%)**	1010	:	**(100%)**
Lithuania	1326	:	**(100%)**	1279	:	**(100%)**
Luxembourg	1156	580	**50%**	838	311	**63%**
Hungary	2664	86	**97%**	2678	99	**96%**
Malta	:	:	**(100%)**	:	:	**(100%)**
Netherlands	10165	3350	**67%**	11511	3150	**73%**
Austria	7681	2320	**70%**	6368	1841	**71%**
Poland	16103	484	**97%**	15123	283	**98%**
Portugal	1628	319	**80%**	2133	:	**(100%)**
Romania	1754	:	**(100%)**	1929	:	**(100%)**
Slovenia	1093	55	**95%**	1215	:	**(100%)**
Slovakia	2122	91	**96%**	2297	56	**98%**
Finland	49	:	**(100%)**	81	:	**(100%)**
Sweden	1242	620	**50%**	1096	464	**58%**
United Kingdom	3515	1081	**69%**	2500	518	**79%**
Total EU-27	96327	23435	**76%**	99483	18649	**81%**
Switzerland	3093	2402	**22%**	1571	1101	**30%**
Russia	1890	:	**(100%)**	87	:	**(100%)**
others	1,466	515	**65%**	1,258	280	**78%**
TOTAL	102776	26352	**74%**	102399	20030	**80%**

: no value () estimate

Table G-3 Volume of sending and receiving of goods by international road freight forwarders in tonne-kilometres for Germany in 2012 [EUROSTAT (2014f)]

German international transport volume (in thousands of tonnes) loaded/ unloaded in:	Sending		Receiving	
	Total	Freq.	Total	Freq.
Belgium	1.951	4%	4.465	8%
Bulgaria	27	0%	32	0%
Czech Republic	3.435	8%	4.228	7%
Denmark	631	1%	84	0%
Estonia	:	:	:	:
Ireland	:	:	:	:
Greece	29	0%	1	0%
Spain	748	2%	785	1%
France	1.865	4%	1.602	3%
Italy	12.071	27%	8.806	15%
Cyprus	:	:	:	:
Latvia	:	:	:	:
Lithuania	2	0%	23	0%
Luxembourg	881	2%	202	0%
Hungary	684	2%	917	2%
Malta	:	:	:	:
Netherlands	2.975	7%	18.029	31%
Austria	8.370	18%	7.083	12%
Poland	2.683	6%	5.285	9%
Portugal	16	0%	26	0%
Romania	131	0%	262	0%
Slovenia	491	1%	199	0%
Slovakia	452	1%	2.124	4%
Finland	8	0%	:	:
Sweden	1.737	4%	2.173	4%
United Kingdom	57	0%	42	0%
Total EU-27	39.244	87%	56.368	97%
Switzerland	5.759	13%	1.704	3%
Russia	87	0%	51	0%
other countries	196	0%	103	0%
TOTAL	45.286		58.226	

: no value

Table G-4 Volume of sending and receiving of goods by rail in tonnes for Germany in 2012 [EUROSTAT (2014e)]

German international transports (in millions of tkm) loaded/ unloaded in:	*Sending*		*Receiving*	
	Total	Freq.	Total	Freq.
Belgium	611	3%	1.359	6%
Bulgaria	11	0%	17	0%
Czech Republic	1.717	8%	1.940	8%
Denmark	300	1%	42	0%
Estonia	:		:	
Ireland	:		:	
Greece	10	0%	1	0%
Spain	193	1%	187	1%
France	672	3%	489	2%
Italy	5.750	28%	4.169	18%
Cyprus	:		:	
Latvia	:		:	
Lithuania	1	0%	9	0%
Luxembourg	262	1%	70	0%
Hungary	286	1%	423	2%
Malta	:		:	
Netherlands	724	4%	5.971	26%
Austria	5.042	25%	4.151	18%
Poland	837	4%	1.429	6%
Portugal	7	0%	10	0%
Romania	65	0%	78	0%
Slovenia	218	1%	70	0%
Slovakia	157	1%	639	3%
Finland	7	0%	:	
Sweden	818	4%	948	4%
United Kingdom	6	0%	5	0%
Total EU-27	17.694	87%	22.007	95%
Switzerland	2.481	12%	1.073	5%
Russia	26	0%	15	0%
other countries	109	1%	64	0%
TOTAL	20.310		23.159	

: no value

Table G-5 Volume of sending and receiving of goods by rail in tonne-kilometres for Germany in 2012 [EUROSTAT (2014e)]

German international transports (in thousands of tonnes) unloaded/ loaded in:	*Sending*			*Receiving*		
	Total of forwarders in EU-27	Transport volume conducted by German forwarders	**Share of foreign forwarders**	Total of forwarders in EU-27	Transport volume conducted by German forwarders	**Share of foreign forwarders**
Belgium	13,122	1,753	**87%**	12,407	3,029	**76%**
Bulgaria	38	6	**84%**	82	7	**91%**
Czech Republic	134	9	**93%**	315	62	**80%**
Denmark	4	:	**100%**	4	3	**25%**
Estonia	0	:	**100%**	0	:	**100%**
Ireland	0	:	**100%**	6	3	**50%**
Greece	0	:	**100%**	0	:	**100%**
Spain	0	:	**100%**	0	:	**100%**
France	1,687	407	**76%**	4,293	1,933	**55%**
Italy	0	:	**100%**	0	:	**100%**
Cyprus	0	0	**100%**	0	0	**100%**
Latvia	0	:	**100%**	6	2	**67%**
Lithuania	0	:	**100%**	9	:	**100%**
Luxembourg	89	14	**84%**	11	7	**36%**
Hungary	44	22	**50%**	756	371	**51%**
Malta	0	0	**100%**	0	0	**100%**
Netherlands	30,082	4,415	**85%**	78,870	15,713	**80%**
Austria	396	240	**39%**	630	484	**23%**
Poland	96	11	**89%**	691	101	**85%**
Portugal	0	:	**100%**	0	:	**100%**
Romania	17	9	**47%**	75	15	**80%**
Slovenia	0	:	**100%**	0	:	**100%**
Slovakia	9	5	**44%**	240	58	**76%**
Finland	2	:	**100%**	8	1	**88%**
Sweden	8	4	**50%**	4	:	**100%**
United Kingdom	265	84	**68%**	52	25	**52%**
Total EU-27	45,993	6,979	**85%**	98,459	21,814	**78%**
Switzerland	1,035	295	**71%**	229	73	**68%**
Russia	0	:	**100%**	32	2	**94%**
other countries	1,220	0	**100%**	0	0	**100%**
TOTAL	48,248	7,274	**85%**	100,041	21,889	**78%**

: no value

Table G-6 Volume of sending and receiving of goods by rail in tonnes for Germany in 2012 [(EUROSTAT, 2014m, 2014o)]

German international transport volume (in millions of tkm) loaded/ unloaded in:	*Sending*			*Receiving*		
	Total of forwarders in EU-27	Transport volume conducted by German forwarders	Share of foreign forwarders	Total of forwarders in EU-27	Transport volume conducted by German forwarders	Share of foreign forwarders
Belgium	3,637	646	82%	3,917	963	75%
Bulgaria	7	4	43%	13	1	92%
Czech Republic	64	5	92%	164	35	79%
Denmark	2	:	100%	2	1	50%
Estonia	0	:	100%	0	:	100%
Ireland	0	:	100%	0	0	100%
Greece	0	:	100%	0	:	100%
Spain	0	:	100%	0	:	100%
France	482	167	65%	1,219	420	66%
Italy	0	:	100%	0	:	100%
Cyprus	0	0	100%	0	0	100%
Latvia	0	:	100%	2	1	50%
Lithuania	0	:	100%	5	:	100%
Luxembourg	41	6	85%	9	6	33%
Hungary	36	18	50%	342	154	55%
Malta	0	0	100%	0	0	100%
Netherlands	7,563	1,474	81%	15,589	3,531	77%
Austria	217	118	46%	207	154	26%
Poland	31	4	87%	194	30	85%
Portugal	0	:	100%	0	:	100%
Romania	16	9	44%	16	2	88%
Slovenia	0	:	100%	0	:	100%
Slovakia	7	3	57%	71	24	66%
Finland	1	:	100%	1	0	100%
Sweden	4	3	25%	3	:	100%
United Kingdom	23	7	70%	5	2	60%
Total EU-27	12,131	2,464	80%	21,759	5,324	76%
Switzerland	266	102	62%	96	32	67%
Russia	0	:	100%	20	1	95%
other countries	291	:	100%	353	:	100%
TOTAL	12,688	2,566	80%	22,228	5,357	76%

: no value

Table G-7 Volume of sending and receiving of goods by IWW in tonne-kilometres for Germany in 2012 [(EUROSTAT, 2014m, 2014o)]

CPA	National					International											TOTAL
						Sending					*Receiving*						
	By German forwarders		By foreign forwarders		IWW total	By German forwarders		By foreign forwarders		IWW total	By German forwarders		By foreign forwarders		IWW total	
	road	rail	road	rail		road	rail	road	rail		road	rail	road	rail		
1	6%	2%	(6%)		12%	(5%)	2%	(6%)		12%	(7%)	1%	(12%)		8%	
2	0%	6%	(0%)		8%	(1%)	0%	(1%)		0%	(0%)	15%	(0%)		27%	
3	10%	14%	(10%)		22%	(7%)	10%	(10%)		19%	(6%)	10%	(3%)		16%	
4	17%	1%	(17%)		5%	(12%)	1%	(10%)		6%	(14%)	1%	(10%)		4%	
5	1%	0%	(1%)		0%	(1%)	0%	(1%)		0%	(1%)	0%	(1%)		0%	
6	7%	4%	(7%)		1%	(8%)	6%	(9%)		1%	(6%)	5%	(10%)		2%	
7	3%	14%	(3%)		20%	(2%)	3%	(3%)		10%	(2%)	5%	(1%)		18%	
8	7%	12%	(7%)		10%	(13%)	7%	(10%)		18%	(14%)	5%	(11%)		12%	
9	9%	4%	(9%)		3%	(8%)	1%	(8%)		3%	(6%)	1%	(5%)		1%	
10	7%	11%	(7%)		5%	(8%)	9%	(10%)		9%	(9%)	12%	(13%)		4%	
11	3%	0%	(3%)		0%	(3%)	0%	(4%)		1%	(3%)	0%	(4%)		0%	
12	4%	6%	(4%)		0%	(4%)	4%	(5%)		2%	(5%)	4%	(6%)		0%	
13	1%	0%	(1%)		1%	(1%)	0%	(2%)		0%	(1%)	0%	(3%)		0%	
1 – 13	77%	72%	(77%)		87%	(73%)	42%	(79%)		81%	(75%)	60%	(80%)		93%	80%
	2,227,164	204,482	(23,541)		47,920	47,595	19,299	97,636		37,310	35,776	39,165	100,268		93,507	2,973,664
14	6%	3%	(6%)		11%	(4%)	3%	(2%)		6%	(5%)	2%	(3%)		2%	
15	3%	0%	(3%)		0%	(1%)	0%	(1%)		0%	(1%)	0%	(0%)		0%	
16	3%	1%	(3%)		0%	(2%)	1%	(3%)		0%	(4%)	1%	(3%)		1%	
17	2%	0%	(2%)		0%	(1%)	0%	(0%)		0%	(1%)	0%	(0%)		0%	
18	9%	1%	(9%)		0%	(7%)	1%	(8%)		0%	(7%)	0%	(8%)		0%	
19	0%	22%	(0%)		1%	(0%)	40%	(2%)		9%	(0%)	32%	(2%)		3%	
20	0%	1%	(0%)		0%	(0%)	0%	(3%)		0%	(0%)	0%	(3%)		0%	
others						(13%)	13%	(2%)		3%	(7%)	5%	(1%)		1%	
Total	245,267	56,326	8,308		10,912	26,352	20,310	76,424		12,688	20,030	23,159	82,369		22,228	3,706,181

Table G-8 Total transport volume per NST commodity division for Germany 2012 in millions of tonne-kilometres

Level	Code	*Description*	*Reference to CPA*
1	01	Products of agriculture, hunting, and forestry; fish and other fishing products	
2	01.1	Cereals	01.11.1, 01.11.2, 01.11.3, 01.11.4, 01.12
2	01.2	Potatoes	01.13.51
2	01.3	Sugar beet	01.13.71
2	1.4	Other fresh fruit and vegetables	01.11.6, 01.11.7, 01.13.1, 01.13.2, 01.13.3, 01.13.4, 01.13.52, 01.13.53, 01.13.59, 01.13.8, 01.13.9, 01.14, 01.21, 01.22, 01.23, 01.24, 01.25.1, 01.25.3, 01.25.9, 01.26, 02.30.4
2	01.5	Products of forestry and logging	02.10.1, 02.10.3, 02.2, 02.30.1, 02.30.2, 02.30.3
2	01.6	Live plants and flowers	01.13.6, 01.13.72, 01.19.2, 01.25.2, 01.30
2	01.7	Other substances of vegetable origin	01.11.5, 01.11.8, 01.11.9, 01.15, 01.16, 01.19.1, 01.19.3, 01.27, 01.28, 01.29
2	01.8	Live animals	01.41.1, 01.42.1, 01.43, 01.44, 01.45.1, 01.46, 01.47.1, 01.49.1
2	01.9	Raw milk from bovine cattle, sheep and goats	01.41.2, 01.45.2
2	01.A	Other raw materials of animal origin	01.42.2, 01.45.3, 01.47.2, 01.49.2, 01.49.3
2	01.B	Fish and other fishing products	03
1	02	Coal and lignite; crude petroleum and natural gas	
2	02.1	Coal and lignite	05
2	02.2	Crude petroleum	06.1
2	02.3	Natural gas	06.2
1	03	Metal ores and other mining and quarrying products; peat; uranium and thorium	
2	03.1	Iron ores	07.1
2	03.2	Non ferrous metal ores (except uranium and thorium ores)	07.29
2	03.3	Chemical and (natural) fertilizer minerals	08.91
2	03.4	Salt	08.93
2	03.5	Stone, sand, gravel, clay, peat and other mining and quarrying products n.e.c.	08.1, 08.92, 08.99
2	03.6	Uranium and thorium ores	07.21
1	04	Food products, beverages and tobacco	
2	04.1	Meat, raw hides and skins and meat products	10.1
2	04.2	Fish and fish products, processed and preserved	10.2
2	04.3	Fruit and vegetables, processed and preserved	10.3
2	04.4	Animal and vegetable oils and fats	10.4
2	04.5	Dairy products and ice cream	10.5
2	04.6	Grain mill products, starches, starch products and prepared animal feeds	10.6, 10.9
2	04.7	Beverages	11
2	04.8	Other food products n.e.c. and tobacco products (except in parcel service or grouped)	10.7, 10.8, 12
2	04.9	Various food products and tobacco products in parcel service or grouped	Various in 10, 11 or 12 / Verschiedene in 10, 11 oder 12 / Divers dans 10, 11 ou 12
1	05	Textiles and textile products; leather and leather products	
2	05.1	Textiles	13
2	05.2	Wearing apparel and articles of fur	14
2	05.3	Leather and leather products	15
1	06	Wood and products of wood and cork (except furniture); articles of straw and plaiting materials; pulp, paper and paper products; printed matter and recorded media	
2	06.1	Products of wood and cork (except furniture)	16
2	06.2	Pulp, paper and paper products	17
2	06.3	Printed matter and recorded media	18, 58, 59
1	07	Coke and refined petroleum products	
2	07.1	Coke oven products; briquettes, ovoids and similar solid fuels	19.1, 19.20.1
2	07.2	Liquid refined petroleum products	19.20.2

Table G-9 NST and referenced CPA codes, p. 1 [EUROSTAT (2009)]

Level	Code	*Description*	*Reference to CPA*
2	07.3	Gaseous, liquefied or compressed petroleum products	19.20.3
2	07.4	Solid or waxy refined petroleum products	19.20.4
1	08	Chemicals, chemical products, and man-made fibers; rubber and plastic products ; nuclear fuel	
2	08.1	Basic mineral chemical products	20.11, 20.12, 20.13.2, 20.13.3, 20.13.4, 20.13.5, 20.13.6
2	08.2	Basic organic chemical products	20.14
2	08.3	Nitrogen compounds and fertilizers (except natural fertilizers)	20.15
2	08.4	Basic plastics and synthetic rubber in primary forms	20.16, 20.17
2	08.5	Pharmaceuticals and parachemicals including pesticides and other agro-chemical products	20.2, 20.3, 20.4, 20.5, 20.6, 21
2	08.6	Rubber or plastic products	22
2	08.7	Nuclear fuel	20.13.1
1	09	Other non metallic mineral products	
2	09.1	Glass and glass products, ceramic and porcelain products	23.1, 23.2, 23.3, 23.4
2	09.2	Cement, lime and plaster	23.5
2	09.3	Other construction materials, manufactures	23.6, 23.7, 23.9
1	10	Basic metals; fabricated metal products, except machinery and equipment	
2	10.1	Basic iron and steel and ferro-alloys and products of the first processing of iron and steel (except tubes)	24.1, 24.3
2	10.2	Non ferrous metals and products thereof	24.4
2	10.3	Tubes, pipes, hollow profiles and related fittings	24.2, 24.5
2	10.4	Structural metal products	25.1
2	10.5	Boilers, hardware, weapons and other fabricated metal products	25.2, 25.3, 25.4, 25.7, 25.9
1	11	Machinery and equipment n.e.c.; office machinery and computers; electrical machinery and apparatus n.e.c.; radio, television and communication equipment and apparatus; medical, precision and optical instruments; watches and clocks	
2	11.1	Agricultural and forestry machinery	28.3
2	11.2	Domestic appliances n.e.c. (White goods)	27.5
2	11.3	Office machinery and computers	26.2, 28.23
2	11.4	Electric machinery and apparatus n.e.c.	27.1, 27.2, 27.3, 27.4, 27.9
2	11.5	Electronic components and emission and transmission appliances	26.1, 26.3
2	11.6	Television and radio receivers; sound or video recording or reproducing apparatus and associated goods (Brown goods)	26.4, 26.8
2	11.7	Medical, precision and optical instruments, watches and clocks	26.5, 26.6, 26.7, 32.5

Table G-10 NST and referenced CPA codes, p. 2 [EUROSTAT (2009)]

Level	Code	*Description*	*Reference to CPA*
2	11.8	Other machines, machine tools and parts	28.1, 28.21, 28.22, 28.24, 28.25, 28.29, 28.4, 28.9
1	12	Transport equipment	
2	12.1	Automobile industry products	29
2	12.2	Other transport equipment	30
1	13	Furniture; other manufactured goods n.e.c.	
2	13.1	Furniture	31
2	13.2	Other manufactured goods	32.1, 32.2, 32.3, 32.4, 32.9
1	14	Secondary raw materials; municipal wastes and other wastes	
2	14.1	Household and municipal waste	38.11.31
2	14.2	Other waste and secondary raw materials	37.00.20, Others 38.11, 38.12, 38.3 / 37.00.20, Andere 38.11, 38.12, 38.3 / 37.00.20, Autres 38.11, 38.12, 38.3
1	15	Mail, parcels	
2	15.1	Mail	Not applicable / Nicht genannt / Sans objet
2	15.2	Parcels, small packages	Not applicable / Nicht genannt / Sans objet
1	16	Equipment and material utilized in the transport of goods	
2	16.1	Containers and swap bodies in service, empty	Not applicable / Nicht genannt / Sans objet
2	16.2	Pallets and other packaging in service, empty	Not applicable / Nicht genannt / Sans objet
1	17	Goods moved in the course of household and office removals; baggage and articles accompanying travellers; motor vehicles being moved for repair; other non market goods n.e.c.	
2	17.1	Household removal	Not applicable / Nicht genannt / Sans objet
2	17.2	Baggage and articles accompanying travellers	Not applicable / Nicht genannt / Sans objet
2	17.3	Vehicles for repair	Not applicable / Nicht genannt / Sans objet
2	17.4	Plant equipment, scaffolding	Not applicable / Nicht genannt / Sans objet
2	17.5	Other non market goods n.e.c.	Not applicable / Nicht genannt / Sans objet
1	18	Grouped goods: a mixture of types of goods which are transported together	
2	18.0	Grouped goods	Not applicable / Nicht genannt / Sans objet
1	19	Unidentifiable goods: goods which for any reason cannot be identified and therefore cannot be assigned to groups 01-16.	
2	19.1	Unidentifiable goods in containers or swap bodies	Not applicable / Nicht genannt / Sans objet
2	19.2	Other unidentifiable goods	Not applicable / Nicht genannt / Sans objet
1	20	Other goods n.e.c.	
2	20.0	Other goods not elsewhere classified	Not applicable / Nicht genannt / Sans objet

Table G-11 NST codes in relation to CPA standard, p.3 [EUROSTAT (2009)]

CPA-2008 two-digit	NST-2007 three-digit							
01	011	012	013	014	016	017	018	01A
02	015							
03	01B							
05	021							
06	022	023						
07	031	032	036					
08	033	034	035					
10	041	042	043	044	045	046	048	
11	047							
12	048							
13	051							
14	052							
15	053							
16	061							
17	062							
18	063							
19	071	072	073	074				
20	081	082	083	084	085	087		
21	085							
22	086							
23	091	092	093					
24	101	102	103					
25	104	105						
26	113	115	116	117				
27	112	114						
28	111	113	118					
29	121							
30	122							
31	131							
32	117	132						
35	081							
37	142							
38	141	142						
58	063							
59	063							
71	192							
74	175							
90	192							
91	175							

Table G-12 Applied correspondence table for conversion from NST-three-digit data to CPA-two-digit information

			CPA							
			01	02	03		96	97	98	99
NUTS			Crop and animal production, hunting and related service activities	Forestry and logging	Fishing and aquaculture		Other personal service activities	Activities of households as employers of domestic personnel	Undifferentiated producing activities of private households	Activities of extraterritorial organisations and bodies
	Total for Germany	4,844,278	91,146	8,039	1,345		161,102	40,760	437	2,702
	thereof in:					...				
1	*01001 Flensburg, City*	5,356	15	*	345		221	32	24	*
2	*01002 Kiel, State capital*	13,339	15	*	555		433	71	*	332
3	*01003 Hanseatic city Lübeck*	11,982	72	*	18		499	90	*	422
4	*01004 Neumünster, City*	4,843	26	*	256		163	32	12	524
...					...					
399	*16074 Saale-Holzland-district*	5,182	87	*	3		161	7	*	23
400	*16075 Saale-Orla-district*	5,674	124	47	8		168	78	*	32
401	*16076 Greiz*	7,064	162	19	34	...	*	5	7	235
402	*16077 County of Altenburg*	5,602	166	*	*		345	8	*	74

table includes random entries

Table G-13 Exemplary format for initial dataset with either a number of companies or a number of employees per company size-class

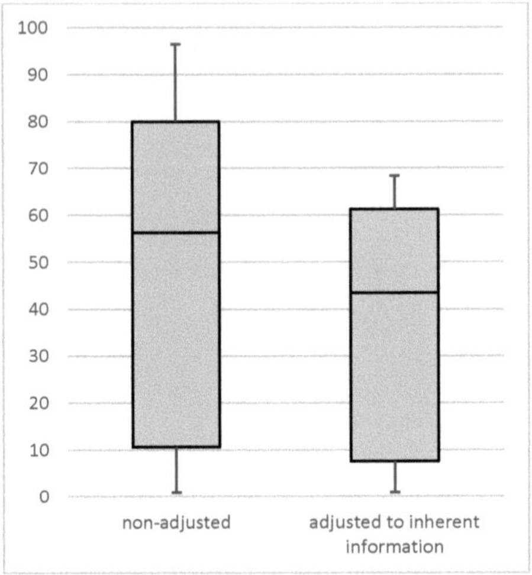

Fig. G-1 Quartiles on the distribution in percentage of missing data on number of firms per region and sector

	t	€	€/t
Cereals	45,397,000	9,166,000,000	202
Potatoes	10,666,000	1,094,000,000	103
Sugar beet	27,687,000	647,000,000	23
Fruit	1,256,000	495,000,000	394
Vegetable	3,830,000	1,375,000,000	359
Wine	90000	1,207,000,000	13,411
Meat seizure	8,508,000	14,603,000,000	1,716
Cow's milk	30,520,000	9,932,000,000	325
Eggs	824,000	825,000,000	1,001
Winter oilseed rape	4,807,000	(6,831,402,561)	1,421

() estimate from weighted average

Table G-14 Annual agricultural production by volume and value for Germany in 2012

Oak	*Beech and other hardwood*	*Pine and larch*	*Spruce, fir, douglas fir and other softwood*	*Total*	
2050000	11824000	13254000	25210000	**52338000**	1000 m³ without bark
0.69	0.74	0.55	0.44		conversion rate t/m³
1414500	8749760	7289700	11092400	**28546360**	t
220	90	50	63		average pricing €/m³
311190000	787478400	364485000	698821200	**2161974600**	€
220	90	50	63	**76**	€/t

Table G-15 Annual timber production by volume and value for Germany in 2012 [incl. data from DESTATIS (2014u); LWKN (2014)]

		t	*€/t*	*€*
Catches		205,384	-	694,512,763
Aquaculture production		26,591	3,382	89,919,976
thereof:				
	Brown trout	657	4,610	3,028,770
	Rainbow trout	8,116	3,210	26,052,360
	Salmon trout	1,278	2,440	3,118,320
	Brook trout	385	4,630	1,782,550
	Alsatian char	1,275	4,350	5,546,250
	Common carp	5,521	2,340	12,919,140
	Tench	160	3,510	561,600
	Pike-perch	50	8,510	425,500
	Pike	49	4,590	224,910
	European eel	706	10,500	7,413,000
	Wels catfish	199	4,490	893,510
	African catfish	430	1,570	675,100
	Siberian sturgeon	294	6,850	2,013,900
	others	7,471	-	-
Total		**231,975**	**3,382**	**784,432,739**

Table G-16 Annual fishing-related production by volume and weight for Germany in 2012 [incl. data from EUROSTAT (2014c); DESTATIS (2013h); DESTATIS (2013c)]

		Thereof, sent via:			
	Total export volume* (in thousands of tonnes)	*German airports allocated*	*German sea-ports reported*	*German sea-ports allocated*	***German seaports allocated as share of reported***
BELGIUM	24,676	19	959	956	100
BULGARIA	540	2	0	0	100
DENMARK	7,802	10	1,775	1,728	97
ESTONIA	647	0	578	395	68
FINLAND	2,691	10	2,322	1,912	82
FRANCE	29,370	17	973	973	100
GREECE	1,065	0	331	290	88
IRELAND	1,082	0	404	404	100
ITALY	16,931	47	746	746	100
LATVIA	418	17	523	317	61
LITHUANIA	680	0	569	411	72
LUXEMBOURG	4,369	0	0	0	100
MALTA	52	2	387	48	12
NETHERLANDS	63,743	0	2,543	2,543	100
AUSTRIA	22,478	0	0	0	100
POLAND	20,040	0	1,319	1,319	100
PORTUGAL	1,147	0	366	357	98
ROMANIA	1,776	2	8	8	100
SWEDEN	7,258	4	3,558	3,447	97
SLOVAKIA	2,526	0	0	0	100
SLOVENIA	969	0	0	0	100
SPAIN	6,313	0	1,039	1,039	100
CZECH REPUBLIC	11,455	0	1	1	100
HUNGARY	3,418	0	0	0	100
UNITED KINGDOM	13,862	0	5,291	5,234	99
CYPRUS	174	6	41	39	97
SWITZERLAND	30,838	4	0	0	100
RUSSIA	8,510	8	3,817	3,254	85
others	79,165	1,438	53,133	48,046	90
Total	363,995	1,587	80,684	73,467	91

Table G-17 Comparison of reported outgoings per country of origin at German seaports with related assignment of German imports

Total export volume to EU-27, CH and RU** (in thousands of tonnes)		Thereof, sent via: German airports allocated	German seaports reported	German seaports allocated	German seaports allocated as share of reported	Port of Rotterdam allocated	Port of Antwerpen allocated
CPA-01	15,252	1	2,410	2,147	89	0	0
CPA-02	2,188	0	182	97	53	0	0
CPA-03	138	0	25	5	19	0	0
CPA-05	6,606	0	66	4	6	0	0
CPA-06	0	0	0	0	100	0	0
CPA-07	567	0	35	14	39	0	0
CPA-08	30,168	0	1,144	1,037	91	0	0
CPA-09	0	0	0	0	100	0	0
CPA-10	26,814	9	3,316	3,097	93	0	0
CPA-11	8,116	1	343	308	90	0	0
CPA-12	78	0	0	0	100	0	0
CPA-13	904	1	330	162	49	0	0
CPA-14	356	7	238	61	26	0	0
CPA-15	117	1	17	11	66	0	0
CPA-16	8,089	1	476	446	94	0	0
CPA-17	12,768	4	696	611	88	0	0
CPA-18	224	2	33	19	58	0	0
CPA-19	29,062	1	2,670	2,639	99	0	0
CPA-20	40,484	13	4,917	4,867	99	0	0
CPA-21	818	5	0	0	100	0	0
CPA-22	6,067	6	509	467	92	0	0
CPA-23	44,166	1	2,062	2,058	100	0	0
CPA-24	18,930	7	2,760	2,748	100	0	0
CPA-25	5,924	5	455	416	91	0	0
CPA-26	835	19	224	110	49	0	0
CPA-27	2,447	13	634	578	91	0	0
CPA-28	7,669	22	1,392	1,291	93	0	0
CPA-29	8,651	23	2,038	1,821	89	0	0
CPA-30	2,537	4	89	88	99	0	0
CPA-31	2,278	1	129	114	88	0	0
CPA-32	556	4	357	144	40	0	0
Total	282,807	149	27,550	25,359	92	0	0

* excluding CPA-06

Table G-18 Allocation of exports to the EU-27, Switzerland and Russia via German seaports, airports and the ports of *Rotterdam* and *Antwerp* [incl. data from EUROSTAT (2014l)]

Total export volume to EU-27, CH and RU** (in thousands of tonnes)		Thereof, sent via: *German airports allocated*	*German seaports reported*	*German seaports allocated*	*German seaports allocated as share of reported*	*Port of Rotterdam allocated*	*Port of Antwerpen allocated*
CPA-01	3,493	7	4,223	3,486	83	0	0
CPA-02	80	0	341	80	23	0	0
CPA-03	1	0	254	1	0	0	0
CPA-05	70	0	25	25	100	27	18
CPA-06	0	0	0	0	100	0	0
CPA-07	593	0	69	69	100	314	210
CPA-08	3,005	0	699	699	100	1,382	923
CPA-09	0	0	0	0	100	0	0
CPA-10	5,853	26	4,779	4,779	100	628	419
CPA-11	1,752	3	1,079	1,079	100	402	269
CPA-12	22	3	0	0	100	11	8
CPA-13	228	8	509	221	43	0	0
CPA-14	107	71	380	36	9	0	0
CPA-15	33	5	83	28	34	0	0
CPA-16	2,305	6	2,296	2,296	100	2	1
CPA-17	3,680	18	4,016	3,663	91	0	0
CPA-18	19	3	81	16	20	0	0
CPA-19	2,524	2	990	990	100	919	614
CPA-20	19,825	82	9,346	9,346	100	6,235	4,163
CPA-21	414	47	0	0	100	220	147
CPA-22	1,752	33	1,904	1,719	90	0	0
CPA-23	11,037	2	1,825	1,825	100	5,523	3,687
CPA-24	8,563	46	4,767	4,767	100	2,249	1,501
CPA-25	2,082	28	1,331	1,331	100	434	290
CPA-26	431	112	1,433	319	22	0	0
CPA-27	1,447	130	1,460	1,317	90	0	0
CPA-28	6,621	346	3,223	3,223	100	1,830	1,222
CPA-29	7,600	343	5,607	5,607	100	989	660
CPA-30	1,731	99	309	309	100	793	530
CPA-31	578	3	1,197	575	48	0	0
CPA-32	208	15	909	193	21	0	0
Total	86,055	1,438	53,133	47,997	90	21,959	14,662

* excluding CPA-06

Table G-19 Allocation of exports to other countries than within the EU-27, Switzerland and Russia via German seaports, airports and the ports of *Rotterdam* and *Antwerp* [incl. data from EUROSTAT (2014l); PORT OF ROTTERDAM AUTHORITY (2013)]

	national transport volume		international transport volume				total	deviation
	incl. imports/ exports	*excl. imports/ exports*	*sending*	*export*	*receiving*	*import*		
Transport volume reported for NST-01 to NST-13	2,503,108	-	201,840	-	268,716	-	2,973,664	-24.8%
Transport volume determined for CPA-01 to CPA-32	-	1,485,846	-	368,862	-	382,099	2,236,807	

Table G-20 Comparison of total transport volumes reported to a calculated counterpart for Germany 2012 in thousands of tonnes [incl. data from Table A-3]

	Road	Rail	IWW	Combined-rail 1			Combined-rail 2		
				pre-haul	*main-haul*	*post-haul*	*pre-haul*	*main-haul*	*post-haul*
EOQ in t					505.38				
Net load in t	21.01	1,966.39	1,058.09	22.87	1075.08	22.87	22.87	1075.08	22.87
EOQ_re in t					525.33				
Number of trip per order	25	1	1	23	1	23	23	1	23
Costs per order in EUR				0.00	9727.47	119.33	83.29	12506.87	119.33
	638	8,860	14,943		9,847			12,709	
Number of trips in total	722	29	29	664	29	664	664	29	664
Total annual costs in EUR	460,715	256,928	433,337		361,333			497,240	

Table G-21 Exemplary application of shipment size evaluation for CPA-20

CPA commodity class	Detail	Cargo handling category	Tractor-trailer combination	Vehicle load factor in %
1	Products of agriculture, hunting and related services	1	1	84.1
2	Products of forestry, logging and related services	4	1	59.3
3	Fish and other fishing products; aquaculture products; support services to fishing	3	1	50.2
5	Coal and lignite	1	1	84.1
6	Crude petroleum and natural gas	-	-	-
7	Metal ores	1	1	84.1
8	Other mining and quarrying products	1	1	84.1
9	Mining support services	-	1	-
10	Food products	3	1	50.2
11	Beverages	3	1	50.2
12	Tobacco products	3	1	50.2
13	Textiles	3	1	50.2
14	Wearing apparel	3	1	50.2
15	Leather and related products	3	1	50.2
16	Wood and of products of wood and cork, except furniture; articles of straw and plaiting materials	4	1	59.3
17	Paper and paper products	4	1	59.3
18	Printing and recording services	3	1	50.2
19	Coke and refined petroleum products	2	3	79.2
20	Chemicals and chemical products	2	3	79.2
21	Basic pharmaceutical products and pharmaceutical preparations	3	1	50.2
22	Rubber and plastic products	4	1	59.3
23	Other non-metallic mineral products	1	1	84.1
24	Basic metals	4	1	59.3
25	Fabricated metal products, except machinery and equipment	4	1	59.3
26	Computer, electronic and optical products	3	1	50.2
27	Electrical equipment	3	1	50.2
28	Machinery and equipment n.e.c.	3	1	50.2
29	Motor vehicles, trailers and semi-trailers	4	1	59.3
30	Other transport equipment	3	1	50.2
31	Furniture	4	1	59.3
32	Other manufactured goods	3	1	50.2
-	all, by container/ swap/ additional transport vehicle	5	2	(36.15)

(*) average for units of <6 meter and >6 meter length

Table G-22 Assignment of modelled commodities to cargo handling types, tractor-trailer combinations and vehicle load factors [incl. data from KBA (2013a)]

NUTS-1-region	Annual wages paid in EUR for 2010 (48 working hours per week, maximum rate)	**Annual wages paid in EUR for 2012** (projection from 2010 to 2012 via labour cost index for CPA-49)	Number of employees in economic segment of *land transport and transport via pipelines* (CPA-49)
Baden Württemberg	33,888	35,311	62,013
Bayern	25,716	26,796	73,830
Berlin	30,636	31,923	35,153
Brandenburg	27,564	28,722	19,633
Bremen	26,916	28,046	6,424
Hamburg	23,844	24,845	19,091
Hessen	26,496	27,609	39,131
Mecklenburg-Vorpommern	19,200	20,006	11,029
Niedersachsen	23,460	24,445	45,692
Nordrhein-Westfalen	26,736	27,859	112,900
Rheinland-Pfalz	24,432	25,458	22,457
Saarland	27,000	28,134	6,655
Sachsen	15,408	16,055	33,824
Sachsen-Anhalt	17,196	17,918	18,445
Schleswig-Holstein	25,812	26,896	18,123
Thüringen	17,016	17,731	15,899
	weighted annual average:	**26,788**	

Table G-23 Calculation of average annual wages for lorry drivers in Germany 2012 [incl. data from BERGRATH (2010, p. 19); DESTATIS (2014s, p. 7) and results from section 12.2]

Country	Labour costs per year in 2008 in EUR	Labour costs per year in 2012 in EUR	**Labour cost index rel. to Germany** (2012 if applicable, else 2008)
Belgium	50269	-	139.57
Bulgaria	4291	5115	13.35
Czech Republic	14714	14804	38.63
Denmark	53818	57980	151.31
Germany	-	38319	100.00
Estonia	11665	12069	31.50
Ireland	47168	49712	129.73
Greece	38801	-	107.73
Spain	32131	35636	93.00
France	-	46958	122.54
Italy	41270	40951	114.59
Cyprus	20694	27094	70.71
Latvia	9324	9497	24.78
Lithuania	11346	11431	29.83
Luxembourg	-	52866	137.96
Hungary	11970	12100	31.58
Malta	18435	-	51.19
Netherlands	47427	52654	137.41
Austria	43089	45801	119.53
Poland	12038	11519	30.06
Portugal	21907	19854	51.81
Romania	7895	7083	18.48
Slovenia	20629	22796	59.49
Slovakia	11669	12929	33.74
Finland	41969	48239	125.89
Sweden	44968	54441	142.07
United Kingdom	42006	45921	119.84
Switzerland	-	88570	231.14
Russia*	4291	5115	13.35
other countries*	4291	5115	13.35

* estimate by values of Bulgaria

Table G-24 International labour costs related to Germany for CPA-49 [incl. data from EUROSTAT (2014r); EUROSTAT (2014s)]

Country	Average price per litre diesel at pump in EUR (incl. excise duties and VAT)	VAT per country in %	Average price 1000l diesel in EUR (incl. excises, excl. VAT)	Fuel price index rel. to Germany
Austria	1.41	20.0	1.18	93.71
Belgium	1.46	21.0	1.21	96.51
Bulgaria	1.27	20.0	1.06	84.59
Cyprus	1.35	17.0	1.15	92.04
Czech Republic	1.45	20.0	1.21	96.52
Germany	1.49	19.0	1.25	100.00
Denmark	1.49	25.0	1.19	95.13
Estonia	1.37	20.0	1.14	91.15
Spain	1.37	18.0	1.16	92.31
Finland	1.55	23.0	1.26	100.78
France	1.40	19.6	1.17	93.12
Greece (EL)	1.54	23.0	1.25	99.54
Hungary	1.50	27.0	1.18	94.46
Ireland	1.55	23.0	1.26	100.57
Italy	1.71	21.0	1.41	112.46
Lithuania	1.33	21.0	1.10	87.69
Luxembourg	1.26	15.0	1.10	87.61
Latvia	1.37	21.0	1.13	90.38
Malta	1.37	18.0	1.16	92.76
Netherlands	1.45	21.0	1.19	95.27
Poland	1.36	23.0	1.10	87.92
Portugal	1.45	23.0	1.18	94.15
Romania	1.32	24.0	1.06	84.87
Sweden	1.67	25.0	1.33	106.31
Slovenia	1.36	20.0	1.13	90.45
Slovakia	1.44	20.0	1.20	95.74
United Kingdom	1.75	20.0	1.46	116.60
Switzerland[2]	1.71	21.0	1.41	112.46
Russia[3]	1.27	20.0	1.06	84.59
others[3]	1.27	20.0	1.06	84.59

[2] estimate by values of Italy

[3] estimate by values of Bulgaria

Table G-25 International diesel fuel prices related to Germany in 2012 [incl. data from EC (2014)]

Country	Infrastructure charges for considered tractor-trailer combinations in €/km	Road infrastructure charge index rel. to GER
Belgium		0.00
Bulgaria		0.00
Czech Republic	0.17	109.68
Denmark		0.00
Germany	0.155	100.00
Estonia		0.00
Ireland		0.00
Greece	0.07	45.16
Spain	0.13	83.87
France	0.216	139.35
Italy	0.167	107.74
Cyprus		0.00
Latvia		0.00
Lithuania		0.00
Luxembourg		0.00
Hungary		0.00
Malta		0.00
Netherlands		0.00
Austria	0.323	208.39
Poland		0.00
Portugal	0.15	96.77
Romania		0.00
Slovenia	0.214	138.06
Slovakia		0.00
Finland		0.00
Sweden		0.00
United Kingdom		0.00
Switzerland	0.641	413.55
Russia		0.00
other countries		0.00

Table G-26 Estimated infrastructure user charges for international road freight transports related to Germany in 2012 [adapted from BMVI (2014a, p. 142)]

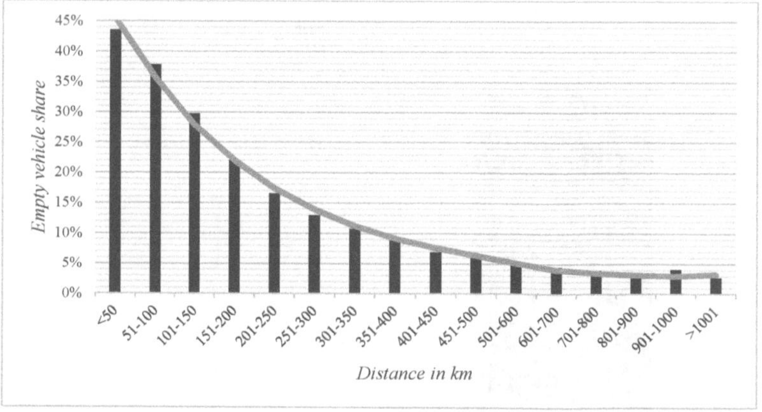

Fig. G-2 Curve fitting for empty vehicle shares per distance class of German road freight vehicles for the year 2012 [based on data from KBA (2013b, p. 60f.)]

	Wagon type	Depiction	Field of application	Average load capacity in t	Average length in m	Wagon weight in t	Picture and technical in-formation from:
1	Zaes 65 m³ Tank wagon		four axle tank wagon	57	14.90	23.50	a)
2	Falns 183		open bulk freight wagon	59	12.54	25.00	b)
3	Eanos-x 059		four -axle open wagon	48	15.74	23.90	c)
4	Roos 639		four-axle flat freight rail-way wagon with hinged sides and stanchions	53	19.90	27.00	d)
5	Sgns 691		four-axle con-tainer and swap body wagon	70	19.74	20.00	e)
6	Faals 151		open hopper car	100	15.05	35.00	f)
7	Habbis 345		covered slid-ing door wagon	47	24.13	29.77	g)
8	Samms 709		six-axle piv-oted bogie flat wagon	56	16.40	30.80	h)
9	Laekks 552		motor vehicle transport wagon	17	26.24	24.30	i)
10	Ucs 908 tank hop-per		tank-wagon for transport of powder	25	8.54	10.80	j)

Table G-27 Depiction of modelled rail freight wagons [see references below]

Pictures and technical data in Table G-27 are referenced as:

- *a)* RAILCO (2014)
- *b)* DB SCHENKER (2012f)
- *c)* DB SCHENKER (2012h)
- *d)* DB SCHENKER (2012c)
- *e)* DYBAS (2008)
- *f)* DB SCHENKER (2012g)
- *g)* DB SCHENKER (2012d)
- *h)* DB SCHENKER (2012b)
- *i)* DB SCHENKER (2012e)
- *j)* DB SCHENKER (2012a).

Annual transport volume	Total	Share of block trains	Share of wagon loads
Total in 1000 t	366,204	270,644	95,560
Share		74	26
Total in mill. tkm	110,797	86,134	24,663
Share		**78**	**22**

Table G-28 Role of block trains and wagon load trains for Germany in 2012 [incl. data from DESTATIS (2014a, p. 13)]

CPA commodity class	Detail	Cargo handling category	Rail freight wagon type
1	Products of agriculture, hunting and related services	1	3
2	Products of forestry, logging and related services	4	4
3	Fish and other fishing products; aquaculture products; support services to fishing	3	7
5	Coal and lignite	1	2
6	Crude petroleum and natural gas	-	-
7	Metal ores	1	6
8	Other mining and quarrying products	1	2
9	Mining support services	-	-
10	Food products	3	7
11	Beverages	3	7
12	Tobacco products	3	7
13	Textiles	3	7
14	Wearing apparel	3	7
15	Leather and related products	3	7
16	Wood and of products of wood and cork, except furniture; articles of straw and plaiting materials	4	7
17	Paper and paper products	4	7
18	Printing and recording services	3	7
19	Coke and refined petroleum products	2	1
20	Chemicals and chemical products	2	1
21	Basic pharmaceutical products and pharmaceutical preparations	3	7
22	Rubber and plastic products	4	7
23	Other non-metallic mineral products	1	10
24	Basic metals	4	8
25	Fabricated metal products, except machinery and equipment	4	8
26	Computer, electronic and optical products	3	7
27	Electrical equipment	3	8
28	Machinery and equipment n.e.c.	3	8
29	Motor vehicles, trailers and semi-trailers	4	9
30	Other transport equipment	3	7
31	Furniture	4	7
32	Other manufactured goods	3	7
-	all, by container/ swap/ additional transport vehicle	5	5

Table G-29 Assignment of modelled commodities to cargo handling types and rail freight wagons

Country	Electricity prices for industrial consumers (500 MWh < Consumption < 2 000 MWh) in €/kWh	Electricity price index rel. to GER
Belgium	0.1092	84.81
Bulgaria	0.0735	57.11
Czech Republic	0.1034	80.34
Denmark	0.0980	76.11
Germany	0.1287	100.00
Estonia	0.0801	62.24
Ireland	0.1357	105.40
Greece	0.1203	93.43
Spain	0.1205	93.63
France	0.0869	67.48
Italy	0.1713	133.10
Cyprus	0.2291	177.97
Latvia	0.1107	85.98
Lithuania	0.1142	88.69
Luxembourg	0.1032	80.15
Hungary	0.0974	75.64
Malta	0.1857	144.29
Netherlands	0.0967	75.10
Austria	0.1109	86.17
Poland	0.0937	72.77
Portugal	0.1146	89.04
Romania	0.0799	62.04
Slovenia	0.0945	73.39
Slovakia	0.1294	100.51
Finland	0.0750	58.24
Sweden	0.0793	61.58
United Kingdom	0.1168	90.75
Switzerland	0.1032	80.15
Russia	0.0735	57.11
other countries	0.1134	88.12

*) estimated (by Luxemburg); [2]) estimated (by Bulgaria); [3]) estimated (EU-27 average)
prices excluding VAT and other recoverable taxes and levies

Table G-30 International electricity prices related to Germany in 2012 [incl. data from EUROSTAT (2014d)]

Country	Rail infrastructure charges for 2012 in €/km	**Rail infrastructure charge index rel. to GER**
Belgium	1.87	66.04
Bulgaria	9.04	320.05
Czech Republic	7.61	269.33
Denmark	0.30	10.52
Germany	2.83	100.00
Estonia	10.73	379.99
Ireland	3.86	136.80
Greece	3.86	136.80
Spain	0.30	10.64
France	2.02	71.60
Italy	2.72	96.32
Cyprus	3.86	136.80
Latvia	7.43	262.96
Lithuania	12.79	452.79
Luxembourg	3.86	136.80
Hungary	2.26	80.00
Malta	3.86	136.80
Netherlands	4.43	156.64
Austria	4.25	150.44
Poland	6.23	220.65
Portugal	1.76	62.40
Romania	4.44	157.04
Slovenia	1.76	62.44
Slovakia	11.30	400.00
Finland	5.00	176.97
Sweden	0.81	28.53
United Kingdom	3.86	136.80
Switzerland	5.85	207.00
Russia	9.04	320.05
other countries	4.56	161.34

*) estimated (by UK); ²) estimated (by Bulgaria); ³) estimated (EU-27 average)

Table G-31 Average rail infrastructure charges in EU-27 [incl. data from OECD (2008); BNETZA (2013, p. 27)]

Transport connection	Annual number of containers in 1000 TEU	Annual net load of containerised transport volume in 1000 tonnes	Net load per TEU in tonnes
National	2,465	24,172	9.81
International - Sending	1,042	12,324	11.83
International - Receiving	900	9,937	11.04
		Weighted average:	10.54

Table G-32 Average load factor per rail freight container for Germany in 2012 [incl. data from DESTATIS (2014q, p. 9)]

	Annual transport volume in containers in tkm			**Transport volume in vehicles in tkm**			**Share of containers**
Transport connection	20-ft ISO container	30-ft ISO container	40-ft ISO container	unaccompanied	accompanied	share of accompanied	
National	5798	475	5412	435	-	0.00	96.41
International - Sending	2753	791	1925	2223	51	2.29	70.63
International - Receiving	1888	598	2163	2053	60	2.92	68.75

Table G-33 Share of different transport units for intermodal transports by rail for Germany in 2012 [incl. data from DESTATIS (2014q, p. 12 ff.)]

				Distance and travel time related factors				Time dependent cost factors		Mileage dependent cost factors			Additional cost factors				Comparison of results		
Origin	***Destination***	***Object***	Data from:	*Trip length, speed, total travel time*	*Net load*	*Number of locomotives/ wagons*	*Train length, net weight*	*Personnel*	*Maintenance a)*	*Maintenance b)*	*Infrastructure*	*Energy*	*Train composition*	*Transhipment*	*Administrative*	*other*	*Total costs*	***Costs per tonne** (avg.)*	***Dev. in %***
Port of Rotterdam	*Duisburg*	*containerised load*	[a]	2x250 km (48.00 km/h) (2x3.45h; 23 h)	840 t 740 t	1 42	720 m 1,620 t 1,520 t	340€	2,064€		1,337€	1,063€	1,145€	6,605€	1,883€	1) 2) 3)	15,060€	**9.53€**	54.93
			[b]	2x255 km 65.00 km/h 16 h	720 t	1 24	670 m 1,479 t	634€	2,340€		1,170€	1,351€	1,200€	-	916€	-	7,611€	**5.29 €**	-14.09
			own away	255 km 40.00 km/h 7.21 h	1,075 t (1,075 t)	1 34	2,090 t 2,090 t	353€	692€	241€	922€	498€	1,112€	1,833€	848€	-	6,498€	**6.15€**	
			own return	255 km 40.00 km/h 7.21 h	1,075 t (999 t)	1 34	2,090 t 2,090 t	379€	744€	259€	990€	535€	1,112€	1,833€	878€	-	6,729€		
Hamburg	*Salzgitter*	*Coal and lignite*	[a]	2x259 km 48.00 km/h 26.6 h	2,207 t 0 t	2 (1) 42	700 m 3,310 t 1,103 t	556€	3,749€		1,388€	1,979€	1,373€	-	1,588€	4) 5)	11,476€	**5.20€**	77.31
			own	211 km 40.00 km/h 12.21 h	2,454 t 0 t	1 35	580 m 3,634 t 1,180 t	514€	1,260€	427€	1,192€	1,439€	1,439€	-	939€		7,197€	**2.93 €**	
Port of Rotterdam	*Hamburg*	*containerised load*	[b]	2x562 km 65.00 km/h 25 h	720 t	1 24	670 m 1,479 t	1,012 €	4,680€		2,970€	2,947€	-	1,699€	-	-	14,508€	**10.08€**	24.92
			own away	586 km 40.00 km/h 15.48 h	1,075 t (0 t)*	1 34	2,090 t 2,090 t*	725€	1,654€	625€	1,939€	1,485€	0€	1,131€	-	0€	8,670€	**8.07€**	
			own return	586 km 40.00 km/h 15.48 h	1,075 t (0 t)*	1 34	2,090 t 2,090 t*	725€	1,654€	625€	1,939€	1,485€	0€	1,131€	-	0€	8,670€		

Table G-34 Comparison of calculated rail freight costs with related references, p.1

				Distance and travel time related factors				Time dependent cost factors		Mileage dependent cost factors			Additional cost factors				Comparison of results		
Origin	***Desti-nation***	***Object***	Data from:	*Trip length, speed, total travel time*	*Net load*	*Number of loco-motives/ wagons*	*Train length, net weight*	*Per-sonnel*	*Mainte nance a)*	*Mainte nance b)*	*Infra-struc-ture*	*Energy*	*Train compo-sition*	*Tran-ship-ment*	*Admin-istra-tive*	*other*	*Total costs*	***Costs per tonne*** *(avg.)*	***Dev. in %***
Port of Rotter-dam	*Mann-heim*	*contai-nerised load*	[b]	566 km 65.00 km/h 25 h	720 t	1 24	670 m 1,479 t	1,017€	4,680€		2,518€	2,966€	1,200€	-	1,857€	-	14,238€	**9.89€**	3.35
			own away	480 km 40.00 km/h 12.83 h*	1,075 t 1,075 t	1 34	2,090 t 2,090 t*	614€	1,371€	512€	1,639€	1,194€	1,112€	1,833€	1,241€	-	9,516€	**9.57€**	
			own re- turn	480 km 40.00 km/h 12.83 h*	1,075 t 269 t	1 34	2,090 t 2,090 t*	768€	1,714€	640€	2,050€	1,494€	1,112€	1,833€	1,442€	-	1,112€		
Port of Rotter-dam	*Mün-chen*	*con-tainer-ised load*	[b]	958 km 65.00 km/h 37 h	720 t	1 24	670 m 1479 t	1,499€	4,680€		4,040€	4,964€	1,200€	-	2,457€	-	18,840€	**13.08€**	-10.55
			own away	838 km 40.00 km/h 21.78 h*	1,075 t 1,075 t	1 34	2,090 t 2,090 t*	989€	2,327€	893€	2,651€	2,175€	1,112€	1,833€	1,797€	-	13,777€	**14.63€**	
			own re- turn	838 km 40.00 km/h 21.78 h*	1,075 t 677 t	1 34	2090 t 2090 t*	1,360€	3,200€	1,228€	3,644€	2,991€	1,112€	1,833€	2,305€	-	17,671€		

Table G-35 Comparison of calculated rail freight costs with related references, p.2

Reference cost calculation data in Table G-34 and Table G-35 are indicated as:

- *a)* BMVI (2014a)
- *b)* DEUTSCH (2013).

Additional cost factors therein are indicated as:

- *1)* for a last mile surcharge of 560€
- *2)* for an additional shunting surcharge of 560€
- *3)* for surcharges for driver changes of 674€
- *4)* for a last mile surcharge of 672€
- *5)* for an additional shunting surcharge of 170€.

These additional factors within BMVI (2014a), among others, give reason to identified link cost calculation differences.

German national transports in millions of tkm enrolled by vessels from:	Total	Freq.	Modelled crew labour costs
Belgium	265	2.4%	111
Bulgaria	4	0.0%	14
Czech Republic	83	0.8%	16
Denmark	3	0.0%	96
Germany	7,498	68.7%	(69.69)
Estonia	4	0.0%	22
Ireland	0	0.0%	90
Greece	0	0.0%	59
Spain	0	0.0%	49
France	13	0.1%	68
Italy	0	0.0%	61
Cyprus	0	0.0%	18
Latvia	2	0.0%	20
Lithuania	0	0.0%	22
Luxembourg	60	0.5%	53
Hungary	7	0.1%	13
Malta	0	0.0%	90
Netherlands	2,403	22.0%	72
Austria	4	0.0%	33
Poland	450	4.1%	24
Portugal	0	0.0%	29
Romania	7	0.1%	12
Slovenia	0	0.0%	49
Slovakia	0	0.0%	20
Finland	1	0.0%	56
Sweden	0	0.0%	86
United Kingdom	0	0.0%	58
Switzerland	80	0.7%	168
Russia	0	0.0%	14
others	28	0.3%	0
Weighted German average			69.33

() German reference labour costs of 69.69 €/h without cabotage

Table G-36 Share of German and international forwarders operating on national IWW freight transport network link relations [incl. data from EUROSTAT (2014n); Table G-41]

CPA commodity class	*Dominant vessel type for non-containerised IWW transports*	*Vessel configuration for non-containerised IWW transports*	*Reported annual average net load per trip for non-containerised IWW transports in tonnes*	*Modelled net load per trip for non-containerised IWW transports in tonnes*	*Average net-load-specific fuel consumption in litres per kilometre for non-containerised IWW transports*
1	1	1	1,169	1,169	12.95
2	1	2	317	317	3.51
3	1	3	1,365	1,365	15.13
5	1	4	1,861	1,826	20.23
6	-	-	-	-	-
7	1	5	2,177	1,826	20.23
8	1	6	1,301	1,301	14.42
9	-	-	-	-	-
10	1	7	799	799	8.85
11	1	8	161	161	1.78
12	1	9	342	342	3.79
13	1	10	18	18	0.20
14	1	11	25	25	0.27
15	1	12	33	33	0.36
16	1	13	377	377	4.17
17	1	14	800	800	8.87
18	1	15	34	34	0.38
19	2	16	1,544	1,544	17.47
20	2	17	1,058	1,058	11.97
21	1	18	558	558	6.18
22	1	19	74	74	0.82
23	1	20	648	648	7.18
24	1	21	1,147	1,147	12.71
25	1	22	170	170	1.89
26	1	23	197	197	2.18
27	1	24	103	103	1.14
28	1	25	108	108	1.20
29	1	26	222	222	2.46
30	1	27	140	140	1.55
31	1	28	32	32	0.36
32	1	29	(377)	377	4.17

Table G-37 Assignment of non-containerised commodities to freight vessel types and specific configurations, with net load and fuel consumption factors [incl. data from EUROSTAT (2015b)]

Type of vessel	Self-propelled barge	Self-propelled tanker barge
Number of movements	94,847	33,396
Load capacity	173,203	62,256
Payload reported in 1,000 tonnes	120,491	47,720
Payload reported in 1,000 TEU	1,048	-
Average load capacity in tonnes	1,826	1,864
average payload in tonnes (for tonnes and TEU, with 7,49 tonnes per TEU)	1,353	1,429
Average load factor	0.74	0.77

Table G-38 IWW transport by number of movements, load capacity and payload for selected freight vessel types for Germany in 2012 [incl. data from DESTATIS (2014o, p. 24)]

	Ascending river	Descending river	Average for river	IWW channel
Exemplary vessel	*average diesel fuel consumption in litres per 1000 tkm*			
European vessel measure: 85 x 9,5 metres load capacity: 1300 tonnes or 90 TEU	16.8	8.3	12.55	11.8
Large motor vessel measure: 110 x 11,4 metres load capacity: 2300 tonnes or 208 TEU	7.5	3.9	5.70	4.6
Jowi-class vessel measure: 135 x 17,2 metres load capacity: 5200 tonnes or 500 TEU	5.2	2.7	4.45	-

Table G-39 Specific diesel fuel consumption rates and size measures for exemplary IWW vessels [adopted from KRANKE (2011)]

IWW crew by function	Nb. of crew members for 16 operating hours	Basic hourly wage for 1. - 14. hour in €	Nb. of hours	Shift surcharges for 10. - 14. Hour in €	Nb. of hours	Overtime wage for 14. - 16. hour in €	Nb. of hours	Night-work surcharges of 2 hours in €	Nb. of hours	Hourly wage in €	Hourly labour costs in €
Skipper/ master	1	16.44	14	7.97	4	20.55	2	7.97	2	19.94	27.96
Helmsman	1	13.36	14	6.47	4	16.70	2	6.47	2	16.20	22.72
Sailor	1	11.17	14	5.43	4	13.96	2	5.43	2	13.56	19.01
											Σ= 69.69

With a number of crew members for self-propelled barges and self-propelled tank barges and night-work surcharges for exceeding working time from 20 - 6 o'clock.

Table G-40 Determination of hourly wages for German IWW crews for selected freight vessel types for Germany in 2012

	Average annual labour costs for the year 2008	Average labour costs per year in 2012	**Labour cost index related to Germany (2012 data if applicable, else 2008)**	Modelled international labour costs (for reference of German labour costs of 69.69 €/h)
Belgium	85,651	-	**159.0%**	111
Bulgaria	8,375	12,373	**20.1%**	14
Czech Republic	12,269	14,401	**23.4%**	16
Denmark	74,059	85,337	**138.4%**	96
Germany	-	61,667		
Estonia[1]	-	-	**30.9%**	22
Ireland	59,582	79,535	**129.0%**	90
Greece	45,360	-	**84.2%**	59
Spain	44,667	43,764	**71.0%**	49
France	-	60,189	**97.6%**	68
Italy	46,863	63,185	**87.0%**	61
Cyprus	31,250	16,125	**26.1%**	18
Latvia	26,690	17,707	**28.7%**	20
Lithuania	19,610	19,071	**30.9%**	22
Luxembourg	-	47,004	**76.2%**	53
Hungary	14,002	11,636	**18.9%**	13
Malta[2]	-	-	**129.0%**	90
Netherlands	59,092	64,117	**104.0%**	72
Austria	:	29,156	**47.3%**	33
Poland	21,564	21,357	**34.6%**	24
Portugal	:	25,754	**41.8%**	29
Romania	9,398	10,335	**16.8%**	12
Slovenia	38,021	:	**70.6%**	49
Slovakia	13,740	17,296	**28.0%**	20
Finland	36,826	49,363	**80.0%**	56
Sweden	60,183	76,524	**124.1%**	86
United Kingdom	47,692	51,360	**83.3%**	58
Averge EU-27	37,745	39,875	68.5%	
Switzerland	-	148,636	**241.0%**	168
Russia[3]	-	12,373	**20.1%**	14
AVERAGE	37,745	43,261	72.9%	

[1] estimate from Lithuania value; [2] estimate from Ireland value; [3] estimate from Bulgaria value

Table G-41 International labour cost index related to Germany in 2012 for employees in the segment of *water transport* (CPA-50) [incl. data from EUROSTAT (2014r); EUROSTAT (2014s)]

Country	Average price per litre heating oil in EUR (incl. excise duties and VAT)	VAT per country in %	Average price 1000l heating oil in EUR (incl. excises, excl. VAT)	Heating oil price index rel. to Germany
Austria	999.97	20.00	833.31	106.62%
Belgium	895.94	21.00	740.44	94.74%
Bulgaria	930.04	20.00	775.04	99.16%
Cyprus	1067.10	17.00	912.05	116.69%
Czech Republic	963.47	20.00	802.89	102.73%
Germany	930.09	19.00	781.59	100.00%
Denmark	1497.41	25.00	1197.93	153.27%
Estonia	1029.90	20.00	858.25	109.81%
Spain	950.74	18.00	805.71	103.09%
Finland	1138.51	23.00	925.62	118.43%
France	969.19	19.60	810.36	103.68%
Greece (EL)	1273.80	23.00	1035.61	132.50%
Hungary	1504.85	27.00	1184.92	151.60%
Ireland	1101.55	23.00	895.57	114.58%
Italy	1458.29	21.00	1205.20	154.20%
Lithuania	876.14	21.00	724.08	92.64%
Luxembourg	819.62	15.00	712.71	91.19%
Latvia	1003.07	21.00	828.98	106.06%
Malta	1030.82	18.00	873.57	111.77%
Netherlands	879.04	21.00	726.48	92.95%
Poland	955.95	23.00	777.20	99.44%
Portugal	1296.65	23.00	1054.19	134.88%
Romania	1193.82	24.00	962.76	123.18%
Sweden	1433.04	25.00	1146.43	146.68%
Slovenia	1019.47	20.00	849.56	108.70%
Slovakia	963.47	20.00	802.89	102.73%
United Kingdom	876.39	20.00	730.32	93.44%
Switzerland[2]	1458.29	21.00	1205.20	154.20%
Russia[3]	930.04	20.00	775.04	99.16%
others[3]	930.04	20.00	775.04	99.16%

[2] estimate by values of Italy

[3] estimate by values of Bulgaria

Table G-42 International heating oil prices related to Germany in 2012 [incl. data from EC (2014)]

CPA- 2008	Detail	Infrastructure usage cost group (NSTR)	Infrastructure usage costs per 1000 tkm or unit in Eurocents
1	Products of agriculture, hunting and related services	I	0,910
2	Products of forestry, logging and related services	V	0,777
3	Fish and other fishing products; aquaculture products; support services to fishing	II	0,910
5	Coal and lignite	VI	0,708
6	Crude petroleum and natural gas	-	-
7	Metal ores	VI	0,708
8	Other mining and quarrying products	VI	0,708
9	Mining support services	-	-
10	Food products	I	0,910
11	Beverages	I	0,910
12	Tobacco products	I	0,910
13	Textiles	I	0,910
14	Wearing apparel	I	0,910
15	Leather and related products	I	0,910
16	Wood and of products of wood and cork, except furniture; articles of straw and plaiting materials	III	0,844
17	Paper and paper products	I	0,910
18	Printing and recording services	I	0,910
19	Coke and refined petroleum products	VI	0,708
20	Chemicals and chemical products	V	0,777
21	Basic pharmaceutical products and pharmaceutical preparations	I	0,910
22	Rubber and plastic products	III	0,844
23	Other non-metallic mineral products	V	0,777
24	Basic metals	III	0,844
25	Fabricated metal products, except machinery and equipment	I	0,910
26	Computer, electronic and optical products	I	0,910
27	Electrical equipment	I	0,910
28	Machinery and equipment n.e.c.	I	0,910
29	Motor vehicles, trailers and semi-trailers	I	0,910
30	Other transport equipment	I	0,910
31	Furniture	I	0,910
32	Other manufactured goods	I	0,910
-	all, by container/ swap/ additional transport vehicle		3.3333[1]

[1]weighted average for containers (2/3 for container up to 20-foot and 1/3 for 40-foot)

Table G-43 Assignment of modelled commodities to IWW infrastructure charging groups [incl. data from WSDW (2012)]

CPA-2008	*Dominant vessel type for containerised IWW transports*	*Vessel configuration for containerised IWW transports*	*Reported annual average net load per trip for containerised IWW transports in tonnes*	*Modelled net load per trip for containerised IWW transports in tonnes*	*Average net-load-specific fuel consumption in litres per kilometre for containerised IWW transports*	*Modelled number of TEU for containerised IWW transports*
1	1	30	975	974.95	10.80	93
2	1	30	317	317.06	3.51	31
3	1	30	0	0.00	0.00	0
5	1	30	1,861	1,507.22	16.70	143
6	-	-	-	-	-	-
7	1	30	1,872	1,507.22	16.70	143
8	1	30	1,258	1,257.79	13.94	120
9	-	-	-	-	-	-
10	1	30	612	611.60	6.78	59
11	1	30	161	161.03	1.78	16
12	1	30	342	341.93	3.79	33
13	1	30	18	18.18	0.20	2
14	1	30	25	24.73	0.27	3
15	1	30	33	32.53	0.36	4
16	1	30	377	376.67	4.17	36
17	1	30	800	800.38	8.87	76
18	1	30	34	34.05	0.38	4
19	1	30	1,585	1,507.22	16.70	143
20	1	30	890	889.59	9.86	85
21	1	30	558	558.11	6.18	53
22	1	30	74	73.95	0.82	8
23	1	30	310	310.39	3.44	30
24	1	30	976	976.26	10.82	93
25	1	30	133	133.17	1.48	13
26	1	30	194	193.83	2.15	19
27	1	30	90	90.48	1.00	9
28	1	30	109	108.74	1.20	11
29	1	30	222	222.29	2.46	22
30	1	30	140	139.82	1.55	14
31	1	30	32	32.13	0.36	4
32	1	30	(30.00)	30.00	0.33	3
Average*	**1**	**30**	**633.35**	**633.35**	**7.02**	**61**

*weighted by reported total annual transport volume

Table G-44 Assignment of containerised commodities to freight vessel types and average configurations, with average net load and related fuel consumption factors [incl. data from EUROSTAT (2015b)]

CPA-2008	*Dominant vessel type for containerised IWW transports*	*Vessel configuration for containerised IWW transports*	*Modelled number of TEU for containerised IWW transports*	*Hourly container provision costs in € per hour*[1]	*Transhipment time in hours*[2]	*Transhipment activity costs in Euro*[3]
1	1	30	93	9.3	3.72	155.03
2	1	30	31	3.1	1.24	51.68
3	1	30	0	0	0	0.00
5	1	30	143	14.3	5.72	238.38
6	-	-	-	-	-	-
7	1	30	143	14.3	5.72	238.38
8	1	30	120	12	4.8	200.04
9	-	-	-	-	-	-
10	1	30	59	5.9	2.36	98.35
11	1	30	16	1.6	0.64	26.67
12	1	30	33	3.3	1.32	55.01
13	1	30	2	0.2	0.08	3.33
14	1	30	3	0.3	0.12	5.00
15	1	30	4	0.4	0.16	6.67
16	1	30	36	3.6	1.44	60.01
17	1	30	76	7.6	3.04	126.69
18	1	30	4	0.4	0.16	6.67
19	1	30	143	14.3	5.72	238.38
20	1	30	85	8.5	3.4	141.70
21	1	30	53	5.3	2.12	88.35
22	1	30	8	0.8	0.32	13.34
23	1	30	30	3	1.2	50.01
24	1	30	93	9.3	3.72	155.03
25	1	30	13	1.3	0.52	21.67
26	1	30	19	1.9	0.76	31.67
27	1	30	9	0.9	0.36	15.00
28	1	30	11	1.1	0.44	18.34
29	1	30	22	2.2	0.88	36.67
30	1	30	14	1.4	0.56	23.34
31	1	30	4	0.4	0.16	6.67
32	1	30	3	0.3	0.12	5.00
Average	1	30	61	6.10	2.44	101.69

[1]with 0.1 €/h per TEU; [2]with 0.04 hours per TEU; [3]with 16.67 € per TEU

Table G-45 Determination of average hourly container provision costs, average transhipment times and related activity costs for containerised freight vessel loads [incl. data from Table G-44]

	NUTS-3 ID within model	NUTS-3 Name within model	NUTS-3 ID Eurostat	Latitude within model	Longitude within model
1	1001	Flensburg, Stadt	DEF01	54.79779816	9.477919579
2	1002	Kiel, Landeshauptstadt	DEF02	54.32324982	10.1322403
3	1003	Lübeck, Hansestadt	DEF03	53.86626816	10.68087959
4	1004	Neumünster, Stadt	DEF04	54.07014	9.98783
5	1051	Dithmarschen	DEF05	54.17990112	9.095509529
6	1053	Herzogtum Lauenburg	DEF06	53.69739914	10.77250004
7	1054	Nordfriesland	DEF07	54.47034836	9.05163002
8	1055	Ostholstein	DEF08	54.13729858	10.6086998
9	1056	Pinneberg	DEF09	53.64479828	9.797780037
10	1057	Plön	DEF0A	54.16133	10.42595
11	1058	Rendsburg-Eckernförde	DEF0B	54.29779816	9.666950226
12	1059	Schleswig-Flensburg	DEF0C	54.53390121	9.562609673
13	1060	Segeberg	DEF0D	53.9292984	10.32530022
14	1061	Steinburg	DEF0E	53.93130112	9.518050194
15	1062	Stormarn	DEF0F	53.81840134	10.38759995
16	2000	Hamburg, Freie und Hansestadt	DE600	53.55334091	9.992469788
17	3101	Braunschweig, Stadt	DE911	52.23550034	10.54189968
18	3102	Salzgitter, Stadt	DE912	52.15327835	10.33312035
19	3103	Wolfsburg, Stadt	DE913	52.42440033	10.79749966
20	3151	Gifhorn	DE914	52.46289825	10.55729961
21	3152	Göttingen	DE915	51.55989838	9.931920052
22	3153	Goslar	DE916	51.91469955	10.42930031
23	3154	Helmstedt	DE917	52.22919846	10.99839973
24	3155	Northeim	DE918	51.70349884	9.992620468
25	3156	Osterode am Harz	DE919	51.7276001	10.24779987
26	3157	Peine	DE91A	52.32279968	10.24569988
27	3158	Wolfenbüttel	DE91B	52.1589	10.5591
28	3241	Region Hannover	DE929	52.37226868	9.738149643
29	3251	Diepholz	DE922	52.60565	8.37079
30	3252	Hameln-Pyrmont	DE923	52.1012001	9.348750114
31	3254	Hildesheim	DE925	52.16080093	9.978560448
32	3255	Holzminden	DE926	51.82381	9.45589
33	3256	Nienburg (Weser)	DE927	52.63742447	9.220729828
34	3257	Schaumburg	DE928	52.31370163	9.213179588
35	3351	Celle	DE931	52.62189865	10.0763998
36	3352	Cuxhaven	DE932	53.84429932	8.687970161
37	3353	Harburg	DE933	53.36230087	10.20860004
38	3354	Lüchow-Dannenberg	DE934	52.96630096	11.15699959
39	3355	Lüneburg	DE935	53.25270081	10.41180038
40	3356	Osterholz	DE936	53.2358017	8.785050392
41	3357	Rotenburg (Wümme)	DE937	53.09880066	9.394630432

Table G-46 List of modelled regions, p. 1

	NUTS-3 ID within model	NUTS-3 Name within model	NUTS-3 ID Eurostat	Latitude within model	Longitude within model
42	3358	Heidekreis	DE938	52.86597061	9.694399834
43	3359	Stade	DE939	53.60229874	9.475330353
44	3360	Uelzen	DE93A	52.96582	10.55803
45	3361	Verden	DE93B	52.92451859	9.241938591
46	3401	Delmenhorst, Stadt	DE941	53.04919815	8.625289917
47	3402	Emden, Stadt	DE942	53.364705	7.172326
48	3403	Oldenburg (Oldenburg), Stadt	DE943	53.14345	8.21455
49	3404	Osnabrück, Stadt	DE944	52.29130173	8.054650307
50	3405	Wilhelmshaven, Stadt	DE945	53.52959824	8.099840164
51	3451	Ammerland	DE946	53.25260162	7.904419899
52	3452	Aurich	DE947	53.46720123	7.488900185
53	3453	Cloppenburg	DE948	52.8404007	8.043840408
54	3454	Emsland	DE949	52.69329834	7.277890205
55	3455	Friesland	DE94A	53.57289886	7.895339966
56	3456	Grafschaft Bentheim	DE94B	52.4457016	7.065020084
57	3457	Leer	DE94C	53.23751068	7.461763859
58	3458	Oldenburg	DE94D	53.14345	8.21455
59	3459	Osnabrück	DE94E	52.29130173	8.054650307
60	3460	Vechta	DE94F	52.7356987	8.281640053
61	3461	Wesermarsch	DE94G	53.32632828	8.478639603
62	3462	Wittmund	DE94H	53.5727005	7.764949799
63	4011	Bremen, Stadt	DE501	53.07509995	8.804690361
64	4012	Bremerhaven, Stadt	DE502	53.52610016	8.623869896
65	5111	Düsseldorf, Stadt	DEA11	51.21562958	6.776040077
66	5112	Duisburg, Stadt	DEA12	51.43143082	6.763929844
67	5113	Essen, Stadt	DEA13	51.45180893	7.01060009
68	5114	Krefeld, Stadt	DEA14	51.33390045	6.562300205
69	5116	Mönchengladbach, Stadt	DEA15	51.1955	6.44268
70	5117	Mülheim an der Ruhr, Stadt	DEA16	51.41857	6.88452
71	5119	Oberhausen, Stadt	DEA17	51.4694519	6.855090141
72	5120	Remscheid, Stadt	DEA18	51.18249893	7.212669849
73	5122	Solingen, Klingenstadt	DEA19	51.16759872	7.085100174
74	5124	Wuppertal, Stadt	DEA1A	51.083403	7.024221
75	5154	Kleve	DEA1B	51.7820015	6.117949963
76	5158	Mettmann	DEA1C	51.24489975	6.96216011
77	5162	Rhein-Kreis Neuss	DEA1D	51.2042	6.68795
78	5166	Viersen	DEA1E	51.2555	6.39652
79	5170	Wesel	DEA1F	51.66431	6.62957
80	5314	Bonn, Stadt	DEA22	50.73242188	7.101860046
81	5315	Köln, Stadt	DEA23	50.93753	6.96028
82	5316	Leverkusen, Stadt	DEA24	51.03319168	6.981739998

Table G-47 List of modelled regions, p. 2

	NUTS-3 ID within model	*NUTS-3 Name within model*	*NUTS-3 ID Eurostat*	*Latitude within model*	*Longitude within model*
83	**5334**	**Städteregion Aachen**	DEA2D	50.77822876	6.088640213
84	**5358**	**Düren**	DEA26	50.80059814	6.499529839
85	**5362**	**Rhein-Erft-Kreis**	DEA27	50.94889832	6.648990154
86	**5366**	**Euskirchen**	DEA28	50.65769958	6.787320137
87	**5370**	**Heinsberg**	DEA29	51.06729889	6.099219799
88	**5374**	**Oberbergischer Kreis**	DEA2A	51.02640152	7.557400227
89	**5378**	**Rheinisch-Bergischer Kreis**	DEA2B	50.98949814	7.123159885
90	**5382**	**Rhein-Sieg-Kreis**	DEA2C	50.79985	7.20745
91	**5512**	**Bottrop, Stadt**	DEA31	51.51819992	6.932169914
92	**5513**	**Gelsenkirchen, Stadt**	DEA32	51.51774	7.08572
93	**5515**	**Münster, Stadt**	DEA33	51.96300888	7.617809772
94	**5554**	**Borken**	DEA34	51.8404007	6.874750137
95	**5558**	**Coesfeld**	DEA35	51.9294014	7.165850163
96	**5562**	**Recklinghausen**	DEA36	51.61949921	7.213409901
97	**5566**	**Steinfurt**	DEA37	52.15008926	7.338950157
98	**5570**	**Warendorf**	DEA38	51.94269943	7.981440067
99	**5711**	**Bielefeld, Stadt**	DEA41	52.01092148	8.540869713
100	**5754**	**Gütersloh**	DEA42	51.90390015	8.425290108
101	**5758**	**Herford**	DEA43	52.1167984	8.655059814
102	**5762**	**Höxter**	DEA44	51.77149963	9.378410339
103	**5766**	**Lippe**	DEA45	51.94300079	8.876379967
104	**5770**	**Minden-Lübbecke**	DEA46	52.28649902	8.945340157
105	**5774**	**Paderborn**	DEA47	51.71813965	8.752039909
106	**5911**	**Bochum, Stadt**	DEA51	51.48199844	7.212639809
107	**5913**	**Dortmund, Stadt**	DEA52	51.51660919	7.4582901
108	**5914**	**Hagen, Stadt der FernUniversi.**	DEA53	51.35960007	7.460989952
109	**5915**	**Hamm, Stadt**	DEA54	51.6711998	7.774710178
110	**5916**	**Herne, Stadt**	DEA55	51.54030991	7.21987009
111	**5954**	**Ennepe-Ruhr-Kreis**	DEA56	51.29460144	7.299389839
112	**5958**	**Hochsauerlandkreis**	DEA57	51.35279846	8.276539803
113	**5962**	**Märkischer Kreis**	DEA58	51.2108	7.58629
114	**5966**	**Olpe**	DEA59	51.02600098	7.847869873
115	**5970**	**Siegen-Wittgenstein**	DEA5A	50.86700058	7.986810207
116	**5974**	**Soest**	DEA5B	51.55879974	8.119050026
117	**5978**	**Unna**	DEA5C	51.52619934	7.695569992
118	**6411**	**Darmstadt, Wissenschaftsstadt**	DE711	49.85340118	8.646519661
119	**6412**	**Frankfurt am Main, Stadt**	DE712	50.11207962	8.683409691
120	**6413**	**Offenbach am Main, Stadt**	DE713	50.10329819	8.777469635
121	**6414**	**Wiesbaden, Landeshauptstadt**	DE714	50.08459854	8.242289543
122	**6431**	**Bergstraße**	DE715	49.64260101	8.637519836
123	**6432**	**Darmstadt-Dieburg**	DE716	49.85340118	8.646519661

Table G-48 List of modelled regions, p. 3

	NUTS-3 ID within model	*NUTS-3 Name within model*	*NUTS-3 ID Eurostat*	*Latitude within model*	*Longitude within model*
124	6433	Groß-Gerau	DE717	49.91279984	8.485799789
125	6434	Hochtaunuskreis	DE718	50.2216	8.61318
126	6435	Main-Kinzig-Kreis	DE719	50.13539886	8.917329788
127	6436	Main-Taunus-Kreis	DE71A	50.08720016	8.447389603
128	6437	Odenwaldkreis	DE71B	49.66559982	8.998209953
129	6438	Offenbach	DE71C	50.10329819	8.777469635
130	6439	Rheingau-Taunus-Kreis	DE71D	50.14089966	8.067540169
131	6440	Wetteraukreis	DE71E	50.31829834	8.756230354
132	6531	Gießen	DE721	50.58409882	8.678359985
133	6532	Lahn-Dill-Kreis	DE722	50.55469894	8.513469696
134	6533	Limburg-Weilburg	DE723	50.39020157	8.055130005
135	6534	Marburg-Biedenkopf	DE724	50.80550003	8.777580261
136	6535	Vogelsbergkreis	DE725	50.63650131	9.395389557
137	6611	Kassel, documenta-Stadt	DE731	51.31610107	9.468520164
138	6631	Fulda	DE732	50.54769897	9.671759605
139	6632	Hersfeld-Rotenburg	DE733	50.86500168	9.702589989
140	6633	Kassel	DE734	51.31610107	9.468520164
141	6634	Schwalm-Eder-Kreis	DE735	51.03170013	9.403120041
142	6635	Waldeck-Frankenberg	DE736	51.27819824	8.857460022
143	6636	Werra-Meißner-Kreis	DE737	51.20059967	10.00809956
144	7111	Koblenz, kreisfreie Stadt	DEB11	50.35694	7.589
145	7131	Ahrweiler	DEB12	50.54500961	7.101709843
146	7132	Altenkirchen (Westerwald)	DEB13	50.68669891	7.655059814
147	7133	Bad Kreuznach	DEB14	49.82559967	7.868330002
148	7134	Birkenfeld	DEB15	49.6507988	7.162360191
149	7135	Cochem-Zell	DEB16	50.14419937	7.163030148
150	7137	Mayen-Koblenz	DEB17	50.35694	7.589
151	7138	Neuwied	DEB18	50.44189835	7.474229813
152	7140	Rhein-Hunsrück-Kreis	DEB19	49.98839951	7.541600227
153	7141	Rhein-Lahn-Kreis	DEB1A	50.32759857	7.728079796
154	7143	Westerwaldkreis	DEB1B	50.44430161	7.830430031
155	7211	Trier, kreisfreie Stadt	DEB21	49.75733948	6.636199951
156	7231	Bernkastel-Wittlich	DEB22	50.00059891	6.913899899
157	7232	Eifelkreis Bitburg-Prüm	DEB23	49.96641	6.52986
158	7233	Vulkaneifel	DEB24	50.19620132	6.836969852
159	7235	Trier-Saarburg	DEB25	49.75733948	6.636199951
160	7311	Frankenthal (Pfalz), kr.f. St.	DEB31	49.52529907	8.354550362
161	7312	Kaiserslautern, kreisfr. Stadt	DEB32	49.44329834	7.755539894
162	7313	Landau in der Pfalz, kr.f. St.	DEB33	49.1957016	8.131930351
163	7314	Ludwigshafen am Rhein, Stadt	DEB34	49.48094177	8.447299957
164	7315	Mainz, kreisfreie Stadt	DEB35	50.00061035	8.272279739

Table G-49 List of modelled regions, p. 4

	NUTS-3 ID within model	*NUTS-3 Name within model*	*NUTS-3 ID Eurostat*	*Latitude within model*	*Longitude within model*
165	7316	Neustadt an der Weinstraße,St.	DEB36	49.35019	8.14869
166	7317	Pirmasens, kreisfreie Stadt	DEB37	49.19839859	7.6073699
167	7318	Speyer, kreisfreie Stadt	DEB38	49.31000137	8.425359726
168	7319	Worms, kreisfreie Stadt	DEB39	49.62779999	8.356789589
169	7320	Zweibrücken, kreisfreie Stadt	DEB3A	49.2471	7.36348
170	7331	Alzey-Worms	DEB3B	49.75941849	8.157179832
171	7332	Bad Dürkheim	DEB3C	49.45999908	8.164690018
172	7333	Donnersbergkreis	DEB3D	49.53739929	8.185250282
173	7334	Germersheim	DEB3E	49.21969986	8.370750427
174	7335	Kaiserslautern	DEB3F	49.44329834	7.755539894
175	7336	Kusel	DEB3G	49.53630066	7.397469997
176	7337	Südliche Weinstraße	DEB3H	49.1957016	8.131930351
177	7338	Rhein-Pfalz-Kreis	DEB3I	49.48094	8.4473
178	7339	Mainz-Bingen	DEB3J	50.00061035	8.272279739
179	7340	Südwestpfalz	DEB3K	49.19839859	7.6073699
180	8111	Stuttgart, Landeshauptstadt	DE111	48.76766968	9.171919823
181	8115	Böblingen	DE112	48.66899872	8.994930267
182	8116	Esslingen	DE113	48.73920059	9.30904007
183	8117	Göppingen	DE114	48.7052002	9.644949913
184	8118	Ludwigsburg	DE115	48.89319992	9.194270134
185	8119	Rems-Murr-Kreis	DE116	48.83810043	9.337120056
186	8121	Heilbronn, Stadt	DE117	49.13830185	9.237910271
187	8125	Heilbronn	DE118	49.13830185	9.237910271
188	8126	Hohenlohekreis	DE119	49.27669907	9.688099861
189	8127	Schwäbisch Hall	DE11A	49.1191	9.74863
190	8128	Main-Tauber-Kreis	DE11B	49.62319946	9.665280342
191	8135	Heidenheim	DE11C	48.62592	10.17231
192	8136	Ostalbkreis	DE11D	48.8370018	10.10229969
193	8211	Baden-Baden, Stadt	DE121	48.78150177	8.203570366
194	8212	Karlsruhe, Stadt	DE122	49.00540161	8.386870384
195	8215	Karlsruhe	DE123	49.00540161	8.386870384
196	8216	Rastatt	DE124	48.86629868	8.199620247
197	8221	Heidelberg, Stadt	DE125	49.4134903	8.708069801
198	8222	Mannheim, Universitätsstadt	DE126	49.48120117	8.482709885
199	8225	Neckar-Odenwald-Kreis	DE127	49.35070038	9.154430389
200	8226	Rhein-Neckar-Kreis	DE128	49.4134903	8.708069801
201	8231	Pforzheim, Stadt	DE129	48.89350128	8.706979752
202	8235	Calw	DE12A	48.71289825	8.743470192
203	8236	Enzkreis	DE12B	48.89350128	8.706979752
204	8237	Freudenstadt	DE12C	48.46699905	8.424779892
205	8311	Freiburg im Breisgau, Stadt	DE131	47.99853134	7.849649906

Table G-50 List of modelled regions, p. 5

	NUTS-3 ID within model	NUTS-3 Name within model	NUTS-3 ID Eurostat	Latitude within model	Longitude within model
206	8315	Breisgau-Hochschwarzwald	DE132	47.99853134	7.849649906
207	8316	Emmendingen	DE133	48.11909866	7.847770214
208	8317	Ortenaukreis	DE134	48.48149872	7.93956995
209	8325	Rottweil	DE135	48.15999985	8.625590324
210	8326	Schwarzwald-Baar-Kreis	DE136	48.06216	8.53802
211	8327	Tuttlingen	DE137	47.97909927	8.793910027
212	8335	Konstanz	DE138	47.67670059	9.182880402
213	8336	Lörrach	DE139	47.6089	7.6604
214	8337	Waldshut	DE13A	47.62369919	8.214449883
215	8415	Reutlingen	DE141	48.49499893	9.202969551
216	8416	Tübingen	DE142	48.52130127	9.052980423
217	8417	Zollernalbkreis	DE143	48.26979828	8.861720085
218	8421	Ulm, Universitätsstadt	DE144	48.39459991	9.968560219
219	8425	Alb-Donau-Kreis	DE145	48.39459991	9.968560219
220	8426	Biberach	DE146	48.10079956	9.774849892
221	8435	Bodenseekreis	DE147	47.66490173	9.478580475
222	8436	Ravensburg	DE148	47.77659988	9.604559898
223	8437	Sigmaringen	DE149	48.0909996	9.235480309
224	9161	Ingolstadt, Stadt	DE211	48.7737999	11.42010021
225	9162	München, Landeshauptstadt	DE212	48.1391	11.5145
226	9163	Rosenheim, Stadt	DE213	47.8667984	12.11289978
227	9171	Altötting	DE214	48.22629929	12.66399956
228	9172	Berchtesgadener Land	DE215	47.71979904	12.85840034
229	9173	Bad Tölz-Wolfratshausen	DE216	47.75839996	11.55860043
230	9174	Dachau	DE217	48.26010132	11.45339966
231	9175	Ebersberg	DE218	48.09555054	11.9647398
232	9176	Eichstätt	DE219	48.89070129	11.20680046
233	9177	Erding	DE21A	48.30739975	11.90320015
234	9178	Freising	DE21B	48.40859985	11.74880028
235	9179	Fürstenfeldbruck	DE21C	48.1721	11.2369
236	9180	Garmisch-Partenkirchen	DE21D	47.49219894	11.08279991
237	9181	Landsberg am Lech	DE21E	48.0573	10.8641
238	9182	Miesbach	DE21F	47.78939819	11.83570004
239	9183	Mühldorf a.Inn	DE21G	47.94292831	12.25841999
240	9184	München	DE21H	48.1391	11.5145
241	9185	Neuburg-Schrobenhausen	DE21I	48.7336998	11.17099953
242	9186	Pfaffenhofen a.d.Ilm	DE21J	48.53110123	11.50199986
243	9187	Rosenheim	DE21K	47.8667984	12.11289978
244	9188	Starnberg	DE21L	47.98929977	11.32549953
245	9189	Traunstein	DE21M	47.87139893	12.63239956
246	9190	Weilheim-Schongau	DE21N	47.83890152	11.14050007

Table G-51 List of modelled regions, p. 6

	NUTS-3 ID within model	*NUTS-3 Name within model*	*NUTS-3 ID Eurostat*	*Latitude within model*	*Longitude within model*
247	9261	Landshut, Stadt	DE221	48.54470062	12.1566
248	9262	Passau, Stadt	DE222	48.58459854	13.48340034
249	9263	Straubing, Stadt	DE223	48.88420105	12.59710026
250	9271	Deggendorf	DE224	48.83539963	12.94509983
251	9272	Freyung-Grafenau	DE225	48.80814	13.54843
252	9273	Kelheim	DE226	48.91500092	11.89840031
253	9274	Landshut	DE227	48.54470062	12.1566
254	9275	Passau	DE228	48.58459854	13.48340034
255	9276	Regen	DE229	48.97359848	13.1402998
256	9277	Rottal-Inn	DE22A	48.44010162	12.93719959
257	9278	Straubing-Bogen	DE22B	48.88420105	12.59710026
258	9279	Dingolfing-Landau	DE22C	48.63629913	12.4968996
259	9361	Amberg, Stadt	DE231	49.44919968	11.83320045
260	9362	Regensburg, Stadt	DE232	49.02399826	12.07590008
261	9363	Weiden i.d.OPf., Stadt	DE233	49.67436	12.14893
262	9371	Amberg-Sulzbach	DE234	49.44919968	11.83320045
263	9372	Cham	DE235	49.21910095	12.66699982
264	9373	Neumarkt i.d.OPf.	DE236	49.27726	11.4672
265	9374	Neustadt a.d.Waldnaab	DE237	49.7378006	12.17889977
266	9375	Regensburg	DE238	49.02399826	12.07590008
267	9376	Schwandorf	DE239	49.32659912	12.11540031
268	9377	Tirschenreuth	DE23A	49.87139893	12.33969975
269	9461	Bamberg, Stadt	DE241	49.89030075	10.90579987
270	9462	Bayreuth, Stadt	DE242	49.91949844	11.5600996
271	9463	Coburg, Stadt	DE243	50.26570129	10.93809986
272	9464	Hof, Stadt	DE244	50.32210159	11.92339993
273	9471	Bamberg	DE245	49.89030075	10.90579987
274	9472	Bayreuth	DE246	49.91949844	11.5600996
275	9473	Coburg	DE247	50.26570129	10.93809986
276	9474	Forchheim	DE248	49.71789932	11.06799984
277	9475	Hof	DE249	50.32210159	11.92339993
278	9476	Kronach	DE24A	50.2397995	11.3416996
279	9477	Kulmbach	DE24B	50.09700012	11.45510006
280	9478	Lichtenfels	DE24C	50.14139938	11.05350018
281	9479	Wunsiedel i.Fichtelgebirge	DE24D	50.03639984	12.00559998
282	9561	Ansbach, Stadt	DE251	49.30279922	10.5795002
283	9562	Erlangen, Stadt	DE252	49.58729935	11.02579975
284	9563	Fürth, Stadt	DE253	49.48040009	10.97990036
285	9564	Nürnberg, Stadt	DE254	49.46519852	11.07600021
286	9565	Schwabach, Stadt	DE255	49.3227005	11.0163002
287	9571	Ansbach	DE256	49.30279922	10.5795002

Table G-52 List of modelled regions, p. 7

	NUTS-3 ID within model	*NUTS-3 Name within model*	*NUTS-3 ID Eurostat*	*Latitude within model*	*Longitude within model*
288	9572	Erlangen-Höchstadt	DE257	49.58729935	11.02579975
289	9573	Fürth	DE258	49.48040009	10.97990036
290	9574	Nürnberger Land	DE259	49.51039886	11.2748003
291	9575	Neustadt a.d.Aisch-Bad Windsh.	DE25A	49.58200836	10.62137985
292	9576	Roth	DE25B	49.23419952	11.08710003
293	9577	Weißenburg-Gunzenhausen	DE25C	49.02600098	10.96440029
294	9661	Aschaffenburg, Stadt	DE261	49.98799896	9.109880447
295	9662	Schweinfurt, Stadt	DE262	50.04539871	10.23159981
296	9663	Würzburg, Stadt	DE263	49.80350113	9.944020271
297	9671	Aschaffenburg	DE264	49.98799896	9.109880447
298	9672	Bad Kissingen	DE265	50.19189835	10.06379986
299	9673	Rhön-Grabfeld	DE266	50.32429886	10.19620037
300	9674	Haßberge	DE267	50.03150177	10.50669956
288	9572	Erlangen-Höchstadt	DE257	49.58729935	11.02579975
289	9573	Fürth	DE258	49.48040009	10.97990036
290	9574	Nürnberger Land	DE259	49.51039886	11.2748003
291	9575	Neustadt a.d.Aisch-Bad Windsh.	DE25A	49.58200836	10.62137985
292	9576	Roth	DE25B	49.23419952	11.08710003
293	9577	Weißenburg-Gunzenhausen	DE25C	49.02600098	10.96440029
294	9661	Aschaffenburg, Stadt	DE261	49.98799896	9.109880447
295	9662	Schweinfurt, Stadt	DE262	50.04539871	10.23159981
296	9663	Würzburg, Stadt	DE263	49.80350113	9.944020271
297	9671	Aschaffenburg	DE264	49.98799896	9.109880447
298	9672	Bad Kissingen	DE265	50.19189835	10.06379986
299	9673	Rhön-Grabfeld	DE266	50.32429886	10.19620037
300	9674	Haßberge	DE267	50.03150177	10.50669956
301	9675	Kitzingen	DE268	49.7364006	10.15509987
302	9676	Miltenberg	DE269	49.70220184	9.247810364
303	9677	Main-Spessart	DE26A	49.96340179	9.783969879
304	9678	Schweinfurt	DE26B	50.04539871	10.23159981
305	9679	Würzburg	DE26C	49.80350113	9.944020271
306	9761	Augsburg, Stadt	DE271	48.35509872	10.93309975
307	9762	Kaufbeuren, Stadt	DE272	47.8742981	10.61830044
308	9763	Kempten (Allgäu), Stadt	DE273	47.72930145	10.30869961
309	9764	Memmingen, Stadt	DE274	47.98690033	10.19659996
310	9771	Aichach-Friedberg	DE275	48.45299911	11.13210011
311	9772	Augsburg	DE276	48.35509872	10.93309975
312	9773	Dillingen a.d.Donau	DE277	48.58000183	10.49629974
313	9774	Günzburg	DE278	48.44990158	10.28359985
314	9775	Neu-Ulm	DE279	48.39780045	10.00500011
315	9776	Lindau (Bodensee)	DE27A	47.54600143	9.688750267

Table G-53 List of modelled regions, p. 8

	NUTS-3 ID within model	*NUTS-3 Name within model*	*NUTS-3 ID Eurostat*	*Latitude within model*	*Longitude within model*
316	**9777**	**Ostallgäu**	DE27B	47.77280045	10.61740017
317	**9778**	**Unterallgäu**	DE27C	48.0522995	10.48589993
318	**9779**	**Donau-Ries**	DE27D	48.7357	10.7918
319	**9780**	**Oberallgäu**	DE27E	47.51409912	10.2791996
320	**10041**	**Regionalverband Saarbrücken**	DEC01	49.23149	6.99833
321	**10042**	**Merzig-Wadern**	DEC02	49.43109894	6.634490013
322	**10043**	**Neunkirchen**	DEC03	49.34899902	7.179150105
323	**10044**	**Saarlouis**	DEC04	49.30659866	6.74116993
324	**10045**	**Saarpfalz-Kreis**	DEC05	49.33110046	7.326720238
325	**10046**	**St. Wendel**	DEC06	49.46670151	7.15899992
326	**11000**	**Berlin, Stadt**	DE300	52.50140762	13.40232849
327	**12051**	**Brandenburg an der Havel, St.**	DE401	52.41090012	12.50090027
328	**12052**	**Cottbus, Stadt**	DE402	51.76089859	14.35079956
329	**12053**	**Frankfurt (Oder), Stadt**	DE403	52.34722	14.55057
330	**12054**	**Potsdam, Stadt**	DE404	52.38288879	13.04601002
331	**12060**	**Barnim**	DE405	52.832901	13.75199986
332	**12061**	**Dahme-Spreewald**	DE406	51.9457016	13.87530041
333	**12062**	**Elbe-Elster**	DE407	51.69120026	13.22029972
334	**12063**	**Havelland**	DE408	52.60749817	12.34399986
335	**12064**	**Märkisch-Oderland**	DE409	52.53139877	14.37250042
336	**12065**	**Oberhavel**	DE40A	52.75299835	13.23600006
337	**12066**	**Oberspreewald-Lausitz**	DE40B	51.51369858	14.00150013
338	**12067**	**Oder-Spree**	DE40C	52.17250061	14.25179958
339	**12068**	**Ostprignitz-Ruppin**	DE40D	52.93099976	12.81280041
340	**12069**	**Potsdam-Mittelmark**	DE40E	52.1435318	12.5896101
341	**12070**	**Prignitz**	DE40F	53.07360077	11.8682003
342	**12071**	**Spree-Neiße**	DE40G	51.7439003	14.65270042
343	**12072**	**Teltow-Fläming**	DE40H	52.08269882	13.15620041
344	**12073**	**Uckermark**	DE40I	53.31689835	13.86250019
345	**13003**	**Rostock, Hansestadt**	DE803	54.0790596	12.13216019
346	**13004**	**Schwerin, Landeshauptstadt**	DE804	53.62591171	11.41656971
347	**13071**	**Mecklenburgische Seenplatte**	DE802	53.55918503	13.27782536
348	**13072**	**Landkreis Rostock**	DE803	54.0790596	12.13216019
349	**13073**	**Vorpommern-Rügen**	DE805	54.30881119	13.09450722
350	**13074**	**Nordwestmecklenburg**	DE80E	53.86003	11.19069
351	**13075**	**Vorpommern-Greifswald**	DE801	54.08974457	13.39124393
352	**13076**	**Ludwigslust-Parchim**	DE80G	53.42684937	11.85428047
353	**14511**	**Chemnitz, Stadt**	DED41	50.83620071	12.93700027
354	**14521**	**Erzgebirgskreis**	DED42	50.70918	12.77501
355	**14522**	**Mittelsachsen**	DED43	50.91059875	13.34889984
356	**14523**	**Vogtlandkreis**	DED44	50.49761	12.13687

Table G-54 List of modelled regions, p. 9

	NUTS-3 ID within model	*NUTS-3 Name within model*	*NUTS-3 ID Eurostat*	*Latitude within model*	*Longitude within model*
357	14524	Zwickau	DED45	50.71789932	12.45730019
358	14612	Dresden, Stadt	DED21	51.05363083	13.74081039
359	14625	Bautzen	DED2C	51.17689896	14.42910004
360	14626	Görlitz	DED2D	51.1295	14.9835
361	14627	Meißen	DED2E	51.16178	13.49766
362	14628	Sächs. Schweiz-Osterzgebirge	DED2F	50.96319962	13.94839954
363	14713	Leipzig, Stadt	DED51	51.34519958	12.3859396
364	14729	Leipzig	DED52	51.34519958	12.3859396
365	14730	Nordsachsen	DED53	51.52557	12.3381
366	15001	Dessau-Roßlau, Stadt	DEE01	51.84283	12.23039
367	15002	Halle (Saale), Stadt	DEE02	51.49698	11.9688
368	15003	Magdeburg, Landeshauptstadt	DEE03	52.13185883	11.62777996
369	15081	Altmarkkreis Salzwedel	DEE04	52.84830093	11.16790009
370	15082	Anhalt-Bitterfeld	DEE05	51.7506	11.97119999
371	15083	Börde	DEE07	52.28919983	11.41230011
372	15084	Burgenlandkreis	DEE08	51.15869904	11.80220032
373	15085	Harz	DEE09	51.90999985	11.02680016
374	15086	Jerichower Land	DEE06	52.2709	11.85535
375	15087	Mansfeld-Südharz	DEE0A	51.47510147	11.29430008
376	15088	Saalekreis	DEE0B	51.33229828	11.99250031
377	15089	Salzlandkreis	DEE0C	51.79546738	11.73930645
378	15090	Stendal	DEE0D	52.60530853	11.85071659
379	15091	Wittenberg	DEE0E	51.8676796	12.64865971
380	16051	Erfurt, Stadt	DEG01	50.98477	11.02988
381	16052	Gera, Stadt	DEG02	50.87979889	12.11330032
382	16053	Jena, Stadt	DEG03	50.92039871	11.59070015
383	16054	Suhl, Stadt	DEG04	50.6108017	10.68150043
384	16055	Weimar, Stadt	DEG05	50.98600006	11.31690025
385	16056	Eisenach, Stadt	DEG0N	50.97940063	10.3032999
386	16061	Eichsfeld	DEG06	51.38169861	10.12969971
387	16062	Nordhausen	DEG07	51.50310135	10.80130005
388	16063	Wartburgkreis	DEG0P	50.81470108	10.23550034
389	16064	Unstrut-Hainich-Kreis	DEG09	51.20986	10.4571
390	16065	Kyffhäuserkreis	DEG0A	51.36539841	10.87080002
391	16066	Schmalkalden-Meiningen	DEG0B	50.57391	10.41901
392	16067	Gotha	DEG0C	50.94260025	10.70180035
393	16068	Sömmerda	DEG0D	51.1589	11.1362
394	16069	Hildburghausen	DEG0E	50.42850113	10.72420025
395	16070	Ilm-Kreis	DEG0F	50.84019852	10.9527998
396	16071	Weimarer Land	DEG0G	51.02429962	11.52970028
397	16072	Sonneberg	DEG0H	50.36018	11.17099

Table G-55 List of modelled regions, p. 10

	NUTS-3 ID within model	*NUTS-3 Name within model*	*NUTS-3 ID Eurostat*	*Latitude within model*	*Longitude within model*
398	16073	Saalfeld-Rudolstadt	DEG0I	50.65190125	11.34500027
399	16074	Saale-Holzland-Kreis	DEG0J	50.96861	11.89265
400	16075	Saale-Orla-Kreis	DEG0K	50.57160187	11.79640007
401	16076	Greiz	DEG0L	50.65200043	12.22900009
402	16077	Altenburger Land	DEG0M	50.98659897	12.43980026
403	[20001]	Non-EU27	-	-	-
404	[20002]	Switzerland	(CH)	46.7999	8.1999
405	[20003]	Russia	(RU)	55.75583	37.6173
406	[20004]	Belgium	(BE)	50.71240775	4.4769405
407	[20005]	Bulgaria	(BG)	42.46454462	25.48232175
408	[20006]	Denmark	(DK)	56.16146	10.18982
409	[20007]	Estonia	(EST)	58.72372488	25.80699675
410	[20008]	Finland	(FI)	59.32932	18.06858
411	[20009]	France	(FR)	46.77364085	1.71856675
412	[20010]	Greece	(GR)	38.2645263	23.321999
413	[20011]	Ireland	(IE)	53.4678	-3.0111
414	[20012]	Italy	(IT)	42.79350857	12.573787
415	[20013]	Latvia	(LV)	57.14453714	24.60612575
416	[20014]	Lithuania	(LT)	55.14391098	23.942214
417	[20015]	Luxembourg	(LU)	49.75286485	6.13335075
418	[20016]	Malta	(MT)	38.11569	13.36127
419	[20017]	Netherlands	(NL)	51.86481775	5.33029275
420	[20018]	Austria	(AT)	47.45185582	13.34538825
421	[20019]	Poland	(PL)	51.87840787	19.13433375
422	[20020]	Portugal	(PT)	39.42396643	-7.844941
423	[20021]	Romania	(RO)	45.72811721	24.98862725
424	[20022]	Sweden	(SE)	59.06269683	17.634344
425	[20023]	Slovakia	(SK)	48.80153205	19.700049
426	[20024]	Slovenia	(SI)	46.05095387	14.986138
427	[20025]	Spain	(ES)	40.06895221	-2.98829575
428	[20026]	Czech Republic	(CZ)	49.87109509	15.474998
429	[20027]	Hungary	(HU)	47.14191	19.3634
430	[20028]	United Kingdom	(UK)	53.27380755	-2.232612
431	[20029]	Cyprus	(CY)	37.94299	23.64698
432	[01]	Port of Hamburg	DE600	53.55334091	9.992469788
433	[02]	Port of Wilhelmshaven	DE945	53.52959824	8.099840164
434	[03]	Port of Bremen	DE501	53.07509995	8.804690361
435	[04]	Port of Bremerhaven	DE502	53.52610016	8.623869896
436	[05]	Port of Brunsbüttel	DEF05	54.17990112	9.095509529

Table G-56 List of modelled regions, p. 11

	NUTS-3 ID within model	*NUTS-3 Name within model*	*NUTS-3 ID Eurostat*	*Latitude within model*	*Longitude within model*
437	[06]	Port of Brake	DE94G	53.32632828	8.478639603
438	[07]	Port of Bützfleth	DE939	53.60229874	9.475330353
439	[08]	Port of Emden	DE942	53.364705	7.172326
440	[09]	Port of Nordenham	DE94G	53.32632828	8.478639603
441	[10]	Port of Cuxhaven	DE932	53.84429932	8.687970161
442	[11]	Port of Leer	DE94C	53.23751068	7.461763859
443	[12]	Port of Papenburg	DE94C	53.23751068	7.461763859
444	[13]	Port of Husum	DEF07	54.47034836	9.05163002
445	[14]	Port of Rostock	DE803	54.0790596	12.13216019
446	[15]	Port of Lübeck	DEF03	53.86626816	10.68087959
447	[16]	Port of Puttgarden	DEF08	54.13729858	10.6086998
448	[17]	Port of Kiel	DEF02	54.32324982	10.1322403
449	[18]	Port of Wismar	DE80E	53.86003	11.19069
450	[19]	Port of Saßnitz	DE805	54.30881119	13.09450722
451	[20]	Port of Lubmin	DE801	54.08974457	13.39124393
452	[21]	Port of Wolgast	DE801	54.08974457	13.39124393
453	[22]	Port of Stralsund	DE805	54.30881119	13.09450722
454	[23]	Port of Flensburg	DEF01	54.79779816	9.477919579
455	[24]	Port of Rendsburg	DEF0B	54.29779816	9.666950226
456	[25]	Port of Rotterdam	(NL)	51.92442	4.47773
457	[26]	Port of Antwerp	(BE)	51.21945	4.40246
458	[27]	Airport of Berlin-Schönefeld	DE300	52.50140762	13.40232849
459	[28]	Airport of Berlin-Tegel	DE300	52.50140762	13.40232849
460	[29]	Airport of Dresden	DED21	51.05363083	13.74081039
461	[30]	Airport of Düsseldorf	DEA11	51.21562958	6.776040077
462	[31]	Airport of Frankfurt/Main	DE712	50.11207962	8.683409691
463	[32]	Airport of Hahn	DEB19	49.98839951	7.541600227
464	[33]	Airport of Hamburg	DE600	53.55334091	9.992469788
465	[34]	Airport of Hannover	DE929	52.37226868	9.738149643
466	[35]	Airport of Köln/Bonn	DEA23	50.93753	6.96028
467	[36]	Airport of Leipzig/Halle	DED53	51.52557	12.3381
468	[37]	Airport of Memmingen	DE27C	48.0522995	10.48589993
469	[38]	Airport of München	DE212	48.1391	11.5145
470	[39]	Airport of Nürnberg	DE254	49.46519852	11.07600021
471	[40]	Airport of Rostock-Laage	DE803	54.0790596	12.13216019
472	[41]	Airport of Stuttgart	DE111	48.76766968	9.171919823
473	[42]	Airport of Zweibrücken	DEB3A	49.2471	7.36348

Table G-57 List of modelled regions, p. 12

	NUTS-3 ID within model	*NUTS-3 Name within model*	*NUTS-3 ID Eurostat*	*Federal German waterway ID*	*Federal German waterway station-km*
1	[01]	Seaport of Hamburg	DE600	0792	631
2	[03]	Seaport of Bremen	DE501	5298	4
3	[04]	Seaport of Bremerhaven	DE502	5298	65.8
4	[05]	Seaport of Brunsbüttel	DEF05	3401	2.6
5	[06]	Seaport of Brake	DE94G	5298	40.6
6	[07]	Seaport of Bützfleth	DE939	701	658
7	[08]	Seaport of Emden	DE942	1191	37
8	[09]	Seaport of Nordenham	DE94G	5298	57
9	[10]	Seaport of Cuxhaven	DE932	701	724
10	[11]	Seaport of Leer	DE94C	2501	22
11	[12]	Seaport of Papenburg	DE94C	513	76
12	[15]	Seaport of Lübeck	DEF03	0801	4.7
13	[17]	Seaport of Kiel	DEF02	3401	96.8
14	[24]	Seaport of Rendsburg	DEF0B	3401	62
15	03101	Inland port of Braunschweig, Stadt	DE911	3101	219.7
16	03102	Inland port of Salzgitter, Stadt	DE912	3114	16
17	03157	Inland port of Peine	DE91A	3101	192.3
18	03241	Inland port of Region Hannover	DE929	3101	162
19	03257	Inland port of Schaumburg	DE928	5201	164.4
20	03359	Inland port of Stade	DE939	4401	0.2
21	03360	Inland port of Uelzen	DE93A	0901	63.4
22	03403	Inland port of Oldenburg (Oldenburg), Stadt	DE943	1901	0
23	03454	Inland port of Emsland	DE949	0501	145
24	05111	Inland port of Düsseldorf, Stadt	DEA11	3901	743.1
25	05112	Inland port of Duisburg, Stadt	DEA12	3901	775.7
26	05114	Inland port of Krefeld, Stadt	DEA14	3901	764.7
27	05154	Inland port of Kleve	DEA1B	3901	851.6
28	05162	Inland port of Rhein-Kreis Neuss	DEA1D	3901	740.2
29	05170	Inland port of Wesel	DEA1F	5101	0.5
30	05314	Inland port of Bonn, Stadt	DEA22	3901	654
31	05315	Inland port of Köln, Stadt	DEA23	3901	690
32	05316	Inland port of Leverkusen, Stadt	DEA24	3901	702
33	05362	Inland port of Rhein-Erft-Kreis	DEA27	3901	668
34	05512	Inland port of Bottrop, Stadt	DEA31	4001	16.4
35	05513	Inland port of Gelsenkirchen, Stadt	DEA32	4001	20.6
36	05562	Inland port of Recklinghausen	DEA36	5101	36.2
37	05770	Inland port of Minden-Lübbecke	DEA46	5201	204.4
38	05913	Inland port of Dortmund, Stadt	DEA52	501	2
39	05915	Inland port of Hamm, Stadt	DEA54	0301	34.9

Table G-58 List of modelled ports, p.1 [incl. data from WSV (2014b)]

	NUTS-3 ID within model	*NUTS-3 Name within model*	*NUTS-3 ID Eurostat*	*Federal German waterway ID*	*Federal German waterway station-km*
40	05916	Inland port of Herne, Stadt	DEA55	4001	32.7
41	05978	Inland port of Unna	DEA5C	0301	14
42	06412	Inland port of Frankfurt am Main, Stadt	DE712	2901	32.9
43	06433	Inland port of Groß-Gerau	DE717	3901	462
44	06435	Inland port of Main-Kinzig-Kreis	DE719	2901	61.9
45	07111	Inland port of Koblenz, kreisfreie Stadt	DEB11	3901	591.5
46	07137	Inland port of Mayen-Koblenz	DEB17	3901	616.1
47	07314	Inland port of Ludwigshafen am Rhein, Stadt	DEB34	3901	420
48	07315	Inland port of Mainz, kreisfreie Stadt	DEB35	3901	498.3
49	07319	Inland port of Worms, kreisfreie Stadt	DEB39	3901	444.5
50	07334	Inland port of Germersheim	DEB3E	3901	385.4
51	08111	Inland port of Stuttgart, Landes-hauptstadt	DE111	3301	186
52	08121	Inland port of Heilbronn, Stadt	DE117	3301	110
53	08212	Inland port of Karlsruhe, Stadt	DE122	3901	359.9
54	08222	Inland port of Mannheim, Univer-sitätsstadt	DE126	3901	415
55	08317	Inland port of Ortenaukreis	DE134	3901	297.7
56	09262	Inland port of Passau, Stadt	DE222	0401	2225.95
57	09263	Inland port of Straubing, Stadt	DE223	0401	2312
58	09271	Inland port of Deggendorf	DE224	0401	2284
59	09273	Inland port of Kelheim	DE226	0401	2411
60	09362	Inland port of Regensburg, Stadt	DE232	0401	2373
61	09461	Inland port of Bamberg, Stadt	DE241	3001	2.1
62	09564	Inland port of Nürnberg, Stadt	DE254	3001	71.2
63	09661	Inland port of Aschaffenburg, Stadt	DE261	2901	83.3
64	11000	Inland port of Berlin, Stadt	DE300	6601	33
65	12051	Inland port of Brandenburg an der Havel, St.	DE401	6701	57.6
66	12060	Inland port of Barnim	DE405	5801	66
67	12061	Inland port of Dahme-Spreewald	DE406	5522	0
68	12063	Inland port of Havelland	DE408	5701	23
69	12065	Inland port of Oberhavel	DE40A	5801	12
70	12067	Inland port of Oder-Spree	DE40C	6501	125.95
71	12069	Inland port of Potsdam-Mit-telmark	DE40E	6601	12
72	12073	Inland port of Uckermark	DE40I	5801	121
73	15003	Inland port of Magdeburg, Lan-deshauptstadt	DEE03	0701	329.9
74	15082	Inland port of Anhalt-Bitterfeld	DEE05	0701	277.4
75	15083	Inland port of Börde	DEE07	3101	297.5
76	[20002]	Inland port of SWITZERLAND	(CH)	-	-
77	[20004]	Inland port of BELGIUM	(BE)	-	-

Table G-59 List of modelled ports, p.2 [incl. data from WSV (2014b)]

	NUTS-3 ID within model	*NUTS-3 Name within model*	*NUTS-3 ID Eurostat*	*Federal German waterway ID*	*Federal German waterway station-km*
78	[20005]	Inland port of BULGARIA	(BG)	-	-
79	[20009]	Inland port of FRANCE	(FR)	-	-
80	[20015]	Inland port of LUXEMBOURG	(LU)	-	-
81	[20017]	Inland port of NETHERLANDS	(NL)	-	-
82	[20018]	Inland port of AUSTRIA	(AT)	-	-
83	[20019]	Inland port of POLAND	(PL)	-	-
84	[20021]	Inland port of ROMANIA	(RO)	-	-
85	[20023]	Inland port of SLOVAKIA	(SK)	-	-
86	[20026]	Inland port of CZECH REPUB-LIC	(CZ)	-	-
87	[20027]	Inland port of HUNGARY	(HU)	-	-
88	[25]	Inland port of NETHERLANDS	(NL)	-	-
89	[26]	Inland port of BELGIUM	(BE)	-	-

Table G-60 List of modelled ports, p.3 [incl. data from WSV (2014b)]

	NUTS-3 ID within model	*NUTS-3 Name within model*	*NUTS-3 ID Eurostat*	*DB DIUM ID within model*
1	05562	Railyard of Recklinghausen	DEA36	100156
2	05512	Railyard of Bottrop	DEA31	102509
3	05358	Railyard of Dueren	DEA26	153312
4	06632	Railyard of Bad Hersfeld	DE733	151076
5	12071	Railyard of Forst-Lausitz	DE40G	040444
6	05566	Railyard of Steinfurt	DEA37	572966
7	15084	Railyard of Naumburg	DEE08	230011
8	05170	Railyard of Wesel	DEA1F	102756
9	07314	Railyard of Ludwigshafen am Rhein	DEB34	190793
10	15083	Railyard of Haldensleben	DEE07	240465
11	05315	Railyard of Köln	DEA23	154799
12	05158	Railyard of Mettmann	DEA1C	505057
13	08212	Railyard of Karlsruhe	DE122	142448
14	03241	Railyard of Hannover	DE929	132357
15	05362	Railyard of Bergheim	DEA27	154799
16	12052	Railyard of Cottbus	DE402	040196
17	16063	Railyard of Bad Salzungen	DEG0P	161299
18	08121	Railyard of Heilbronn	DE117	717769
19	09371	Railyard of Amberg	DE234	261958
20	05334	Railyard of Aachen	DEA2D	153692
21	12073	Railyard of Prenzlau	DE40I	280701
22	05513	Railyard of Gelsenkirchen	DEA32	100446
23	06631	Railyard of Fulda	DE732	056382
24	03154	Railyard of Helmstedt	DE917	133447
25	10044	Railyard of Saarlouis	DEC04	253617
26	09162	Railyard of München	DE212	205179
27	05112	Railyard of Duisburg	DEA12	100008
28	08222	Railyard of Mannheim	DE126	140004
29	06611	Railyard of Kassel	DE731	838268
30	09171	Railyard of Altötting	DE214	200345
31	10041	Railyard of Saarbrücken	DEC01	253930
32	03454	Railyard of Meppen	DE949	500090
33	14524	Railyard of Zwickau	DED45	066407
34	09273	Railyard of Kelheim	DE226	355115
35	15089	Railyard of Bernburg	DEE0C	241844
36	01051	Railyard of Heide	DEF05	870444
37	07133	Railyard of Bad Kreuznach	DEB14	197749
38	07138	Railyard of Neuwied	DEB18	194902
39	07315	Railyard of Mainz	DEB35	250829
40	09176	Railyard of Eichstätt	DE219	159632

Table G-61 List of modelled freight yards, p.1 [incl. data from DB CARGO (2015b)]

	NUTS-3 ID within model	*NUTS-3 Name within model*	*NUTS-3 ID Eurostat*	*DB DIUM ID within model*
41	15088	Railyard of Merseburg	DEE0B	230011
42	12066	Railyard of Senftenberg	DE40B	042234
43	15087	Railyard of Sangerhausen	DEE0A	208488
44	15085	Railyard of Halberstadt	DEE09	208488
45	07339	Railyard of Mainz	DEB3J	250829
46	08128	Railyard of Tauberbischofsheim	DE11B	291880
47	09184	Railyard of München	DE21H	205179
48	05111	Railyard of Düsseldorf	DEA11	080986
49	03103	Railyard of Wolfsburg	DE913	373100
50	05962	Railyard of Lüdenscheid	DEA58	342766
51	14625	Railyard of Bautzen	DED2C	060996
52	09362	Railyard of Regensburg	DE232	629485
53	08117	Railyard of Göppingen	DE114	190843
54	05913	Railyard of Dortmund	DEA52	434068
55	05316	Railyard of Leverkusen	DEA24	505057
56	09577	Railyard of Weissenburg in Bayern	DE25C	224345
57	07131	Railyard of Bad Neuenahr-Ahrweiler	DEB12	154864
58	09780	Railyard of Sonthofen	DE27E	023085
59	08135	Railyard of Heidenheim an der Br	DE11C	297036
60	05116	Railyard of Mönchengladbach	DEA15	152181
61	14729	Railyard of Leipzig	DED52	659870
62	06433	Railyard of Gross-Gerau	DE717	190678
63	07338	Railyard of Ludwigshafen a. Rhein	DEB3I	190793
64	06533	Railyard of Limburg an der Lahn	DE723	114546
65	09675	Railyard of Kitzingen	DE268	220434
66	08115	Railyard of Böblingen	DE112	293837
67	09274	Railyard of Landshut	DE227	264994
68	09186	Railyard of Pfaffenhofen	DE21J	204479
69	05117	Railyard of Mülheim	DEA16	102624
70	16062	Railyard of Nordhausen	DEG07	164533
71	03453	Railyard of Cloppenburg	DE948	116038
72	09161	Railyard of Ingolstadt	DE211	200980
73	08317	Railyard of Offenburg	DE134	143107
74	07333	Railyard of Kirchheim-Bolanden	DEB3D	197749
75	09376	Railyard of Schwandorf	DE239	262261
76	10042	Railyard of Merzig	DEC02	253492
77	05124	Railyard of Wuppertal	DEA1A	082503
78	01061	Railyard of Itzehoe	DEF0E	014951
79	14521	Railyard of Stollberg	DED42	064121
80	06412	Railyard of Frankfurt am Main	DE712	110692

Table G-62 List of modelled freight yards, p.2 [incl. data from DB CARGO (2015b)]

	NUTS-3 ID within model	NUTS-3 Name within model	NUTS-3 ID Eurostat	DB DIUM ID within model
81	09775	Railyard of Neu-Ulm	DE279	020891
82	03152	Railyard of Göttingen	DE915	131847
83	08116	Railyard of Esslingen am Neckar	DE113	290551
84	08118	Railyard of Ludwigsburg	DE115	290189
85	03102	Railyard of Salzgitter	DE912	588012
86	08336	Railyard of Lörrach	DE139	144410
87	14523	Railyard of Plauen	DED44	167221
88	08237	Railyard of Freudenstadt	DE12C	193235
89	06438	Railyard of Offenbach	DE71C	113373
90	06532	Railyard of Wetzlar	DE722	114249
91	06636	Railyard of Eschwege	DE737	131847
92	08417	Railyard of Balingen	DE143	293456
93	09476	Railyard of Kronach	DE24A	220350
94	05515	Railyard of Münster	DEA33	572966
95	08437	Railyard of Sigmaringen	DE149	292540
96	05978	Railyard of Unna	DEA5C	341446
97	03254	Railyard of Hildesheim	DE925	134148
98	06535	Railyard of Lauterbach	DE725	110551
99	09178	Railyard of Freising	DE21B	204479
100	05911	Railyard of Bochum	DEA51	101808
101	07143	Railyard of Montabaur	DEB1B	190140
102	09679	Railyard of Würzburg	DE26C	225383
103	16052	Railyard of Gera	DEG02	167056
104	03153	Railyard of Goslar	DE916	588012
105	06634	Railyard of Homberg	DE735	151076
106	08425	Railyard of Ulm	DE145	291047
107	09673	Railyard of Bad Neustadt a. d. S.	DE266	333260
108	13076	Railyard of Parchim	DE80G	273573
109	12067	Railyard of Beeskow	DE40C	032995
110	09271	Railyard of Deggendorf	DE224	264515
111	05114	Railyard of Krefeld	DEA14	150409
112	09676	Railyard of Miltenberg	DE269	226480
113	09373	Railyard of Neumarkt i.d. Opf.	DE236	223537
114	16076	Railyard of Greiz	DEG0L	167221
115	03460	Railyard of Vechta	DE94F	210278
116	09477	Railyard of Kulmbach	DE24B	220178
117	07232	Railyard of Bitburg	DEB23	251801
118	08111	Railyard of Stuttgart	DE111	290676
119	11000	Railyard of Berlin	DE300	344168
120	07134	Railyard of Birkenfeld	DEB15	254581

Table G-63 List of modelled freight yards, p.3 [incl. data from DB CARGO (2015b)]

NUTS-3 ID within model		*NUTS-3 Name within model*	*NUTS-3 ID Eurostat*	*DB DIUM ID within model*
121	08211	Railyard of Baden-Baden	DE121	142455
122	14730	Railyard of Delitzsch	DED53	231720
123	15003	Railyard of Magdeburg	DEE03	240242
124	14627	Railyard of Meißen	DED2E	060541
125	09479	Railyard of Wunsiedel	DE24D	260034
126	03357	Railyard of Rotenburg	DE937	566026
127	14522	Railyard of Freiberg	DED43	063495
128	05162	Railyard of Neuss	DEA1D	509620
129	03359	Railyard of Stade	DE939	332668
130	01004	Railyard of Neumünster	DEF04	012963
131	05958	Railyard of Meschede	DEA57	053017
132	09763	Railyard of Kempten	DE273	023085
133	06435	Railyard of Hanau	DE719	110551
134	08221	Railyard of Heidelberg	DE125	803791
135	03155	Railyard of Northeim	DE918	130567
136	09190	Railyard of Weilheim	DE21N	202721
137	05119	Railyard of Oberhausen	DEA17	102640
138	06432	Railyard of Darmstadt	DE716	113373
139	05974	Railyard of Soest	DEA5B	100065
140	05916	Railyard of Herne	DEA55	100396
141	06534	Railyard of Marburg	DE724	110353
142	03255	Railyard of Holzminden	DE926	133306
143	09777	Railyard of Marktoberdorf	DE27B	023085
144	05914	Railyard of Hagen	DEA53	342766
145	07137	Railyard of Koblenz	DEB17	190140
146	15086	Railyard of Burg	DEE06	240085
147	09473	Railyard of Coburg	DE247	228148
148	08436	Railyard of Ravensburg	DE148	291534
149	01059	Railyard of Schleswig	DEF0C	014951
150	09279	Railyard of Dingolfing	DE22C	264424
151	07111	Railyard of Koblenz	DEB11	190140
152	03251	Railyard of Diepholz	DE922	210419
153	09574	Railyard of Lauf a.d. Pegnitz	DE259	224535
154	09275	Railyard of Passau	DE228	0460
155	08421	Railyard of Ulm	DE144	291047
156	09261	Railyard of Landshut	DE221	264994
157	05382	Railyard of Siegburg	DEA2C	135335
158	09276	Railyard of Regen	DE229	264655
159	03151	Railyard of Gifhorn	DE914	373100
160	08125	Railyard of Heilbronn	DE118	717769

Table G-64 List of modelled freight yards, p.4 [incl. data from DB CARGO (2015b)]

	NUTS-3 ID within model	NUTS-3 Name within model	NUTS-3 ID Eurostat	DB DIUM ID within model
161	05774	Railyard of Paderborn	DEA47	103598
162	08426	Railyard of Biberach	DE146	291260
163	09172	Railyard of Bad Reichenhall	DE215	201053
164	16072	Railyard of Sonneberg	DEG0H	163329
165	09562	Railyard of Erlangen	DE252	221630
166	09471	Railyard of Bamberg	DE245	220939
167	05954	Railyard of Schwelm	DEA56	101808
168	03256	Railyard of Nienburg	DE927	137331
169	09672	Railyard of Bad Kissingen	DE265	220434
170	09363	Railyard of Weiden i.d. Opf	DE233	261370
171	06431	Railyard of Heppenheim-Bergstrasse	DE715	113373
172	06440	Railyard of Friedberg Hessen	DE71E	113811
173	08315	Railyard of Freiburg	DE132	143529
174	03404	Railyard of Osnabrück	DE944	210278
175	08415	Railyard of Reutlingen	DE141	292540
176	06414	Railyard of Wiesbaden	DE714	729194
177	07233	Railyard of Daun	DEB24	154278
178	03156	Railyard of Osterode	DE919	588012
179	07336	Railyard of Kusel	DEB3G	197749
180	01060	Railyard of Bad Segeberg	DEF0D	013391
181	09662	Railyard of Schweinfurt	DE262	220434
182	03459	Railyard of Osnabrück	DE94E	210278
183	05166	Railyard of Viersen	DEA1E	151225
184	06633	Railyard of Kassel	DE734	838268
185	07334	Railyard of Germersheim	DEB3E	190975
186	08335	Railyard of Konstanz	DE138	145862
187	05570	Railyard of Warendorf	DEA38	100065
188	09187	Railyard of Rosenheim	DE21K	201756
189	01056	Railyard of Pinneberg	DEF09	014951
190	12060	Railyard of Eberswalde	DE405	282277
191	06531	Railyard of Giessen	DE721	110353
192	09179	Railyard of Fürstenfeldbruck	DE21C	205179
193	08215	Railyard of Karlsruhe	DE123	142448
194	09774	Railyard of Günzburg	DE278	020842
195	09472	Railyard of Bayreuth	DE246	220178
196	09779	Railyard of Donauwörth	DE27D	020784
197	05754	Railyard of Gütersloh	DEA42	136077
198	01055	Railyard of Eutin	DEF08	013540
199	16077	Railyard of Altenburg	DEG0M	148445
200	15082	Railyard of Köthen	DEE05	241372

Table G-65 List of modelled freight yards, p.5 [incl. data from DB CARGO (2015b)]

	NUTS-3 ID within model	NUTS-3 Name within model	NUTS-3 ID Eurostat	DB DIUM ID within model
201	05766	Railyard of Detmold	DEA45	134619
202	09572	Railyard of Erlangen	DE257	221630
203	07331	Railyard of Alzey-Worms	DEB3B	250829
204	08311	Railyard of Freiburg im Breisgau	DE131	143529
205	16053	Railyard of Jena	DEG03	160218
206	14612	Railyard of Dresden	DED21	060541
207	07231	Railyard of Wittlich	DEB22	251801
208	08216	Railyard of Rastatt	DE124	142455
209	09773	Railyard of Dillingen a. d. Donau	DE277	020784
210	07318	Railyard of Speyer	DEB38	190926
211	05970	Railyard of Siegen	DEA5A	082479
212	09189	Railyard of Traunstein	DE21M	201178
214	14713	Railyard of Leipzig	DED51	659870
215	09677	Railyard of Karlstadt	DE26A	225466
216	09772	Railyard of Augsburg	DE276	024000
217	05711	Railyard of Bielefeld	DEA41	135996
218	09262	Railyard of Passau	DE222	0460
219	09177	Railyard of Erding	DE21A	264994
220	05113	Railyard of Essen	DEA13	101857
221	09173	Railyard of Bad Tölz	DE216	202721
222	10046	Railyard of St. Wendel	DEC06	254458
223	09272	Railyard of Freyung	DE225	0460
224	03361	Railyard of Verden	DE93B	566026
225	03257	Railyard of Stadthagen	DE928	537019
226	03458	Railyard of Oldenburg	DE94D	016501
227	13071	Railyard of Neubrandenburg	DE802	281261
228	05366	Railyard of Euskirchen	DEA28	154278
229	01058	Railyard of Rendsburg	DEF0B	546051
230	12070	Railyard of Perleberg	DE40F	240622
231	16070	Railyard of Arnstadt	DEG0F	161737
232	09462	Railyard of Bayreuth	DE242	220178
233	07141	Railyard of Bad Ems	DEB1A	190140
234	01053	Railyard of Ratzeburg	DEF06	011742
235	05370	Railyard of Heinsberg	DEA29	152181
236	05374	Railyard of Gummersbach	DEA2A	195289
237	15081	Railyard of Salzwedel	DEE04	240622
238	16073	Railyard of Saalfeld/Saale	DEG0I	163329
239	16074	Railyard of Eisenberg	DEG0J	167221
240	16061	Railyard of Heiligenstadt	DEG06	160770
241	09183	Railyard of Mühldorf amInn	DE21G	200345

Table G-66 List of modelled freight yards, p.6 [incl. data from DB CARGO (2015b)]

	NUTS-3 ID within model	*NUTS-3 Name within model*	*NUTS-3 ID Eurostat*	*DB DIUM ID within model*
242	05314	Railyard of Bonn	DEA22	154864
243	05554	Railyard of Borken	DEA34	053769
244	05154	Railyard of Kleve	DEA1B	151225
245	08235	Railyard of Calw	DE12A	295014
246	16065	Railyard of Sondershausen	DEG0A	208488
247	16071	Railyard of Apolda	DEG0G	160218
248	09374	Railyard of Neustadt a.d. Waldnaab	DE237	261370
249	08416	Railyard of Tübingen	DE142	293183
250	05378	Railyard of Bergisch-Gladbach	DEA2B	084541
251	03462	Railyard of Wittmund	DE94H	214387
252	16068	Railyard of Sömmerda	DEG0D	160218
253	06411	Railyard of Darmstadt	DE711	113373
254	05558	Railyard of Coesfeld	DEA35	572966
255	08225	Railyard of Mosbach	DE127	182923
257	09185	Railyard of Neuburg a.d. Donau	DE21I	200980
258	12062	Railyard of Herzberg-Elster	DE407	042234
259	14628	Railyard of Pirna	DED2F	060541
260	08226	Railyard of Heidelberg	DE128	803791
261	09461	Railyard of Bamberg	DE241	220939
262	08236	Railyard of Pforzheim	DE12B	295014
263	09372	Railyard of Cham	DE235	263970
264	03101	Railyard of Braunschweig	DE911	132605
265	03356	Railyard of Osterholz-Scharmbeck	DE936	585638
266	07140	Railyard of Simmern(Hunsrück)	DEB19	190140
267	08327	Railyard of Tuttlingen	DE137	145862
268	09188	Railyard of Starnberg	DE21L	205179
269	09674	Railyard of Hassfurt	DE267	220939
270	09776	Railyard of Lindau	DE27A	0467
271	12072	Railyard of Luckenwalde	DE40H	344168
272	03358	Railyard of Fallingbostel	DE938	136705
273	03456	Railyard of Nordhorn	DE94B	410134
274	05762	Railyard of Höxter	DEA44	103598
275	07235	Railyard of Trier	DEB25	251801
276	07319	Railyard of Worms	DEB39	190678
277	09478	Railyard of Lichtenfels	DE24C	220566
278	12064	Railyard of Seelow	DE409	282277
279	15091	Railyard of Wittenberg	DEE0E	273177
280	09377	Railyard of Tirschenreuth	DE23A	261370
281	08136	Railyard of Aalen	DE11D	295493
282	01057	Railyard of Plön	DEF0A	546051

Table G-67 List of modelled freight yards, p.7 [incl. data from DB CARGO (2015b)]

	NUTS-3 ID within model	*NUTS-3 Name within model*	*NUTS-3 ID Eurostat*	*DB DIUM ID within model*
283	03158	Railyard of Wolfenbüttel	DE91B	133447
284	03252	Railyard of Hameln	DE923	134288
285	09175	Railyard of Ebersberg	DE218	205179
286	09764	Railyard of Memmingen	DE274	022533
287	13075	Railyard of Greifswald	DE801	213819
288	16067	Railyard of Gotha	DEG0C	160440
289	14626	Railyard of Görlitz	DED2D	060871
290	03360	Railyard of Ülzen	DE93A	517003
291	09576	Railyard of Roth	DE25B	224535
292	16051	Railyard of Erfurt	DEG01	160440
293	16075	Railyard of Schleiz	DEG0K	260034
294	09671	Railyard of Aschaffenburg	DE264	226126
295	09761	Railyard of Augsburg	DE271	024000
296	07312	Railyard of Kaiserslautern	DEB32	197749
297	05758	Railyard of Herford	DEA43	237321
298	09475	Railyard of Hof	DE249	260034
300	09571	Railyard of Ansbach	DE256	227280
301	01062	Railyard of Bad Oldesloe	DEF0F	013268
302	14511	Railyard of Chemnitz	DED41	065581
303	09174	Railyard of Dachau	DE217	205179
304	08127	Railyard of Schwõbisch Hall	DE11A	295493
305	08337	Railyard of Waldshut-Tiengen	DE13A	145219
306	12068	Railyard of Neuruppin	DE40D	242909
307	07211	Railyard of Trier	DEB21	251801
308	03351	Railyard of Celle	DE931	136325
309	16069	Railyard of Hildburghausen	DEG0E	228148
310	15002	Railyard of Halle an der Saale	DEE02	230011
311	09771	Railyard of Aichach	DE275	024000
312	06436	Railyard of Hofheim am Taunus	DE71A	110692
313	03353	Railyard of Winsen	DE933	010660
314	05966	Railyard of Olpe	DEA59	082479
315	16064	Railyard of Mühlhausen/Th.	DEG09	160770
316	16066	Railyard of Meiningen	DEG0B	333260
317	03451	Railyard of Westerstede	DE946	016501
318	08231	Railyard of Pforzheim	DE129	295014
319	08119	Railyard of Waiblingen	DE116	290551
320	08126	Railyard of Künzelsau	DE119	717769
321	06439	Railyard of Bad Schwalbach	DE71D	729194
322	09277	Railyard of Pfarrkirchen	DE22A	264424
323	09263	Railyard of Straubing	DE223	263970

Table G-68 List of modelled freight yards, p.8 [incl. data from DB CARGO (2015b)]

	NUTS-3 ID within model	NUTS-3 Name within model	NUTS-3 ID Eurostat	DB DIUM ID within model
324	10045	Railyard of Homburg	DEC05	254581
325	05770	Railyard of Minden	DEA46	135780
326	07135	Railyard of Cochem	DEB16	190140
327	03355	Railyard of Lüneburg	DE935	011742
328	09778	Railyard of Mindelheim	DE27C	332429
329	12061	Railyard of Löbben-Spreewald	DE406	042234
330	16055	Railyard of Weimar	DEG05	160218
331	09575	Railyard of Neustadt a. d. Aisch	DE25A	227280
332	09563	Railyard of Fürth	DE253	224345
333	06413	Railyard of Offenbach am Main	DE713	110692
334	10043	Railyard of Neunkirchen	DEC03	254458
335	06635	Railyard of Korbach	DE736	053017
336	12051	Railyard of Brandenburg a. d. Havel	DE401	242909
337	09564	Railyard of Nürnberg	DE254	224535
338	12069	Railyard of Belzig	DE40E	242909
339	09663	Railyard of Würzburg	DE263	225383
340	05915	Railyard of Hamm	DEA54	100065
341	12065	Railyard of Oranienburg	DE40A	281261
343	05122	Railyard of Solingen	DEA19	081828
344	07335	Railyard of Kaiserslautern	DEB3F	197749
345	09361	Railyard of Amberg	DE231	261958
346	09182	Railyard of Miesbach	DE21F	201756
347	07317	Railyard of Pirmasens	DEB37	197749
348	09678	Railyard of Schweinfurt	DE26B	220434
349	09573	Railyard of Fürth	DE258	224345
350	07332	Railyard of Bad Dürkheim	DEB3C	194043
351	08325	Railyard of Rottweil	DE135	193235
352	09181	Railyard of Landsberg a.Lech	DE21E	024000
353	07132	Railyard of Altenkirchen	DEB13	195289
354	12063	Railyard of Rathenow	DE408	242909
355	03401	Railyard of Delmenhorst	DE941	432021
356	03403	Railyard of Oldenburg	DE943	016501
357	03452	Railyard of Aurich	DE947	214387
358	03455	Railyard of Jever	DE94A	212233
359	15090	Railyard of Stendal	DEE0D	240622
360	05120	Railyard of Remscheid	DEA18	084541
361	09561	Railyard of Ansbach	DE251	227280
362	09661	Railyard of Aschaffenburg	DE261	226126
363	08316	Railyard of Emmendingen	DE133	143529
364	09375	Railyard of Regensburg	DE238	629485

Table G-69 List of modelled freight yards, p.9 [incl. data from DB CARGO (2015b)]

	NUTS-3 ID within model	*NUTS-3 Name within model*	*NUTS-3 ID Eurostat*	*DB DIUM ID within model*
365	06437	Railyard of Erbach	DE71B	226480
366	07337	Railyard of Landau i. d. Pfalz	DEB3H	194266
367	06434	Railyard of Bad Homburg v. d. Höh	DE718	110692
368	09474	Railyard of Forchheim	DE248	221630
369	[01]	Railyard of Hamburg	DE600	010660
370	[02]	Railyard of Wilhelmshaven	DE945	212233
371	[03]	Railyard of Bremen	DE501	137893
372	[04]	Railyard of Bremerhaven	DE502	137968
373	[05]	Railyard of Brunsbüttel	DEF05	015024
374	[06]	Railyard of Brake	DE94G	138404
375	[07]	Railyard of Bützfleth	DE939	011932
376	[08]	Railyard of Emden	DE942	214569
377	[09]	Railyard of Nordenham	DE94G	138495
378	[10]	Railyard of Cuxhaven	DE932	012054
379	[11]	Railyard of Leer	DE94C	210914
380	[12]	Railyard of Papenburg	DE94C	210864
381	[13]	Railyard of Husum	DEF07	870444
382	[14]	Railyard of Rostock	DE803	270611
383	[15]	Railyard of Lübeck	DEF03	016881
384	[17]	Railyard of Kiel	DEF02	546051
385	[18]	Railyard of Wismar	DE80E	278317
386	[19]	Railyard of Saßnitz	DE805	285510
387	[20]	Railyard of Lubmin	DE801	285734
388	[22]	Railyard of Stralsund	DE805	213819
389	[23]	Railyard of Flensburg	DEF01	0406
390	[20002]	Railyard of Rothenburg	CH	020206
391	[20003]	Railyard of Noginsk	RU	-
392	[20004]	Railyard of Schaerbeek	BE	110072
393	[20005]	Railyard of Stara Zagora	BG	480103
394	[20006]	Railyard of Arhus	DK	000554
395	[20007]	Railyard of Tallinn	EST	-
396	[20008]	Railyard of Stockholm	FI	037507
397	[20009]	Railyard of St-Pierre-Des-Corps	FR	571240
398	[20010]	Railyard of Athine	GR	001057
399	[20011]	Railyard of Liverpool	IE	216200
400	[20012]	Railyard of Arezzo	IT	069161
401	[20013]	Railyard of Riga	LV	-
402	[20014]	Railyard of Sestokai	LT	123802

Table G-70 List of modelled freight yards, p.10 [incl. data from DB CARGO (2015b)]

	NUTS-3 ID within model	NUTS-3 Name within model	NUTS-3 ID Eurostat	Latitude within model	Longitude within model
403	[20015]	Railyard of Bettembourg	LU	300004	
404	[20016]	Railyard of Palermo Brancaccio	MT	120550	
405	[20017]	Railyard of Oss	NL	004952	
406	[20019]	Railyard of Lodz	PL	046490	
407	[20020]	Railyard of Entronacamento	PT	340091	
408	[20021]	Railyard of Brasov	RO	306910	
409	[20022]	Railyard of Södertälje	SE	039768	
410	[20023]	Railyard of Zilina	SK	179150	
411	[20024]	Railyard of Ljubljana	SI	423517	
412	[20025]	Railyard of Madrid	ES	951046	
413	[20026]	Railyard of Pardubice	CZ	536136	
414	[20027]	Railyard of Ferencvaros	HU	100255	
415	[20028]	Railyard of Manchester	UK	296939	
416	[25]	Railyard of Rotterdam	NL	005579	
417	[26]	Railyard of Antwerpen	BE	240671	

Table G-71 List of modelled freight yards, p.11 [incl. data from DB CARGO (2015b)]

1	15082	Aken	DEE05	51.7506	11.97119999
2	9661	Aschaffenburg	DE261	49.98799896	9.109880447
3	11000	Berlin	DE300	52.50140762	13.40232849
4	12063	Wustermark	DE408	52.60749817	12.34399986
5	5314	Bonn	DEA22	50.73242188	7.101860046
6	3101	Braunschweig	DE911	52.23550034	10.54189968
7	5913	Dortmund	DEA52	51.51660919	7.4582901
8	7137	Andernach	DEB17	50.35694	7.589
9	3454	Dörpen	DE949	52.69329834	7.277890205
10	5111	Düsseldorf	DEA11	51.21562958	6.776040077
11	5162	Dormagen	DEA1D	51.2042	6.68795
12	5112	Duisburg	DEA12	51.43143082	6.763929844
13	5170	Emmelsum	DEA1F	51.66431	6.62957
14	8317	Kehl	DE134	48.48149872	7.93956995
15	6412	Frankfurt	DE712	50.11207962	8.683409691
16	7334	Germersheim	DEB3E	49.21969986	8.370750427
17	6433	Gernsheim	DE717	49.91279984	8.485799789

Table G-72 List of modelled national intermodal IWW terminals, p. 1 [incl. data from AGORA (2014)]

	NUTS-3 ID within model	*NUTS-3 Name within model*	*NUTS-3 ID Eurostat*	*Latitude within model*	*Longitude within model*
18	3241	Hannover	DE929	52.37226868	9.738149643
19	8212	Karlsruhe	DE122	49.00540161	8.386870384
20	5315	Köln	DEA23	50.93753	6.96028
21	7111	Koblenz	DEB11	50.35694	7.589
22	5114	Krefeld	DEA14	51.33390045	6.562300205
23	7314	Ludwigshafen	DEB34	49.48094177	8.447299957
24	15003	Magdeburg	DEE03	52.13185883	11.62777996
25	7315	Mainz	DEB35	50.00061035	8.272279739
26	8222	Mannheim	DE126	49.48120117	8.482709885
27	5770	Minden	DEA46	52.28649902	8.945340157
28	9564	Nürnberg	DE254	49.46519852	11.07600021
29	5154	Emmerich	DEA1B	51.7820015	6.117949963
30	8111	Stuttgart	DE111	48.76766968	9.171919823
31	15083	Haldensleben	DEE07	52.28919983	11.41230011
32	7319	Worms	DEB39	49.62779999	8.356789589
33	[01]	Hamburg	DE600	53.55334091	9.992469788
34	[03]	Bremen	DE501	53.07509995	8.804690361
35	[04]	Bremerhaven	DE502	53.52610016	8.623869896
36	[05]	Brunsbüttel	DEF05	54.17990112	9.095509529
37	[06]	Brake	DE94G	53.32632828	8.478639603
38	[07]	Bützfleth	DE939	53.60229874	9.475330353
39	[08]	Emden	DE942	53.364705	7.172326
40	[10]	Cuxhaven	DE932	53.84429932	8.687970161
41	[11]	Leer	DE94C	53.23751068	7.461763859
42	[15]	Lübeck	DEF03	53.86626816	10.68087959
43	[17]	Kiel	DEF02	54.32324982	10.1322403
44	[24]	Rendsburg	DEF0B	54.29779816	9.666950226

Table G-73 List of modelled national intermodal IWW terminals, p. 2 [incl. data from AGORA (2014)]

	NUTS-3 ID within model	*NUTS-3 Name within model*	*Service*	*NUTS-3 ID Eurostat*	*Latitude within model*	*Longitude within model*
1	[20002]	Birsfelden	Rail, IWW	CH	47.55377	7.6279
2	[20019]	Gdansk	Rail, IWW	PL	54.15606	19.40449
3	[20004]	Brussels	Rail, IWW	BE	50.85034	4.35171
4	[20005]	Ruse	Rail, IWW	BG	43.67423	26.02386
5	[17]	Kiel	Rail, IWW	DEF02	54.32324982	10.1322403
6	[20009]	Nancy-Frouard	Rail, IWW	FR	48.75038	6.14658
7	[20009]	Dunkerque	Rail, IWW	FR	51.03437	2.37678
8	[20002]	Basel	Rail, IWW	CH	47.56744	7.59755
9	[20015]	Mertert	Rail, IWW	LU	49.70035	6.48123
10	[20017]	Tiel	IWW	NL	51.88762	5.42788
11	[20018]	Linz	Rail, IWW	AT	48.30694	14.28583
12	[20019]	Gliwice	Rail, IWW	PL	50.29449	18.67138
13	[20023]	Komárno	Rail, IWW	SK	47.76258	18.12941
14	[20026]	Ústí nad Labem	Rail, IWW	CZ	50.66112	14.05315
15	[20027]	Budapest	Rail, IWW	HU	47.49791	19.04023
16	[20021]	Constanta	Rail, IWW	RO	44.17333	28.63833
17	[20004]	Vilvoorde	Rail, IWW	BE	50.92725	4.42579
18	-	(Belgrade)	Rail, IWW	-	44.78657	20.44892
19	[05]	Brunsbüttel	Rail, IWW	DEF05	53.89889	9.13389
20	[20019]	Swinoujscie	Rail, IWW	PL	53.91003	14.24758
21	[20009]	Metz	Rail, IWW	FR	49.11931	6.17572
22	-	(Belgrade)	Rail, IWW	-	44.78657	20.44892
23	[20004]	Oostende	Rail, IWW	BE	51.21543	2.92866
24	[20009]	Thionville	Rail, IWW	FR	49.35757	6.16843
25	[20017]	Oss	Rail, IWW	NL	51.76118	5.51405
26	9275	Passau	Rail, IWW	DE228	48.56674	13.43195
27	16056	Eisenhüttenstadt	Rail, IWW	DEG0N	52.14366	14.6419
28	[20027]	Győr-Gőnyű	Rail, IWW	HU	47.73051	17.76812
29	[20018]	Vienna	Rail, IWW	AT	48.20817	16.37382
30	[20026]	Decin	Rail, IWW	CZ	50.77256	14.21276
31	8336	Weil am Rhein	Rail, IWW	DE139	47.59344	7.61981
32	[20004]	Garocentre	Rail, IWW	BE	50.49493	4.18809
33	-	(Novi Sad)	Rail, IWW	-	45.26714	19.83355
34	[10]	Cuxhaven	Rail, IWW	DE932	53.85934	8.68791
35	[20019]	Szczecin	Rail, IWW	PL	53.42854	14.55281
36	[20004]	Zeebrugge	Rail, IWW	BE	51.33333	3.2
37	[20017]	Utrecht	IWW	NL	52.09074	5.12142
38	9271	Deggendorf	Rail, IWW	DE224	48.84085	12.95748
39	[20023]	Bratislava	Rail, IWW	SK	48.14589	17.10714

Table G-74 List of considered international intermodal IWW and rail terminals [incl. data from AGORA (2014); DB CARGO (2012)]

	NUTS-3 ID within model	*NUTS-3 Name within model*	*NUTS-3 ID Eurostat*	*Latitude within model*	*Longitude within model*
1	15082	Aken (Elbe)	DEE05	51.7506	11.97119999
2	9671	Aschaffenburg Hafen	DE264	49.98799896	9.109880447
3	9772	Augsburg Oberhausen Ubf	DE276	48.35509872	10.93309975
4	8336	Basel - Weil am Rhein	DE139	47.6089	7.6604
5	5978	Bönen (ABX Bahntrans)	DEA5C	51.52619934	7.695569992
6	6634	Beiseförth Ubf	DE735	51.03170013	9.403120041
7	11000	Berlin Westhafen KV Terminal	DE300	52.50140762	13.40232849
8	3101	Braunschweig Hgbf Ubf	DE911	52.23550034	10.54189968
9	5913	Dortmund Westerholz Ubf	DEA52	51.51660919	7.4582901
10	15002	Halle (Saale) Gbf	DEE02	51.49698	11.9688
11	5916	Herne	DEA55	51.54030991	7.21987009
12	5970	Kreuztal	DEA5A	50.86700058	7.986810207
13	7137	Andernach Hafen	DEB17	50.35694	7.589
14	3454	Dörpen Ubf	DE949	52.69329834	7.277890205
15	5111	Düsseldorf Hafen	DEA11	51.21562958	6.776040077
16	5112	Duisburg-Ruhrort Hafen Ubf	DEA12	51.43143082	6.763929844
17	10044	Dillingen (Saar)	DEC04	49.30659866	6.74116993
18	5162	Dormagen	DEA1D	51.2042	6.68795
19	14612	Dresden Neustadt Gbf Ubf	DED21	51.05363083	13.74081039
20	16056	Eisenach	DEG0N	50.97940063	10.3032999
21	12062	Elsterwerda	DE407	51.69120026	13.22029972
22	5170	Wesel	DEA1F	51.66431	6.62957
23	16051	Erfurt-Vieselbach	DEG01	50.98477	11.02988
24	8317	Kehl	DE134	48.48149872	7.93956995
25	12053	Frankfurt (Oder) Ubf	DE403	52.34722	14.55057
26	6412	Frankfurt/M Ost Ubf	DE712	50.11207962	8.683409691
27	3152	Göttingen Ubf	DE915	51.55989838	9.931920052
28	5754	Gütersloh Hbf	DEA42	51.90390015	8.425290108
29	7334	Germersheim	DEB3E	49.21969986	8.370750427
30	6433	Gernsheim	DE717	49.91279984	8.485799789
31	7233	Gerolstein	DEB24	50.19620132	6.836969852
32	14524	Glauchau (Sachs) KV	DED45	50.71789932	12.45730019
33	12072	Großbeeren Ubf	DE40H	52.08269882	13.15620041
34	3241	Hannover Linden Hafen Ubf	DE929	52.37226868	9.738149643
35	5362	Hürth-Knapsack	DEA27	50.94889832	6.648990154
36	9475	Hof Hbf	DE249	50.32210159	11.92339993
37	9161	Ingolstadt-Nord Ubf	DE211	48.7737999	11.42010021
38	8212	Karlsruhe Hbf Ubf	DE122	49.00540161	8.386870384
39	6611	Kassel GVZ Ubf	DE731	51.31610107	9.468520164
40	5315	Köln-Eifeltor Ubf	DEA23	50.93753	6.96028

Table G-75 List of modelled national intermodal rail terminals, p. 1 [incl. data from AGORA (2014); DB CARGO (2012)]

	NUTS-3 ID within model	*NUTS-3 Name within model*	*NUTS-3 ID Eurostat*	*Latitude within model*	*Longitude within model*
41	8118	Kornwestheim Ubf	DE115	48.89319992	9.194270134
42	5114	Krefeld Gbf	DEA14	51.33390045	6.562300205
43	9274	Landshut (Bay) Hbf Ubf	DE227	48.54470062	12.1566
44	14729	Leipzig-Wahren Ubf	DED52	51.34519958	12.3859396
45	7314	Ludwigshafen (Rhein) BASF	DEB34	49.48094177	8.447299957
46	15003	Magdeburg Hafen	DEE03	52.13185883	11.62777996
47	7315	Mainz-Hafen Ubf	DEB35	50.00061035	8.272279739
48	8222	Mannheim Hgbf Ubf	DE126	49.48120117	8.482709885
49	9162	München- Riem Ubf	DE212	48.1391	11.5145
50	5770	Minden (Westf)	DEA46	52.28649902	8.945340157
51	12063	Wustermark-Nord Ubf	DE408	52.60749817	12.34399986
52	9564	Nürnberg Hgbf Ubf	DE254	49.46519852	11.07600021
53	9362	Regensburg Ost Ubf	DE232	49.02399826	12.07590008
54	5154	Emmerich Grenze	DEA1B	51.7820015	6.117949963
55	13072	Rostock Seehafen	DE803	54.0790596	12.13216019
56	3102	Salzgitter Hütte Nord	DE912	52.15327835	10.33312035
57	15088	Buna Werke	DEE0B	51.33229828	11.99250031
58	9662	Schweinfurt Hafen	DE262	50.04539871	10.23159981
59	8335	Singen (Htw.) Ubf	DE138	47.67670059	9.182880402
60	8111	Stuttgart Hafen Ubf	DE111	48.76766968	9.171919823
61	15083	Haldensleben	DEE07	52.28919983	11.41230011
62	8425	Ulm Hbf	DE145	48.39459991	9.968560219
63	6632	Philippsthal (Werra) KV	DE733	50.86500168	9.702589989
64	3103	Wolfsburg Ubf	DE913	52.42440033	10.79749966
65	7319	Worms Hafen	DEB39	49.62779999	8.356789589
66	5124	Wuppertal-Langerfeld Ubf	DEA1A	51.083403	7.024221
67	5914	Hagen Hbf Ubf	DEA53	51.35960007	7.460989952
68	9479	Marktredwitz Ubf	DE24D	50.03639984	12.00559998
69	5562	Marl (UTM Marl Ubf)	DEA36	51.61949921	7.213409901
70	9775	Neu Ulm Ubf	DE279	48.39780045	10.00500011
71	10041	Saarbrücken Rbf Ubf	DEC01	49.23149	6.99833
72	9172	Salzburg Hbf	DE215	47.71979904	12.85840034
73	[01]	Hamburg-Billwerder Ubf	DE600	53.55334091	9.992469788
74	[02]	Wilhelmshaven	DE945	53.52959824	8.099840164
75	[03]	Bremen-Roland Ubf	DE501	53.07509995	8.804690361
76	[04]	Bremerhaven Kaiserhafen	DE502	53.52610016	8.623869896
77	[05]	Brunsbüttel	DEF05	54.17990112	9.095509529
78	[06]	Brake (Unterweser)	DE94G	53.32632828	8.478639603
79	[07]	Bützfleth	DE939	53.60229874	9.475330353

Table G-76 List of modelled national intermodal rail terminals, p. 2 [incl. data from AGORA (2014); DB CARGO (2012)]

	NUTS-3 ID within model	*NUTS-3 Name within model*	*NUTS-3 ID Eurostat*	*Latitude within model*	*Longitude within model*
80	[08]	Emden-Außenhafen	DE942	53.364705	7.172326
81	[10]	Cuxhaven	DE932	53.84429932	8.687970161
82	[11]	Leer (Ostfriesl)	DE94C	53.23751068	7.461763859
83	[13]	Husum CFL	DEF07	54.47034836	9.05163002
84	[14]	Rostock Seehafen	DE803	54.0790596	12.13216019
85	[15]	Lübeck Hgbf	DEF03	53.86626816	10.68087959
86	[17]	Kiel Nordhafen	DEF02	54.32324982	10.1322403
87	[18]	Wismar Hafen	DE80E	53.86003	11.19069
88	[19]	Sassnitz-Mukran	DE805	54.30881119	13.09450722
89	[20]	Lubmin Werkbahnhof	DE801	54.08974457	13.39124393
90	[23]	Flensburg Grenze	DEF01	54.79779816	9.477919579

Table G-77 List of modelled national intermodal rail terminals, p. 3 [incl. data from AGORA (2014); DB CARGO (2012)]

MIX
Papier aus verantwortungsvollen Quellen
Paper from responsible sources
FSC® C105338

If you have any concerns about our products,
you can contact us on
ProductSafety@springernature.com

In case Publisher is established outside the EU,
the EU authorized representative is:
Springer Nature Customer Service Center GmbH
Europaplatz 3, 69115 Heidelberg, Germany

Printed by Libri Plureos GmbH
in Hamburg, Germany